Ecological Studies, Vol. 193

Analysis and Synthesis

Ecological Studies

Volumes published since 2002 are listed at the end of this book.

W. Nentwig (Ed.)

Biological Invasions

With 38 Figures and 19 Tables

Springer

Prof. Dr. Wolfgang Nentwig
Community Ecology
Zoological Institute
University of Bern
Baltzerstrasse 6
3012 Bern
Switzerland

Cover illustration: The cover drawing shows the global dimensions of three invasion histories. (1) Imports of the first potato (*Solanum tuberosum*) from the Andean highlands to Europe began around 1550. In the eighteenth century potatoes were brought to the east coast of North America from where potato cultivation expanded to the west coast, until potatoes came into contact with the native weed *Solanum rostratum*. From this plant, the Colorado potato beetle (*Leptinotarsa decemlineata*) switched in 1859 to *Solanum tuberosum*, reached via potato fields the east coast of North America, crossed the Atlantic in 1874, and invaded Europe to East Asia. (2) The common water hyacinth (*Eichhornia crassipes*), native to a small area in tropical South America, has been spread into the tropical and subtropical areas of all continents. (3) In Brazil, cultivated European honey bees (*Apis mellifera*) and African wild forms of the honey bee, were interbred. Around 1956, colonies of the so-called Africanized honey bee escaped and these are currently spreading into North America.

ISSN 0070-8356
ISBN-10 3-540-36919-8 Springer Berlin Heidelberg New York
ISBN-13 978-3-540-36919-6 Springer Berlin Heidelberg New York

Library of Congress Control Number: 2006934733

Springer is a part of Springer Science+Business Media
springer.com

Editor: Dr. Dieter Czeschlik, Heidelberg, Germany
Desk editor: Dr. Andrea Schlitzberger, Heidelberg, Germany
Cover design: WMXDesign GmbH, Heidelberg, Germany
Typesetting and production: Friedmut Kröner, Heidelberg, Germany

31/3152 YK – 5 4 3 2 1 0 – Printed on acid free paper

Preface

Yet another book on "Biological Invasions"? It is true, the market already provides several recent works on this topic and, in the next few years, probably many more will follow. There are, however, two important points which argue in favour of the relevance and need of exactly this book.

Most books on biological invasions treat only a small part of the subject. They cover either invasive plants or invasive pest arthropods, address invasive species of a country, an island or a habitat, discuss the impact of alien species on economy or evolution, or gather an impressive number of case studies. This book is clearly different insofar that it attempts to cover all (or at least most) of these undoubtedly very important topics. A joint effort of 42 specialists, it deals with plants and animals, includes both the terrestrial and the aquatic environment, guides us from ecology via economy to socio-economy, and comprises also administrative and management aspects. Our intention is a strong focus on mechanisms and so, in the opening chapters we analyse the main pathways of biological invasions and discuss the traits of good invaders. The patterns of invasion and invasibility point to central aspects such as land management, nitrogen pollution or climate change. A presentation of the ecological impact of invasive species, based on striking case studies from major ecosystems worldwide, also tackles the key question whether genetically modified organisms may become invasive. This all includes relevant economic and socio-economic facets. The closing chapters claim an enormous current lack of preventive means, and demand more administrative and control measures as well as eradication programs.

This all leads to my second main point, of urgent need. We already live in a global world, in which the globalizing process has started with full power only a few decades ago. Still, the pace will increase considerably, and there will be ever more people and goods moving from one point of the world to another. This complete loss of biogeographical borders will lead to much more alien species everywhere and an increasing number of these will become invasive. Invasion biology, until recently known only to a few experts, is becoming ever more topical in newspapers, keeping governments and administrations more than busy. The public awareness of biohazard due to invasive species is rising. Still, this process, reaching from the public opinion to politicians and a growing scientific community, needs to be intensified even more to face biological invasions as the most serious threat to biodiversity.

This work is based on the joint effort of all authors and co-authors involved, and I wish to thank these all for their state-of-the-art contributions. It was truly an enjoyable task producing this book together. Warmest thanks also go to the many people behind the scenes, helping us in many ways, and to the publisher who invited me to edit this book.

Bern, September 2006 *Wolfgang Nentwig*

Contents

Section III Patterns of Invasion and Invasibility

Section V Economy and Socio-Economy of Biological Invasions

Contributors

BABENDREIER, DIRK

Biosafety Group, Federal Department of Economic Affairs DEA, Agroscope Reckenholz-Tänikon Research Station ART, Reckenholzstrasse 191, 8046 Zürich, Switzerland, e-mail: dirk.babendreier@art.admin.ch

BAUR, BRUNO

Department of Environmental Sciences, Section of Conservation Biology, University of Basel, St. Johanns-Vorstadt 10, 4056 Basel, Switzerland, e-mail: bruno.baur@unibas.ch

BAYHA, KEITH M.

Dauphin Island Sea Lab and University of South Alabama, 101 Bienville Blvd., Dauphin Island, AL 36528, USA

BINIMELIS I ADELL, ROSA

Institut de Ciència i Tecnologia Ambientals (ICTA), Edifici Ciències, Torre Àrea 9, 4a planta, Universitat Autònoma de Barcelona, 08193 Bellaterra (Barcelona), Spain, e-mail: rosa.binimelis@gmx.net

BORN, WANDA

UFZ – Centre for Environmental Research Leipzig-Halle, Department of Economics, Permoserstr. 15, 04318 Leipzig, Germany

BROOKS, MATT

US Geological Survey, Western Ecological Research Center, Las Vegas Field Station, 160 N. Stephanie St., Henderson, NV 89074, USA, e-mail: matt_brooks@usgs.gov

BUSCHMANN, HOLGER

Albrecht-von-Haller Institute for Plant Sciences, University of Göttingen, Grisebachstraße 1a, 37077 Göttingen, Germany

CHARLES, HEATHER

Department of Biology, University of Massachusetts, 100 Morrissey Blvd., Boston, MA 02125, USA

DEHNEN-SCHMUTZ, KATHARINA

Department of Biology, University of York, York YO10 5DD, UK

DUKES, JEFFREY S.

Department of Biology, University of Massachusetts, 100 Morrissey Blvd., Boston, MA 02125, USA, e-mail: jeffrey.dukes@umb.edu

GALIL, BELLA

National Institute of Oceanography, P.O. Box 8030, Haifa 31080, Israel, e-mail: bella@ocean.org.il

GENOVESI, PIERO

Chair European Section IUCN SSC ISSG, INFS (National Wildlife Institute), Via Ca Fornacetta 9, 40064 Ozzano Emilia BO, Italy, e-mail: piero.genovesi@infs.it

GOLLASCH, STEPHAN

GoConsult, Bahrenfelder Str. 73a, 22765 Hamburg, Germany, e-mail: sgollasch@aol.com

GRAHAM, WILLIAM

Dauphin Island Sea Lab and University of South Alabama, 101 Bienville Blvd., Dauphin Island, AL 36528, USA, e-mail: mgraham@disl.org

HAILS, ROSIE

Pathogen Population Ecology, Centre for Ecology and Hydrology, Mansfield Rd., Oxford OX1 3SR, UK, e-mail: RHA@ceh.ac.uk

HALLMAN, GUY

USDA-ARS, 2413 E. Highway 83, Weslaco, TX 78596, USA, e-mail: ghallman@weslaco.ars.usda.gov

Hufbauer, Ruth A.

Department of Bioagricultural Sciences and Pest Management, Colorado State University, Fort Collins, CO 80523-1177, USA, e-mail: hufbauer@lamar.colostate.edu

Jones, Glyn

Environment Department, University of York, York YO10 5DD, UK

Klotz, Stefan

UFZ – Centre for Environmental Research Leipzig-Halle, Department of Community Ecology, Theodor-Lieser-Strasse 4, 06120 Halle, Germany

Kowarik, Ingo

Institute of Ecology, Technical University Berlin, Rothenburgstrasse 12, 12165 Berlin, Germany, e-mail: kowarik@tu-berlin.de

Kühn, Ingolf

UFZ – Centre for Environmental Research Leipzig-Halle, Department of Community Ecology, Theodor-Lieser-Strasse 4, 06120 Halle, Germany, e-mail: ingolf.kuehn@ufz.de

Largiadèr, Carlo R.

Institute of Clinical Chemistry, Inselspital, University Hospital, University of Bern, 3010 Bern, Switzerland, e-mail: carlo.largiader@insel.ch

Midgley, Guy F.

Climate Change Research Group, Ecology and Conservation, Kirstenbosch Research Center, National Biodiversity Institute, P/Bag X7 Claremont 7735; Cape Town, South Africa

Monterroso, Iliana

Institut de Ciència i Tecnologia Ambientals (ICTA), Edifici Ciències, Torre Àrea 9, 4a planta, Universitat Autònoma de Barcelona, 08193 Bellaterra (Barcelona), Spain

Nehring, Stefan

AeT umweltplanung, Bismarckstraße 19, 56068 Koblenz, Germany e-mail: nehring@aet-umweltplanung.de

NENTWIG, WOLFGANG

Community Ecology, Zoological Institute, University of Bern, Baltzerstrasse 6, 3012 Bern, Switzerland, e-mail: wolfgang.nentwig@zos.unibe.ch

OLDE VENTERINK, HARRY

Institute of Integrative Biology, ETH Zurich, Universitaetsstrasse 16, 8092 Zurich, Switzerland

PANOV, VADIM E.

Zoological Institute of the Russian Academy of Sciences, Universitetskaya Nab.1, 199034 St. Petersburg, Russia

PIMENTEL, DAVID

College of Agriculture and Life Sciences Cornell University, Ithaca, NY 14853, USA, e-mail: dp18@cornell.edu

PIMENTEL, MARCIA

Division of Nutritional Sciences, Cornell University, Ithaca, NY 14853, USA

PYŠEK, PETR

Institute of Botany, Academy of Sciences of the Czech Republic, 252 43 Pruhonice, Czech Republic, e-mail: pysek@ibot.cas.cz

RICHARDSON, DAVID M.

Centre for Invasion Biology, University of Stellenbosch, Stellenbosch, South Africa

RODRÍGUEZ-LABAJOS, BEATRIZ

Institut de Ciència i Tecnologia Ambientals (ICTA), Edifici Ciències, Torre Àrea 9, 4a planta, Universitat Autònoma de Barcelona, 08193 Bellaterra (Barcelona), Spain

SCHERER-LORENZEN, MICHAEL

Institute of Plant Sciences, ETH Zurich, Universitaetsstrasse 2, 8092 Zurich, Switzerland, e-mail: michael.scherer@ipw.agrl.ethz.ch

SCHMIDLIN, STEPHANIE

Department of Environmetal Sciences, Section of Conservation Biology, University of Basel, St. Johanns-Vorstadt 10, 4056 Basel, Switzerland

Sol, Daniel

CREAF (Center for Ecological Research and Applied Forestries), Autonomous University of Barcelona, 08193 Bellaterra, Catalonia, Spain, e-mail: d.sol@creaf.uab.es

Thuiller, Wilfried

Laboratoire d'Ecologie Alpine, CNRS, Université Joseph Fourier, B.P. 53, 38041 Grenoble Cedex 9, France, e-mail: wilfried.thuiller@ujf-grenoble.fr

Timms-Wilson, Tracey

Molecular Microbial Ecology, Centre for Ecology and Hydrology, Mansfield Rd., Oxford OX1 3SR, UK

Torchin, Mark E.

Smithsonian Tropical Research Institute, Apartado 0843–03092, Balboa, Ancon, Panama, Republic of Panama

Touza, Julia

UFZ, Centre for Environmental Research Leipzig-Halle, Department of Ecological Modelling, Permoserstr. 15, 04318 Leipzig, Germany

von der Lippe, Moritz

Institute of Ecology, Technical University Berlin, Rothenburgstrasse 12, 12165 Berlin, Germany

Wilson, Anne

Research Assistant, Cornell University, Ithaca, NY 14853, USA

1 Biological Invasions: why it Matters

WOLFGANG NENTWIG

The dispersal of organisms is a natural process, important for the distribution of life on earth. It is also important for the appearance and expression of biodiversity, strengthening the multiple forms and functions of diversity in living organisms. We, too, profit from this process and its dynamics, and are even dependent on it. On a longer time scale, dispersal is one of the drivers of evolution, responsible for life itself.

Dispersal is limited by multiple barriers, among which geographical barriers are the most evident. However, human dispersal has overcome all biogeographical barriers, and humans now inhabit all parts of the world. Our activities are on a global scale, and we have been working intensively for centuries to connect all parts of the world ever closer together. In human history, roads were the first expression of these interconnections, as well as shipping. Today, we can reach any spot on earth by plane within 24 h, and vessels transport cargo around the globe within a few weeks. In addition, new connections between water drainage systems, lakes and oceans have been constructed.

Man did not conquer the world alone. He was always accompanied by his domesticated animals, crop plants, pets, pathogens and parasites. In addition to this, we gather and transport ornamental plants and animals or parts of these. Already our ancestors collected feathers and fur, shells and seeds for multiple purposes. Wilson called this search or even love for all types of life expressions "biophilia" (Wilson 1984). From a philosophical point of view, one could state that we obviously need to surround ourselves with nature because we are part of it. From a more ecological (and also medical) point of view, one could state that we need our natural surrounding because it enables our survival, personally and as a species.

The dark side of this entourage of many species which always surrounds us is the spread of exactly these species to every spot where we stay and to which we move. Consequently, we distribute hundreds and thousands of species intentionally and unintentionally worldwide. Bringing a given species into a new habitat is not neutral to the environment because it interacts with resident species or abiotic parameters or energy and matter fluxes.

Ecological Studies, Vol. 193
W. Nentwig (Ed.)
Biological Invasions

Usually, the result of an additional species is neither an enrichment of the ecosystem nor any amelioration whatsoever. The acclimatisation societies (which I would call pseudoscientific, rather than scientific), founded in the second half of the 19th century to "vaccinate", professionally and on a large scale, new world colonies with European species in order to "ameliorate" their "inferior" species richness, made a fatal error. It is meanwhile widely known, at least to the scientific community, that alien species are among the major dangers to the well-functioning of our environment.

The consequences of introducing alien species can be manifold: though the majority of these species has no (or no visible) immediate negative effect, some do exhibit ecological impacts directly or after a short time of adapting to the new environment. The more obvious cases are invasive aliens competing with native species, dominating their new environment or even replacing native residents. Quite often, this is done with the complicity of newly introduced pathogens or parasites to which the alien species is adapted but to which a close resident relative is not. Thus, alien species regularly lead to a loss of biodiversity and to a homogenization of the invaded habitat. Secondary damage may be chemical pollution and erosion. Alien species consume and contaminate water, thereby reducing its quantity and quality. Many alien species themselves are pathogens or parasites which threaten the indigenous species community. Weeds and animal pests are of high importance to humans because they impede on agriculture. Human pathogens such as bacteria and viruses attack humans directly and may cost millions of lives. On a global scale, this ever-increasing mixture of species originating from all over the world and being spread everywhere results in one gigantic muddle, initiating a homogenization of world ecosystems towards a least-common denominator.

One of the earliest invasions of a dangerous microorganism led to "the plague" or Black Death which, in medieval times, reached Europe from East Asia and killed about one third of the European population. Caused by the pathogen *Yersinia pestis*, with fleas, mice and rats as vectors, this disease has been spreading all over the world until recently. The potato blight *Phytophthora infestans*, a fungal pathogen of the potato, was introduced in 1840 from North America into Europe, causing mass famine in many countries the following years. Most severely affected was the Irish population, which was reduced to half its size not only by starvation as such but also by an associated mass emigration (Nentwig 2005).

To date, the most serious disease to have attacked mankind has been the pandemic influenza which spread from North America via Europe to all parts of the world at the end of the First World War. Recent estimates indicate that, over the period 1918–1920, about one third of the world population had been infected and 100 million people lost their lives – up to 20 % victims. The most recent pandemic is AIDS. Since the 1960s, the human immunodeficiency virus has spread from West or Central Africa via North America to the rest of the world. So far, more than 40 million people have been infected, some 5 million people are newly infected each year and 3–4 million, mostly in sub-Saharan Africa, have died. Though HIV has as yet infected only 1 % of the world population, it has the potential of becoming a very serious threat to mankind (Nentwig 2005).

Meanwhile, the economic costs associated with alien species are known for some countries, some species, some time periods as well as for some processes. There are increasingly good data which facilitate some generalizations and extrapolations. Such data (albeit incomplete) already indicate the various economic damages associated with invasive alien species in several nations of the world to amount to about 5 % of the world GNP. Including the countries, species and processes still unaccounted for, this value would certainly be much higher (cf. Chap. 18).

Fifty years ago, the British ecologist Charles Elton published his *Ecology of invasions by animals and plants*, already then clearly stating that our world's new mix of native and alien species has unfavourable and dangerous aspects: "The whole matter goes far wider than any technological discussion of pest control, though many of the examples are taken from applied ecology. The real thing is that we are living in a period of the world's history when the mingling of thousands of kinds of organisms from different parts of the world is setting up terrific dislocations in nature. We are seeing huge changes in the natural population balance of the world" (Elton 1958). Elton was among the first to realize the typical pattern of a biological invasion which he also called biological explosion. He asked pertinent questions: why and how are species dispersed by human activities? What is the negative impact of species in a new environment? How can this be prevented? Elton is rightly considered as one of the founders not only of ecology but also of the so-called invasion biology.

The discovery of America by Columbus in 1492 is usually set as the zero point of our definition of biological invasions. This is arguably rather an arbitrary date but it indeed marks the start of a new era of fast global population movements and trade. Thus, it does not really matter that already the Romans – and other earlier cultures, too – had imported new species into their empire. Rather, it was approximately 500 years ago when the main process began which today is called globalization, and its basic principles have not changed in the last centuries. The speed, however, is accelerating from year to year.

A new development of the last few decades concerns the self-conception of globalization, and the ease with which global trade proceeds. Global regulatory concepts such as the General Agreement on Tariffs and Trade GATT and its successor, the World Trade Organization WTO, are intended to facilitate exchange between all nations. These treaties reduce tariffs, export subsidies, protective measures, any kind of import limits, and quotas. On a worldwide basis, the goal is to eliminate all obstructions for free trade, which is seen as a basic right for people, nations, industries and trading companies.

In principle, this trend is certainly positive and could promote growth in less-developed countries. So far, however, the main profit has gone to industrialized countries. A very important side-effect of the new WTO-world is that the controls of goods involving alien species can be easily denounced as trade obstruction. It could become more difficult to set up stricter quarantine measures or large-scale controls for pests. It may also become less easy to strengthen the prohibition of trade dealing with potentially invasive alien species classified as ornamentals or pets. Strong efforts will be necessary to prevent the development of the WTO-world from turning fully in the direction initiated by early capitalistic societies. Especially politicians but also decision-makers at all hierarchical levels, including opinion-makers such as journalists, need intensive furthering of education with respect to the need of preventing the spread of alien species.

Even a WTO-world is not free of regulations and of responsibility for its own activities. Since the problem of invasive alien species is primarily economic, it is open for economic solutions (Perrings et al. 2002). One promising solution could be for each trading partner to contract an obligatory insurance covering any hazards of alien species caused by international trade. Such education – to take responsibility for one's activities – needs to be strengthened.

There is increasing concern that strong lobbies may prevent necessary countermeasures to biological invasions. Indeed, urgently needed import and general trade restrictions are becoming ever less enforceable in a WTO-world. Restrictions to trade with ornamental plants will fail on the front set up by gardeners and plant lovers. It is difficult to convince public opinion about the need for eradication programs for squirrels, parakeets, racoons and other charismatic vertebrates. Such beautiful birds and appealing mammals generate much public sympathy, and this despite their alien status, and even some scientists defend exotic species and debate about the purpose of eradication measures, smartly taking advantage of the psychological and sociological impact of such species. The public obviously needs continuing education to convince the majority of the negative aspects of even cuddly aliens!

It is an oft-cited argument that, in case-by-case studies, the negative effect of each and every alien species has to be proven. This "innocent until proven guilty" approach is justified in human jurisdiction but fatal in dealing with alien species because it immediately leads to uncontrolled introductions of the worst pest species. When ecological damage is detected, it is always too late since, once released, an organism can not simply be removed. Certainly, alien species have to be treated differently – to be on the safe side, a general zero-tolerance attitude is the far better position to take. This argumentation additionally shows that our society needs much more information on the ecological hazards of alien species.

Raising public awareness is always a tricky balance between panic and lethargy. Both extremes are usually counterproductive but, indirectly, they have proved useful in the past. When Rachel Carson published her *Silent*

Table 1.1 Numbers of alien species per continent. These values are minimum numbers, empty fields indicating gaps in our knowledge (data combined after Pimentel 2002; DAISIE 2006; other sources)

	Plants	Vertebrates	Invertebrates	Microorganisms
Africa	8,750[a]	83 fish 24 herp[a] 16 mammals[a] 8 birds[a]		
North America	5,000	145 fish 53 herp[b] 20 mammals[b] 97 birds[b]	4,500 arthropods[b] 11 earthworms[b] 91 molluscs[b] 100 aquatic species[b]	20,000[b]
South America	11,605[c]	76 fish[c] 25 mammals[c] 3 birds[c]	25 nematodes[c]	500 fungi[c] 100 viruses[c]
Asia	18,000[d]	300 fish[d] 30 mammals[d] 4 birds[d]	1,100 arthropods[d]	
Australia	3,020	180 fish 20 herp 20 mammals 70 birds	1,000 terrestrial species 250 aquatic species	188
Europe	3,691	140 fish 40 herp 90 mammals 51 birds	1,350 insects 210 arachnids 65 annelids 135 other "worms" 155 crustaceans 201 molluscs 17 cnidarians	
Oceania	2,000[e]	112 fish	2,200[e]	

[a]South Africa, [b]USA, [c]Brazil, [d]India, [e]New Zealand

spring in 1962, she wanted to achieve a more responsible and carefully managed use of environmental chemicals, especially pesticides. Her book caused much concern as well as overreactions but also led to the beginning of the modern environmental conservation movement. Today's use of environmental chemicals has become much more restricted. In another notable example in 1972 when Meadows and colleagues brought out *Limits to growth*, the book suffered from a poor database and insufficient computer software for predictions. Consequently, the critics "tore it to pieces" but, today, the main message of this key work is considered accurate and is generally accepted: the resources of the earth are finite and, thus, are inevitably subject to natural limits. The modern movement of sustainability roots in the ideas of Meadows and co-workers.

Elton (1958) characterized the introduction of alien species as "one of the great convulsions of the world's flora and fauna". Astonishingly, the hazards provoked by alien species did not cause that much concern among scientists, nor did it attract public awareness as much as would have been expected. However, the ultimate reason for the loss of more than 5 % of the world GNP, one main reason for the loss of biodiversity, for millions of human deaths, and for the loss of more than 20 % of the world's food production cannot be ignored.

The simple question as to how many alien species we have worldwide has no precise answer. Per continent or larger geographic area, some estimates indicate up to 10,000 alien plant species, up to 300 alien vertebrates, more than 5,000 alien invertebrates and many 1,000s of alien microorganisms (Table 1.1). Giving a more precise answer is not yet possible. This alarming knowledge gap is indicative of our whole predicament in this field, and clearly points to our urgent need for more activities at all levels to stem against the increasing flood of alien species. This is why biological invasions do matter!

Acknowledgements. My thanks for valuable comments and support on the concept of this book or on individual chapters go to Sven Bacher, Cecily Klingler and Rita Schneider. Also, I gratefully acknowledge the considerable support by the European Union of my studies on invasive alien species within FP 5 (Giant Alien EVK2-CT-2001-00128) and within FP 6 through the projects ALARM (GOCE-CT-2003-506675) and DAISIE (SSPI-CT-2003-511202). Special thanks go to Gerhard Heldmaier (University of Marburg) and to Andrea Schlitzberger (Springer-Verlag, Heidelberg).

References

Carson R (1962) Silent spring. Houghton Mifflin, Boston
DAISIE (2006) Delivering alien invasive inventories for Europe. EU project 6th FP. www.europe-aliens.org
Elton CS (1958) The ecology of invasions by animals and plants. Methuen, London
Meadows DH, Meadows DL, Randers J, Behrens WW III (1972) The limits to growth. University Books, New York
Nentwig W (2005) Humanökologie. Springer, Berlin Heidelberg New York
Perrings C, Williamson M, Barbier EB, Delfino D, Dalmazzone S, Shogren J, Simmons P, Watkinson A (2002) Biological invasion risks and the public good: an economic perspective. Conserv Ecol 6(1):1
Pimentel D (2002) Biological invasions. CRC Press, Boca Raton
Wilson EO (1984) Biophilia. Harvard University Press, Cambridge

Section I

Pathways of Biological Invasions

Short Introduction

Wolfgang Nentwig

The following four chapters present the main pathways for alien species invasions, i.e. they concern the routes by which aliens leave their habitat of origin, how they are distributed, and the ways by which they enter their new, invaded habitat. The subdivision into animals, plants, ships and waterways is admittedly arbitrary but offers useful possibilities to detect parallels and differences between various invasion pathways. Chapters 2 and 3 deal primarily, but not exclusively, with terrestrial species whereas Chaps. 4 and 5 concern the aquatic environment, including both animals and plants (mostly algae).

The drive behind species distributing across biogeographical barriers into new habitats where they assume their role as aliens is based on human activities, by direct migration, by transport of goods or by facilitating dispersal through the elimination of existing barriers. This process probably began already when early man started to conquer the world but it has intensified dramatically over the last few centuries, culminating in the modern phase of globalization.

Many species have been spread deliberately but, for the majority of invaders, spread has occurred accidentally. Parallels may be conspicuous between animals and plants when alien species reach their new habitat as contaminants, hitchhikers and stowaways or when they are transported by humans as ornamentals and pets.

Pathway analysis is a first and important step of curtailing the spread of alien species. Only when the corridors through which species become aliens are known can effective countermeasures be taken. This is most obvious in the examples given in Chaps. 4 and 5 where ballast water, hull fouling, waterways and man-made ocean channels are easily recognizable pathways. By means of international conventions, it is intended that their harmful effect be eliminated or at least reduced.

Short Introduction

2 Pathways in Animal Invasions

Wolfgang Nentwig

2.1 Natural Dispersal Versus more Recent Invasions

Two main ways of dispersal of species can be distinguished: natural dispersal and anthropogenic spread, either indirectly or directly. Natural spread is usually slow and occurs within evolutionary times, it hardly crosses biogeographic borders, and is mostly undirectional. Anthropogenic dispersal is enabled or facilitated directly by human activities. This includes domestication and the worldwide spread of selected species, releases into the wild of suitable game, and escapes from captivity. Humans use animals for nutrition in multiple ways (farming, game, aquaculture and mariculture) and, as humans settle in the world, other species accompany them. More recent motivations to spread species worldwide include the demand for luxury and exotic products (e.g. fur farms), biological control and the pet trade. The main directions of anthropogenic dispersal until the 19th century were from Europe to the European colonies and many other parts of the world. Later, with the increasing independence of numerous countries, with growing world trade, and also with the actual step of globalization, species have been distributed to and from everywhere in the world.

In the past, many introductions occurred intentionally, e.g. as game or as "enrichment" of a new environment, or they were accepted as unavoidable. Looking back today, the lack of even basic ecological knowledge is astonishing and the naive attitude of even scientists is frightening. By contrast, unintentional introductions concern many smaller species such as arthropods, parasites of other species, as contaminants of goods or stowaways on means of transportation.

Ecological Studies, Vol. 193
W. Nentwig (Ed.)
Biological Invasions

2.2 Unintentional Introductions

All means of transport enable alien species to reach new habitats where they may become invasive. Many goods are transported with species as stowaways, and the vehicles themselves may also serve as transportation means. Terrestrial transport can take place in all directions, limited only by the bounds of continental borders. Transport by shipping needs ports and waterways, avian transport is linked to airports. Usually, these three main means of transport are interconnected and, in combination, enable alien species to reach each and every spot on earth within a relatively short time.

2.2.1 Transports

2.2.1.1 Tramps in Vehicles and Planes

Cars, trucks and aircrafts allow fast transportation for many species which would not otherwise survive longer voyages over large distances. Such organisms enter and leave vehicles uncontrolled – much like a tramp – and they are very rapidly transported globally between the main centres of human populations.

Adult mosquitoes can travel by plane from the tropics into the temperate zone. Around airports with frequent connections to tropical countries, these mosquitoes, often vectors of tropical diseases, may infect people. Between 1969 and 1989, there were 87 cases of so-called airport-malaria reported from 12 countries, primarily France, Belgium, Switzerland and the UK – people living in the close vicinity of airports but who have never travelled to the tropics can be infected with malaria. Adult mosquitoes survive only for a short time in the temperate zone and can easily be controlled by insecticidal spraying of the plane's cabin. Eggs, larvae and pupae of mosquitoes, when transported in a small volume of water, can withstand longer transportation times and obviously adapt more easily to temperate climates. The Asian tiger mosquito *Aedes albopictus* was introduced to North and South America, Europe and Australia in used tyres and containers with remnants of rainwater. In its native area, *A. albopictus* is known as a vector of dengue fever and other diseases, in North America it transmits the West Nile virus disease and several other diseases, and in Europe there are concerns for the transmission of a variety of pathogens (Mitchell 1995). It is listed among the 100 world's worst invaders (ISSG 2006).

Another recent tramp species is the Western corn rootworm (*Diabrotica virgifera*), North America's most destructive corn pest, now actually becoming a serious threat to European farmers since it has been displaced to Europe. Within a few years, several obviously independent introductions (at least to

Serbia, northern Italy, Belgium, France and the UK; Miller et al. 2005) have occurred in the vicinity of international airports, from where the beetles have meanwhile spread very rapidly over neighbouring parts of Europe (Fig. 2.1).

Tramp behaviour is typically shown by some subtropical and tropical ants which, in addition, are spread by all kinds of cargo shipment, thus becoming a serious pest worldwide in the last decades. The tiny African pharaoh ant *Monomorium pharaonis* became very abundant in buildings including hospitals, and is now an urban pest which is extremely difficult to control. The Argentine ant *Linepithema humile*, one of the 100 world's worst invaders, has invaded buildings and natural habitats where their huge colonies reduce the native biodiversity considerably.

In a comparable way, many cockroach species have been distributed globally as tramps by vehicles and cargo. Some species of probably tropical or subtropical origin now occur everywhere. *Blatta orientalis*, *Blatella germanica*, *Periplaneta americana*, *P. australasiae* and many more species are serious urban pests very difficult to control, feeding on nearly everything and acting as vectors for many human pathogens.

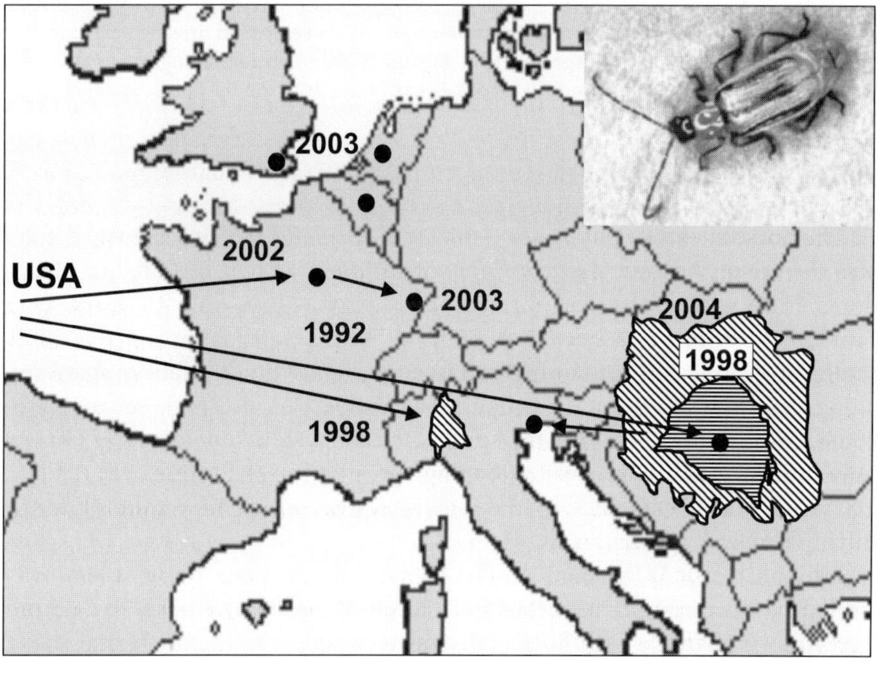

Fig. 2.1 Several independent introductions of the Western corn rootworm, *Diabrotica virgifera*, by plane from the USA to Serbia, north-western Italy, France, Holland, Belgium and England, and subsequent spread within Europe (*hatched area* invaded area; modified from Miller et al. 2005)

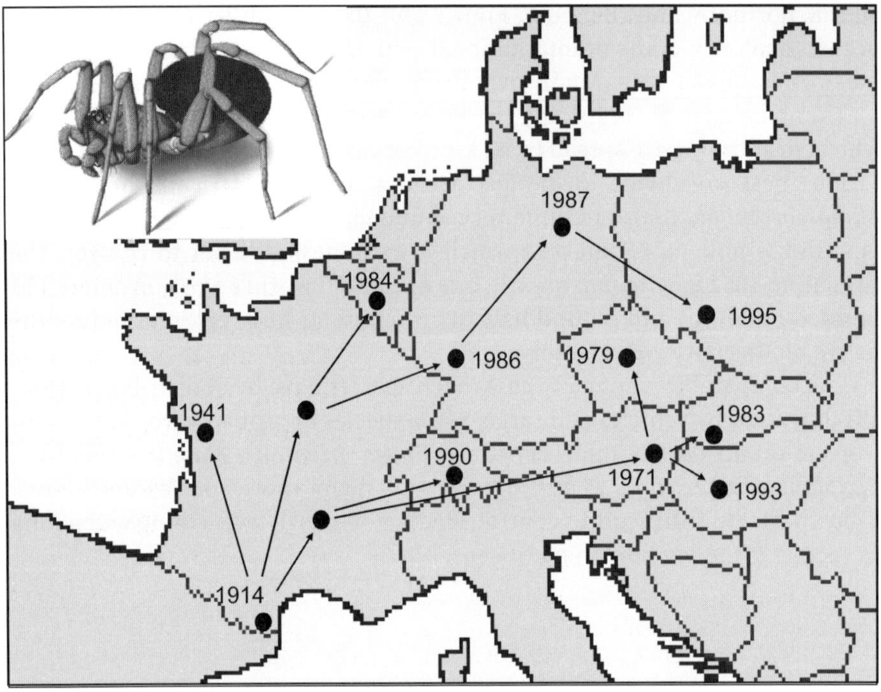

Fig. 2.2 Spread of the spider *Zodarion rubidum* from its restricted area of origin in the eastern French Pyrenees to larger parts of Europe within 80 years along major international railway routes (map modified from Pekar 2002, drawing by Jan Bosselaers, with kind permission)

The horse chestnut leaf miner *Cameraria ohridella*, which seriously defoliates the horse chestnut *Aesculus hippocastanum*, was first detected in Macedonia from where it spread throughout Europe within only 15 years. This rapid distribution has been explained by vehicles transporting the adult moths or dead leaves containing pupae, consistent with distribution patterns along main highways and in urban areas (Gilbert et al. 2004). Pekar (2002) considers that the spread of the spider *Zodarion rubidum*, formerly known only in the French Pyrenees but having spread all over Europe over the last 100 years, can be explained by the railway system which allows the spider to hitchhike over large distances (Fig. 2.2).

Military traffic is "normal" traffic, as mentioned above, but it is less controlled, has its own infrastructure and, therefore, bears more risks. To date, the best-known example with huge consequences concerns the nocturnal snake *Boiga irregularis*. Hiding in containers, it was transported from the Admiralty Islands to Guam and later to other Pacific islands as well. On Guam, *B. irregularis* preyed so intensively on the birds, most of them endemic, that these became extinct or very rare. Today, Guam is an avian desert (Savidge 1987), and *B. irregularis* has been put on the list of the 100 world's worst invaders (ISSG 2006).

2.2.1.2 Waterways and Shipping

For several centuries, humans have connected river systems by canals and cut land bridges to enable shorter shipping routes. Today, particularly Europe has a well-developed waterway system (Chap. 5). These interconnections also offer organisms unique opportunities to cross biogeographical borders by reaching the next drainage system, sea or ocean. The connection of the Caspian Sea and Central Asian waters to Western Europe, the Suez Canal connecting the Red Sea (Indian Ocean) with the Mediterranean Sea (North Atlantic), and the Panama Canal connecting the Atlantic and Pacific oceans are well-known examples (Chap. 5). The Suez Canal enabled hundreds of species to migrate from one sea to the other, often causing considerable ecological damage such as strongly modifying eastern Mediterranean species communities. This phenomenon has been studied in detail and was named the Lessepsian migration (or invasion), in "honour" of the architect of the canal, Ferdinand de Lesseps (Por 1978; Galil 2000; Chap. 5).

Shipping offers unique opportunities for hitchhikers and stowaways to be transported and distributed globally:

1. with cargo: for balance and stability purposes, until the 19th century ships used soil and stones as ballast. In combination with the soil and dirt in the ship itself, and the possibilities which cargo and containers offer in general for stowaways, maritime transport enabled many species to reach far-distant coasts. Notable examples include the brown rat (*Rattus norvegicus*) and the house mouse (*Mus musculus*), also among the 100 world's worst invaders (ISSG 2006) and today globally distributed.
2. as hull fouling: the planktonic larvae of many sessile species regularly colonize the hull of ships and boats. This hull fouling of ships is characterized by crustaceans and bivalves, and may involve more than 100 species (Gollasch 2002) which are transported on a global scale. The Ponto-Caspian zebra mussel *Dreissena polymorpha*, among the 100 world's worst invaders, is one of the best-known examples of an alien species which has invaded Europe and North America via canals and as hull fouling (Chaps. 5 and 15). A *Dreissena* population also contains several parasites, among them, the trematode *Bucephalus polymorphus* which continues its development in several cyprinid species.
3. in ballast water: according to Carlton (1999), about 80 % of the world's commodities is transported aboard a global ship fleet of approximately 35,000 large ships. For more than 100 years, ships have used ballast water for stability purposes. Huge cargo vessels of over 300,000 t carry up to one third of their volume as water when unloaded, amounting to some 12 bil-

lion t of ballast water being transported annually (Chap. 4). This ballast water and its sediment load transport thousands of marine species from uptake to discharge points. This includes virtually any species of plankton or higher organisms, including their planktonic larvae (Chap. 4). Many species have major ecological effects on their new habitats (e.g. the jellyfish *Mnemiopsis leidyi*, one the 100 world's worst invaders, see Chap. 14).

2.2.1.3 Transported Plant Material

The worldwide cultivation of a few important crops has led to a distribution of the main pests (usually insects) of these plants, chiefly through lack of quarantine measures (Chap. 21). In 1874, the Colorado potato beetle (*Leptinotarsa decemlineata*) was transported from North America to France from where it is still spreading to the eastern parts of Eurasia. Many aphid, psyllid, whitefly and coccid species, some dipterans and lepidopteran pests as well as some slugs and snails, originally restricted to smaller areas, now occur in many or most agricultural parts of the world, it often being hardly possible to determine their region of origin.

Harvested plant materials, processed as storable products, are traded worldwide and, consequently, the potential pests of these products are also dispersed globally. Most pest species are insects, especially beetles (primarily Anobiidae, Bostrichidae, Bruchidae, Cucujidae, Curculionidae, Dermestidae, Mycetophagidae, Nitidulidae, Ptinidae, Silvanidae and Tenebrionidae), lepidopterans (e.g. *Plodia interpunctella*, *Ephestia kuehniella* and *Sitatroga cerealella*) and psocids. Though these alien insects are not as spectacular as alien mammals and birds, they represent an enormous number of species (more than 1,000 worldwide), they destroy more than 20 % of the world plant production after harvest, and they cause an economic damage of more than US$ 1.4 trillion per year, amounting to about 5 % of the world GNP (Chap. 18).

Greenhouses are stepping stones for the survival of subtropical and tropical species in otherwise harsh climates. With the transport of plants, soil and equipment, species are moved from and to these greenhouses. Many hemipterans of tropical origin have reached greenhouses all over the world, establishing metapopulation-like colonies and even conquering outdoor habitats when climatic conditions ameliorate (e.g. the aphid *Aphis gossypii*, the whiteflies *Bemisia tabaci* and *Trialeurodes vaporarium*, and the scale *Planococcus citri*). The same is true for thysanopteran species, e.g. *Frankliniella intonsa*, *F. occidentalis* and *Heliothrips haemorroidalis*. Several of these pests are listed among the 100 world's worst invaders (ISSG 2006). Spiders can survive harsh conditions very successfully in the cocoon stage, attached to plants, until more suitable habitats are reached. The pantropical spider *Coleosoma floridanum*, the Mediterranean cribellate spider *Uloborus plumipes*, and several tropical oonopid spider species are widespread in European greenhouses; the Asian giant huntsmen spider *Heteropoda venatoria* is globally distributed (Platnick 2006).

Many xylophagous insects have been dispersed by international movements of timber, timber products and solid wood-packing material. Recent examples include the Japanese scolytid or ambrosia beetle *Xylosandrus germanus* imported to Europe and North America, attacking a variety of native tree species as well as leading to quality loss from direct damage and, in addition, possibly causing tree infections with *Fusarium* fungus. The Asian longhorn beetles *Anoplophora glabripennis* and *A. chinensis* are already major forest pests in Asia. They were introduced to North America and Europe where they attack a wide range of broadleaf tree species. In China, *A. glabripennis* caused the felling of 50 million poplars in one province within 3 years and, in the USA, expensive eradication programs have been organized to prevent the establishment of this serious forest pest.

The transport of soil offers many species an easy way of entering new habitats. A historical example concerns early ship traffic between Europe and the North Atlantic islands (Faroe Islands, Iceland, Greenland) by the Vikings who settled on these islands. In the soil and dirt of their ships (used partly for stability and balance), the Vikings also transported carabid beetles and, today, most carabids of these islands are Palaearctic, rather than Nearctic (Coope 1986). More recently, soil transport has been identified also as one of the major pathways for alien ants. Several soil-dwelling *Trachyzelotes* spiders have been distributed from Europe into the USA, South America and several Pacific islands (Platnick 2006). In the 1960s, the New Zealand flatworm *Arthurdendyus triangulates* was introduced repeatedly on root-balled plants entering the UK where it is now widely spread. The flatworms prey on earthworms and potentially may have an major impact on the soil ecosystem (Cannon et al. 1999).

2.2.2 Escapes

Unintentional escapes occur when species are bred or cultivated outside of their native range under controlled conditions, and escape. This concerns animals from fur farms, pet animals, farm animals, etc. Sometimes, animals may be released deliberately, e.g. to "enrich" the environment. Thus, escapes may be unintentional or intentional (for mammals, see Long 2003).

All over the world, animals have escaped regularly from fur farms and built up feral populations. Escape reasons can be unsafe construction of cages or catastrophic events such as natural disasters or war. Examples include the South American nutria *Myocastor coypus* which escaped in Europe and North America and is now among the 100 world's worst invaders (ISSG 2006), and the East Asian racoon dog *Nyctereutes procyonoides*, spreading now in Europe. Examples of North American mammals escaping from European fur farms comprise the muskrat *Ondathra zibethicus*, the American mink *Mustela vison* and the racoon *Procyon lotor*. The European invasion history of the latter shows that multiple dispersal pathways are involved because the German racoon population has two roots: a deliberate release of four animals for hunting in 1934 (central Germany) and escapes after a racoon farm had been bombed in 1945 (eastern Germany). Additionally, an intentional release by American soldiers in northern France (1966) and farm escapes in Byelorussia and in the Caucasus accelerated further spread in Europe. Today, most racoons in European populations are infected with the racoon roundworm *Baylisascaris procyonis* which causes severe and even fatal encephalitis in a variety of birds and mammals as well as in humans.

The South American nutria (coypu) *M. coypus*, a large semi-aquatic rodent, was bred in fur farms of North America, Europe and Asia. It escaped at numerous locations and built up large feral populations. In addition, intentional releases occurred and, today, the nutria is distributed over all the continents of the world. By their burrowing activity, these rodents damage river banks, dikes and irrigation facilities, associated with high economical follow-up costs; their feeding activity has a destructive effect on the vegetation, e.g. in marshland, reed swamp and other wetlands. Eradication programs were successful in Great Britain (Chap. 22) but failed in most other parts of Europe. The nutria has been classified among of the world's 100 worst invaders.

Closed systems such as fur farms or laboratory animal stations also have the special problem of "liberation" actions by naive, ignorant and/or militant animal lovers. The release of the American mink *Mustela vison* from fur farms in Europe gave rise to an inevitable population spread of this invasive species, with strong negative effects on the European mink *M. lutreola* (replaced in most parts of Europe) and on populations of nesting birds (considerably reduced in numbers; Maran et al. 1998; Nordström et al. 2002). In addition, such actions caused the Aleutian disease to be transmitted to the much more sensitive European mustelid species (Oxenham 1990). It should also be mentioned that fur farmers occasionally released fur animals when market conditions worsened.

2.3 Intentional Introductions

2.3.1 Human Nutrition

2.3.1.1 Global Distribution of Domesticated Animals

As humans spread all over the world, they were always accompanied by their livestock. This became accelerated as the Europeans systematically colonized major parts of the world on all continents, shipping in European horses

(*Equus caballus*), cattle (*Bos taurus*), sheep (*Ovis ammon aries*), asses (*Equus asinus asinus*), goats (*Capra aegagrus hircus*), pigs (*Sus scrofa domestica*) and rabbits (*Oryctolagus cuniculus*) as well as dogs (*Canis lupus familiaris*) and cats (*Felis silvestris catus*). All these species escaped into the new environment or were intentionally released, and goats, pigs, rabbits and cats list now among the 100 world's worst invaders (ISSG 2006). In addition, it was common practice to "vaccinate" many islands with a founding population of domesticated mammals as food for shipwrecked sailors. As a consequence, the actual distribution of these species is a global one (Long 2003).

Domesticated animals always suffer from a variety of parasites and diseases which were then also introduced into the new habitats. The livestock served as a vector for these pests and, in some cases, even enabled the parasites and diseases to spread to wild relatives. In the 1890s and in 1982–1984, rinderpest outbreaks killed major proportions not only of cattle in sub-Saharan Africa but also of wildebeests *Connochaetes taurinus* and buffalos *Syncereos caffer*. Thus, wildlife was not the reservoir of the disease but rather the livestock itself, and consequent vaccination of cattle eliminated the threat to native species (McCallum and Dobson 1995). The domestic pigeon *Columba livia domestica* harbours pathogens and parasites such as the paramyxovirus, *Chlamydophila psittaci*, *Salmonella* sp., *Trichomonas gallinae*, *Eimeria* sp., *Capillaria* sp. and *Ascaridia columbae*. Some of these also infect humans (e.g. psittacosis) and pet birds (Dove et al. 2004).

2.3.1.2 Release of Mammals and Birds for Hunting

Among deer, many species were introduced into most continents as game, especially because of their antlers for trophy hunters. Initially, the deer were usually kept in fenced areas, from which individual animals were subsequently accidentally released, the fences being either not deer-safe or destroyed by natural disasters or wars. In European countries, at least seven non-European species were released: the fallow deer *Dama dama* from the Near East was already introduced during Roman times into most Mediterranean countries and by the Normans into England sometime after 1066. The chital *Axis axis*, sika deer *Cervus nippon*, Chinese water deer *Hydropotes inermis* and muntjacs *Muntiacus reevesi* were imported from Southeast Asia, and the wapiti *Cervus canadensis* and white-tailed deer *Odocoileus virginianus* from North America. The result of this species mix is a worldwide distribution of most and hybridisation of, amongst others, *Cervus* species (Chap. 16). In addition, deer are a reservoir for alien parasites. For example, the sika deer *Cervus nippon* transmitted the Asiatic blood-sucking nematode *Asworthius sidemi* into Europe. Meanwhile, this roundworm has affected 100 % of the Polish population of the European bison *Bison bonasus*, a globally threatened species. Roe deer *Capreolus capreolus*, red deer *Cervus ela-*

phus, cattle and sheep are also susceptible to this parasite (Drozdz et al. 2003).

Rabbits (*Oryctolagus cuniculus*) originate from Spain. In mediaeval times, monks and noblemen spread them to France and other European countries and, in the 12th century, to England. Some domestication occurred but escaping specimens naturalized easily and built up large game populations. British colonists brought rabbits into Australia in 1788 and later into New Zealand, primarily to shoot as game. Rabbits have been released on more than 800 islands throughout the world (Thompson and King 1994). When rabbits became a pest in Australia, the *Myxomatosis* virus was introduced into Australia and other countries to control them. Later, the disease also reduced the native rabbit populations in Spain, France and Italy but, as game compensation, the North American Eastern cottontail *Sylvilagus floridanus* was introduced into these countries.

Probably since Roman times, common pheasants (*Phasianus colchicus*), native to Asia, have been introduced to many European countries for game but could not establish self-maintaining populations in most areas due to unfavourable climatic conditions. Since pheasants are a favourite game bird throughout Europe, millions are bred and released each year. This means that the small naturalized populations of mixed genetic origin are maintained only by large releases, also of different origins. In the UK, for example, 20 million pheasants are released annually, 12 million of which are shot (150 birds km^{-2}), the remaining 8 million birds dying probably due to predator pressure, diseases and insufficient food. Hunting for pheasants is an industry which annually generates over £ 300 million and sustains 26,500 jobs (Tapper 1999).

2.3.1.3 Release of Fish and Other Species

It is a widespread habit for anglers to introduce fish into their waters, with the aim of increasing catch volume or improving the attractiveness of their catch. Since fish survival depends primarily on water quality and environmental structure, many blue white fish never got established – despite enormous stocking, higher yields were not achieved. However, repeated restocking of millions of specimens yielded a mix of species and genetic origins throughout the fishing sites, introduced new species or varieties, and offered access of parasites and disease to hitherto healthy populations.

One of the most prominent cases concerned is that of the rainbow trout *Oncorhynchus mykiss*, native to the Pacific coast river systems of North America. It was first introduced to British rivers in 1874, later to all European river systems, and belongs now to the 100 world's worst invaders (ISSG 2006). It has more tolerance to polluted waters, poor diet, and stress, and has consequently replaced the native brown trout *Salmo trutta* which requires higher-quality waters (Chap. 16), and other native species (Drake and Naiman 2000). Similarly, the North American largemouth brass *Micropterus salmoides* was released throughout the world for fishing (also among the 100 world's worst invaders), and the East European pike-perch (*Sander lucioperca*) now also occurs in Western Europe, including England. The American catfish *Ameiurus melas* and *A. nebulosus* were introduced into Germany in the 19th century for aquaculture, were subsequently found to be not valuable as food but were soon spread all over Europe.

In the 1950s, the Nile perch (*Lates niloticus*), native to North and West African river systems, was released into Lake Victoria and other East African lakes, and soon became of great commercial importance as a food item. Being a strong predator on all organisms in its ecosystem, only 20–30 years later, 60 % of the 300 endemic *Haplochromis* cichlid species were extinct. Today, it is considered as one of the world's 100 worst invasive species (Schofield 1999; Verschuren and Johnson 2002). In the early 1980s, the eel parasite *Anguillicola crassus* was introduced with transports of eel (*Anguilla* sp.) from Southeast Asia to Europe and also to North America. This parasite spread in Europe within two decades, and it affects not only the natural European eel populations but also the aquaculture of eels (Peters and Hartmann 1986).

Several crayfish species from North America were introduced into Europe, with the aim of growing these in farms, e.g. *Pacifastacus leniusculus*, *Procambarus clarkii* and *Orconectes limosus*. Some escaped or were released, and are now widespread in Europe, transferring the crayfish plague, the fungus *Aphanomyces astaci*, to native crayfish species which are much more susceptible and often have become locally extinct. The fungus is one of the 100 world's worst invaders.

Comparably, many mussel and oyster species, distributed worldwide for mariculture, escaped. The Mediterranean mussel *Mytilus galloprovincialis* now replaces native species in South Africa and parts of the United States. Also, frogs such as the North American bullfrog, *Rana catesbeiana*, have been released as harvestable game animals for food. The East African giant snail, *Achatina fulica*, became established all over the tropics because it was transported for food purposes, escaped from gardens, was intentionally released or was moved with agricultural products, equipment, cargo, plant or soil matter. *A. fulica* was also introduced into many areas for its use in medicinal remedies. The latter three species belong to the list of the 100 world's worst invaders (ISSG 2006).

2.3.2 Beneficials or Biological Control Agents

2.3.2.1 Vertebrates

Using vertebrates as biological control agents is associated mainly with the dark and pre-scientific period of biological control. All the foxes and weasels, dogs, cats and toads which were released to control pest species soon became pests themselves because they predated on everything, except the target pest. One of many typical examples concerns the ermine *Mustela erminea*, now also ranked among the 100 world's worst invaders (ISSG 2006), native to the Holarctic region north of the 40th parallel. It was introduced into New Zealand and into some small European islands because it was believed that it could control rabbits. As a predator specialised in small mammals and birds, in New Zealand *M. erminea* preys upon a variety of native species, particularly kiwi chicks and hole-nesting forest birds. Ermines spread easily over long distances on land and also reach small off-shore islands unaided.

Vertebrates have also been used as herbivores. The nutria *Myocastor copyus* has been introduced into Texas as a "cure-all" for ponds with dense vegetation. It reduces many kinds of aquatic plants but, when nutrias get established, overpopulation soon results and the animals move into places where they destroy vegetation which is particularly valuable, e.g. for waterfowl. In 1963, the grass carp or white amure *Ctenopharyngodon idella* was imported into the USA from Malaysia as a biological control of aquatic macrophytes. It was introduced into selected ponds and rivers but escaped. Since grass carps are capable of moving well beyond areas intended for plant control within a single season, within a few decades they had spread throughout the USA (Guillory and Gasaway 1978). Grass carps were also widely introduced into Europe and Central Asia. *C. idella* consumes all types of aquatic plants, including reeds, reed sweet-grass, reed-mace, sedges, bulrushes and horsetail, thus destroying essentially all aquatic vegetation. This has also indirectly reduced native invertebrate and fish populations to extinction (Maeceina et al. 1992; Bain 1993; Crivelli 1995). Shipments of grass carps were not always controlled, so that they also contained foreign species, such as the stone moroko (*Pseudorasbora parva*), now widely spread in Europe. In addition, grass carps are hosts of the Asian tapeworm *Bothriocephalus opsarichthydis* which became established in North America, Europe, Asia and New Zealand. It now parasites indigenous fish, has the potential to reduce local biodiversity, and causes considerable economic harm (Köting 1974).

2.3.2.2 Invertebrates

Invertebrates, primarily insects, are the most common agents to be released within a biocontrol project against a target weed or an insect pest. Nowadays, these projects are very carefully performed and agents are tested intensively prior to release (for the pros and cons of biocontrol, see Chap. 23). Nevertheless, these releases add species to an environment where they previously did not occur. A catalogue by Julien and Griffiths (1998) lists the agents released for the control of weeds: 201 beetle species, 132 lepidopteran species, 58 dipteran species, 55 hemipteran species and 37 other organisms. There is no such compilation for released agents against insect pests but a few 100 hymenopteran species can undoubtedly be added to the above list. A rough estimate of 1,000 invertebrate species released worldwide for biocontrol purposes should certainly not be regarded as being too exaggerated.

2.3.3 Ornamental Animals and Pets

For different reasons, ever-increasing numbers of animals are traded as ornamentals and pets. Whereas the worldwide trade of 3,000 species of endangered mammals, birds and reptiles is prohibitively regulated by the appendix lists of CITES since 1975, other vertebrate taxa and most invertebrates are excluded. Especially non-endangered species still have hardly any trade restrictions in most countries. This leads to the paradox situation that many invasive species or potentially invasive species can easily be traded. It should not be ignored that virtually all species in the pet trade occasionally may become released, intentionally or unintentionally, may build up a population, and may cause problems.

The introduction of ornamental wildfowl was once commonplace, and many species have been able to escape. Probably the most famous example concerns the ruddy duck *Oxyura jamaicensis*. It escaped in England from a wildfowl facilities in 1949, invaded the European continent and produced fertile hybrids with the white-headed duck *Oxyura leucocephala*, thereby strongly reducing the number of pure *O. leucocephala*. Formerly widespread in the Mediterranean area, the European populations of *O. leucocephala* are now close to extinction. The most obvious way of safeguarding the European *O. leucocephala* is a rigorous shooting of *O. jamaicensis* in Europe (Hughes et al. 1999). Other examples of escaping waterfowl include the East Asian Mandarin duck *Aix galericulata*, the ruddy shelduck *Tadorna ferruginea*, and the Canada goose *Branta canadensis* which became the most successful avian invader in many regions in Europe.

The import and trade of songbirds are problematical because individuals escape continuously. The more anthropophilic the species are, the more often they are introduced and released (e.g. sparrows, starlings, parakeets), leading to a higher chance to establish in the wild. When the trade of some species was prohibited, many held as pets were illegally released, e.g. the Californian house finch, the most destructive bird pest in California, when first offered on sale in local New York shops (Williamson 1996). On all continents, several parrot and parakeet species have escaped or were released. They survive primarily in city parks with sufficient food resources such as shrubs and trees with fruits and seeds – as well as bird feeders. The Afro-Indian ring-necked parakeet (*Psittacula krameri*) and the South American monk parakeet (*Myiopsitta monachus*) have established large breeding colonies in many urban areas of North America, Europe, Africa and Asia (Lever 1987). Additionally, many introduced birds carry avian malaria and birdpox which usually infect local bird species much more strongly. Some species even went extinct – e.g. several Hawaiian birds. Some infection routes were facilitated by the earlier import of mosquitoes such as *Culex* or *Aedes* species, which actually happens worldwide (Williamson 1996).

Between 1876 and 1929, at least 30 separate introductions to parks and estates in the UK from the pet stock of the North American grey squirrel *Neosciurus carolinensis* established it countrywide, and it soon replaced the native red squirrel *Sciurus vulgaris*. An analogous situation is predicted for Italy (Bertolino and Genovesi 2003) where, independently from the British populations, grey squirrels have spread after having been released at several locations. Additionally, the grey squirrel transmits diseases such as the parapoxvirus to the red squirrel – the latter succumbs whereas the former is not susceptible to this virus (Tompkins et al. 2003). This pathogen transfer may even be the main reason for the replacement of the native red squirrel. Comparably, the Siberian chipmunk, *Tamias sibiricus*, has been released at several locations from The Netherlands into Italy where it has established growing populations.

Many pets cause problems when they are no longer desired by their owners. They may have become too large (e.g. sliders, snakes, alligators), too numerous (e.g. goldfish), too dangerous or expensive, or simply too bothersome, e.g. during the holiday season! Such surplus animals are often simply released into the wild and may establish self-maintaining populations. One notable example is the North American bullfrog *Rana catesbeiana*, released at several locations in Europe. Its large size, high mobility, generalist eating habits, function as a disease vector to other amphibians, and huge reproductive capabilities make this species an extremely successful invader and a major threat to biodiversity (Werner at al. 1995; Daszak et al. 1999).

The red-eared slider, *Trachemys scripta elegans*, now among the 100 world's worst invaders (ISSG 2006), and other species of aquatic turtles are frequently released in high numbers into European waters where they compete with the rare native European pond terrapins (*Emys orbicularis*) for food and sun-basking places. Usually, the alien species have a competitive advantage over the European pond turtle due to their lower age at maturity, higher fecundity, larger adult body size, and more aggressive behaviour (Cady and Joly 2004).

The goldfish (*Carassius auratus*) and related species, native to Central and Eastern Asia, are favourite fish for aquarium and garden ponds, and have reached a global distribution. They regularly escape or are released, have a high reproduction rate, and today populate most European waters. As general predators with highly flexible behaviour, they reduce biodiversity and change ecosystem structure (Lehtonen 2002). Also North American pumpkinseed species (*Lepomis auritius, L. gibbosus*) have been released into European waters because of their colourful appearance and their attractive breeding behaviour.

2.4 Conclusions

The main conclusions of this short presentation focus on the high variety of pathways for the introduction of alien species. All pathways are linked to or caused by human activities and, for many species, several pathways or combinations of pathways are realised. Many introductions show astonishing historical roots but globalization is a main driver of the actual acceleration of invasion processes. When reading the wealth of information dealing with the introduction of invasive alien species, one can argue that basic ecological knowledge or, at least, common sense must have been weakly developed at one point in time, and one would expect that the negative experience of the past can largely be avoided today. However, there are present invasion events which are even worse than those from past experience. The overall conclusion, therefore, can only be that we need much tighter control, stricter quarantine measures and more caution when shipping species and goods around the world. These control costs are obviously well-invested money to prevent much higher ecological and economical follow-up costs.

Acknowledgements. I would like to thank Cecily Klingler for valuable comments on this chapter. My studies on invasive alien species have been considerably supported by funding of the European Union within the FP 6 integrated project ALARM (GOCE-CT-2003-506675) and by the specific targeted research project DAISIE (SSPI-CT-2003-511202), which is gratefully acknowledged.

References

Bain MB (1993) Assessing impacts of introduced aquatic species: grass carp in large systems. Environ Manage 17:211–224

Bertolino S, Genovesi P (2003) Spread and attempted eradication of the grey squirrel (*Sciurus carolinensis*) in Italy, and consequences for the red squirrel (*Sciurus vulgaris*) in Eurasia. Biol Conserv 109:351–358

Cady A, Joly P (2004) Impact of the introduction of the red-eared slider (*Trachemys scripta elegans*) on survival rates of the European pond turtle (*Emys orbicularis*). Biodiversity Conserv 13:2511–2518

Cannon RJC, Baker RHA, Taylor M, Moore JP (1999) A review of the pest status of the New Zealand flatworm in the U.K. Ann Appl Biol 135:597–614

Carlton JT (1999) The scale and ecological consequences of biological invasions in the world's oceans. In: Sandlund OT, Schei PJ, Viken A (eds) Invasive species and biodiversity management. Kluwer, Dordrecht, pp 195–212

CITES (2006) Convention on international trade in endangered species of wild fauna and flora. www.cites.org

Coope GR (1986) The invasion and colonization of the North Atlantic islands: a palaeoecological solution to a biogeographic problem. Philos Trans R Soc B 314:619–635

Crivelli AJ (1995) Are fish introductions a threat to endemic freshwater fishes in the northern Mediterranean region? Biol Conserv 72:311–319

Daszak P, Berger L, Cunningham AA, Hyatt AD, Earl Green D, Speare R (1999) Emerging infectious diseases and amphibian population declines. Emerging Infect Diseases 5:735–748

Dove A, Zorman-Rojs O, Vergles Rataj A, Bole-Hribovsek; Krape U; Dobeic M (2004) Health status of free-living pigeons (*Columba livia domestica*) in the city of Ljubljana. Acta Vet Hung 52:219–216

Drake CD, Naiman JR (2000) An evaluation of restoration efforts in fishless lakes stocked with exotic trout. Conserv Biol 14:1807–1820

Drozdz J, Demiaszkiewicz AW, Lachowicz J (2003) Expansion of the Asiatic parasite *Ashworthius sidemi* (Nematoda, Trichostrongylidae) in wild ruminants in Polish territory. Parasitol Res 89:94–97

Galil B (2000) A sea under siege – Alien species in the Mediterranean. Biol Invasions 2:177–186

Gilbert M, Grégoire JC, Freise JF, Heitland W (2004) Long-distance dispersal and human population density allow the prediction of invasion patterns in the horse chestnut leafminer *Cameraria ohridella*. J Anim Ecol 73:459–468

Gollasch S (2002) The importance of ship fouling as a vector of species introduction into the North Sea. Biofouling 18:105–121

Guillory V, Gasaway RD (1978) Zoogeography of the grass carp in the United States. Trans Am Fish Soc 107:105–112

Hughes B, Kirby J, Rowcliffe JM (1999) Waterbird conflicts in Britain & Ireland: Ruddy ducks, Canada geese and cormorants. Wildfowl 50:70–99

ISSG (2006) 100 of the world's worst invasive alien species. Invasive species specialist group. http://www.issg.org

Julien MH, Griffiths MW (1998) Biological control of weeds. A world catalogue of agents and their target weeds. CABI, Wallingford

Köting W (1974) Bothriocephalosis of the carp. Vet Med Rev 2:165–171

Lehtonen H (2002) Alien freshwater fishes of Europe. In: Leppaekoski E, Gollasch S, Olenin S (eds) Invasive aquatic species in Europe. Kluwer, Dordrecht, pp 153–161

Lever C (1987) Naturalized birds of the world. Longman, London

Long JL (2003) Introduced mammals of the world. Their history, distribution and influence. CABI, Wallingford
Maeceina MJ, Cichra MF, Betsill RK, Bettoli PW (1992) Limnological changes in a large reservoir following vegetation removal by grass carp. J Freshwater Ecol 7:81–95
Maran T, Macdonald DW, Kruuk H, Sidorovich V, Rozhnov VV (1998) The continuing decline of the European mink *Mustela lutreola*: evidence for the intraguild aggression hypothesis. In: Dunstone N, Gorman M (eds) Behaviour and ecology of riparian mammals. Cambridge University Press, Cambridge, Symp Zoological Society of London 71, pp 297–323
McCallum H, Dobson A (1995) Detecting disease and parasite threats to endangered species and ecosystems. Trends Ecol Evol 10:190–194
Miller N, Estoup A, Toepfer S, Bourguet D, Lapchin L, Derridj S, Seok Kim K, Reynaud P, Furlan L, Guillemaud T (2005) Multiple transatlantic introductions of the Western corn rootworm. Science 310:992
Mitchell CJ (1995) Geographic spread of *Aedes albopictus* and potential for involvement in arbovirus cycles in the Mediterranean basin. J Vector Ecol 20:44–58
Nordström M, Högmander J, Nummelin J, Laine J, Laanetu N, Korpimäki E (2002) Variable responses of waterfowl breeding populations to long-term removal of introduced American mink. Ecography 25:385–394
Oxenham M (1990) Aleutian disease in the ferret. Vet Rec 126:585
Pekar S (2002) *Zodarion rubidum* Simon 1914: railroad riders? Newslett Br Arachnol Soc 95:11–12
Peters G, Hartmann F (1986) *Anguillicola*, a parasitic nematode of the swim bladder spreading among eel populations in Europe. Diseases Aquat Organisms 1:229–230
Platnick NI (2006) The world spider catalogue. http://research.amnh.org/entomology/spiders/catalog/
Por FD (1978) Lessepsian migration: the influx of Red Sea biota into the Mediterranean by way of the Suez Canal. Ecological Studies 23. Springer, Berlin Heidelberg New York
Savidge JA (1987) Extinction of an island forest avifauna by an introduced snake. Ecology 68:660–668
Schofield P (1999) Interactions between Nile perch, *Lates niloticus*, and other fishes in Lake Nabugabo, Uganda. Environ Biol Fish 55:343–358
Tapper S (ed) (1999) A question of balance: game animals and their role in the British countryside. The Game Conservancy Trust, Hampshire
Thompson HV, King CM (eds) (1994) The European rabbit, the history and biology of a successful colonizer. Oxford University Press, Oxford
Tompkins DM, White A, Boots M (2003) Ecological replacement of native red squirrels by invasive greys driven by disease. Ecol Lett 6:189–196
Verschuren D, Johnson T (2002) History and timing of human impact on Lake Victoria, East Africa. Proc R Soc Biol Sci Ser B 269:289–294
Werner EE, Wellborn GA, McPeek MA (1995) Diet composition in postmetamorphic bullfrogs and green frogs: implications for interspecific predation and competition. J Herpetol 29:600–607
Williamson M (1996) Biological invasions. Chapman & Hall, London

3 Pathways in Plant Invasions

INGO KOWARIK and MORITZ VON DER LIPPE

3.1 Introduction

At least at a global scale, species transfer through human agency is much more frequent, efficient and effective than through natural mechanisms and has no parallel in evolutionary history (Elton 1958; Mack et al. 2000). As propagule pressure is one of the most powerful bottlenecks in invasions (Williamson 1996), human-mediated dispersal is a key process in the range expansion of non-native plant species.

Due to the role of biological invasions as a major threat to biodiversity, recent research has aimed at identifying pathways in invasions which can be regulated to prevent or, at least, curb negative impacts of non-native species (Carlton and Ruiz 2005). Information on the functioning and effectiveness of different pathways is therefore necessary to set priorities in regulation or management (Mack 2003).

In literature on human-mediated plant dispersal, "pathway" is used in two ways: functionally, to describe why and how species are moved by human-mediated agency and geographically, to describe explicit parts of landscapes where dispersal proceeds. Consequently, Carlton and Ruiz (2005) aimed at a more detailed analysis of pathways and proposed to analyse "causes" as the human motivation for introducing species, "vectors" as physical means or agents by which a species is transported, and "routes" and "corridors" as geographic paths over which a species is transferred.

Pathways of plant dispersal vary conspicuously with time (Poschlod and Bonn 1998; Mack and Lonsdale 2001) and at different spatial scales (Pyšek and Hulme 2005; Pauchard and Shea 2006), and so does the underlying human motivation. We illustrate here the relative importance of different pathways in the accidental and deliberate transfer of species in space and time. In order to do so, we describe the usefulness of differentiating first between processes leading to the introduction of a species to a new range and those which subsequently provide secondary releases of the species within its

Ecological Studies, Vol. 193
W. Nentwig (Ed.)
Biological Invasions

new range. Then, we argue for a detailed analysis of propagule transport and release as two sub-processes of dispersal. Both may be driven by human agency with or without intention for long periods after initial introduction. We thus emphasise the need to analyse "causes" of plant dispersal far beyond the reasons for introducing a species to a new range.

3.2 Introductions to a New Range: Relative Role of Deliberate Versus Accidental Transfer of Species

Since the first human migrations and the beginning of agriculture and keeping of livestock, humans have purposefully or accidentally transferred plants whenever they have moved themselves, together with animals, seeds and other goods (Hodkinson and Thompson 1997). If this occurs at regional to continental scales, then such transfers result in species introductions, defined as a range expansion of a species through human agency by overcoming major geographical barriers (Richardson et al. 2000).

Both deliberate and accidental plant introductions have long been associated to human migrations. Palaeobotanical studies have revealed a significant influx of new species to Central Europe since the Neolithic period, which evidently increased during the Roman period (Willerding 1986). Most early introductions occurred accidentally, probably as impurities in crop seeds or by attachment to animals. Consistently, accidental introductions dominate pre-1500 introductions to Central Europe, as shown by the Czech flora (Table 3.1).

Table 3.1 Role of deliberately and accidentally introduced species in the non-native Czech flora (modified from Pyšek et al. 2002)

Introduction period	Status	Number of species	Mode of introduction Deliberate	Both	Accidental
Archaeophytes (pre-1500)	Casual	74	30	4	40
	Naturalised	237	17	25	195
	Invasive	21	2	4	15
	Total	332	49	33	250
Neophytes (post-1500)	Casual	817	400	47	370
	Naturalised	160	94	18	48
	Invasive	69	45	4	20
	Total	1,046	539	69	438
Total	Casual	891	430	51	410
	Naturalised	397	111	43	243
	Invasive	90	47	8	35
	Total	1,378	588	102	688

The Romans transferred a broad array of cereals and other useful plants within their empire (Franz 1984). *Castanea sativa* may have started spreading at this time, but its dispersal has been promoted mostly by further cultivation since medieval times (Conedera et al. 2004). Charlemagne's decree *Capitulare de villis*, in 812 AD, enhanced the use of introduced, mainly Mediterranean plants, far beyond the northern limits of their ranges (Franz 1984).

In the post-Columbian period, the number and efficiency of deliberate introductions greatly increased due to the magnitude of intercontinental transfer of organisms (Crosby 1986). The temporal sequence of introductions from different donor areas to Central Europe echoes the history of voyages and discoveries, as illustrated by woody plant introductions to Central Europe (Fig. 3.1a). At a regional scale, the sequence of invasions follows the pattern of infra- and intercontinental introductions (Fig. 3.1b), with an exceedingly variable lag phase which averaged 170 years for trees and 131 for shrubs (Kowarik 1995). As global commerce grows, the frequency of introductions of ornamentals to new areas will continue to increase. A huge potential exists, for example, for importing thousands of previously unavailable Chinese species to the United States (Mack 2001).

Ornamentation is the predominant purpose for introducing plant species, as shown by American plant introductions to Europe (Forman 2003) or the non-native species of the Czech flora which were introduced deliberately: 74 % were ornamentals, 22 % nutritional plants, 14 % medicinal plants, 11 % fodder plants, 6 % were used for landscaping, 5 % as bee plants, and all less than 2 % each as forest crops or oil, dye and fibre plants (Pyšek et al. 2002).

In recent decades, the relative importance of intentional introductions has increased at the infra- and intercontinental scale, as advances in seed cleaning and quarantine measures reduced the efficiency of accidental pathways (Mack and Lonsdale 2001).

3.2.1 Introduction Mode and Invasion Success

Global transfer of goods and people has also fostered the unintentional transport of "hitchhiking" plants. As was illustrated early by the seminal *Flore adventice de Montpellier* (Thellung 1912), accidental introductions may dominate the regional non-native species pool but are clearly less important among the naturalised species (Table 3.2). This also holds at a greater spatial-temporal scale, as for post-1500 introductions to the Czech Republic (Table 3.1) and especially for most recent introductions. Among those now-naturalised species first recorded in Australia between 1971 and 1995, only 2 % are known to be accidental introductions (Mack and Lonsdale 2001).

Both accidental and deliberate introductions contribute to the group of non-native species with detrimental effects. At a landscape scale, the importance of the introduction mode varies strongly. About one-half of the 50

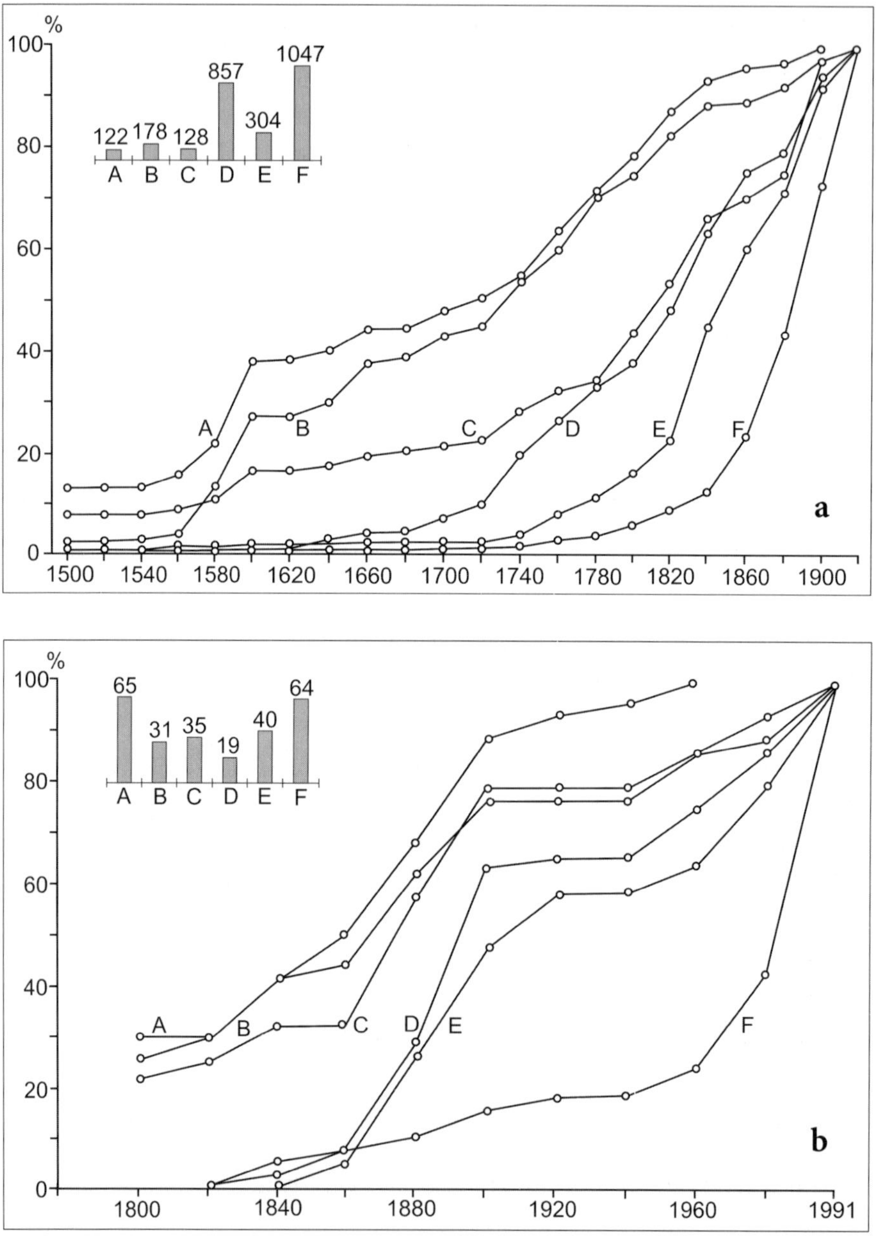

Fig. 3.1a, b Temporal sequence of **a** woody plant introductions from different donor areas to Central Europe and of **b** successive invasions by species from the same donor areas in Brandenburg, Germany, reflecting the history of infra- and intercontinental introductions at a regional scale. The *inserted columns* in **a** show absolute numbers of introductions from *A* parts of Europe excluding the Mediterranean, *B* the Mediterranean, *C* western Asia, *D* North America, *E* Central Asia and *F* East Asia. The *cumulative curves* illustrate the relative importance of introductions from different donor areas for

Table 3.2 Invasion success of accidentally versus deliberately introduced alien plant species to the region of Montpellier (southern France), expressed by the rate of naturalisation of alien species (data from Thellung 1912)

Pathway of introduction	Total of alien species Number	Subset of casual species Number	%	Subset of naturalised species Number	%
Total species number	800	693	86.6	107	13.4
a) Deliberate introductions	148	87	58.8	61	41.2
b) Accidental introductions with	621	575	92.6	46	7.4
Wool	526	507	96.4	19	3.6
Seed and feed grains	40	31	77.5	9	22.5
Grain crops	18	18	100.0	0	0
Ballast materials	19	10	52.6	9	47.4
Transportation vehicles	18	9	50.0	9	50.0
c) Resulting from hybridisation	31	31	100.0	0	0

species which are classified as noxious in Germany were introduced accidentally, but all of these are virtually confined to arable fields. Almost all non-agricultural conflicts due to invasive plant species result from deliberate introductions (Kowarik 2003). Except for agricultural weeds, the majority of invasive species have evidently been introduced on purpose, and most of these as ornamentals (Reichard and White 2001; Kowarik 2005). In dealing with conservation issues, deliberate introduction is thus the most efficient driver in plant invasions for most terrestrial ecosystems. In marine ecosystems, however, accidentally introduced plants prevail, due mainly to aquaculture and ballast water which serve as effective pathways for species transfer (Chap. 4).

3.2.2 Coinciding Pathways of Deliberate and Accidental Introductions

The example of the seed trade illustrates how deliberate and accidental pathways may coincide, thereby increasing vector efficiency. Deliberate transfer of seeds has led to an efficient infra- and intercontinental exchange of arable

20-year time spans over a period of 400 years (data from Kowarik 1992). The *inserted columns* in **b** show absolute numbers of spontaneously occurring non-native woody species in the region of Brandenburg from the same donor areas (*A–F*) as shown in **a**. The *cumulative curves* illustrate the relative importance of spontaneously emerging species from different donor areas for 20-year time spans over a period of 200 years, calculated from data in regional floras (100 %=210 species; species occurring in more than one donor area were calculated repeatedly; data from Kowarik 1992) crops, grasses and legumes. Crops have always been potential weeds, and vice versa (Gressel 2005). De-domestication by endo- or exoferality evidently functions as a prerequisite of potential invasions by many crop species (Gressel 2005). Among the deliberately sown species, mainly perennials have started to spread, such as African grass species which were previously used to establish pastures in tropical America (Williams and Baruch 2000).

Since early times, the deliberate transfer of crop seeds has provided a powerful pathway for accidental introductions through seed impurities. The Roman period, for example, provided a significant influx of Mediterranean species to the flora of arable fields in occupied territories north of the Alps (Willerding 1986). In the 19th century, transport routes of crop seeds could be precisely reconstituted based on the native area of associated species (Thellung 1915). The commercialisation of the seed trade undoubtedly enhanced the spread of associated weeds (Mack 1991). Prior to the establishment of efficient seed-cleaning procedures during the last century, 6 billion seeds per year are believed to have been sown accidentally with clover and grass seeds in Great Britain (Salisbury 1953). On arable fields, the recent sharp decline of many species dispersed with crop seeds indirectly highlights the key role of seed transfer in accidental introductions (Kornas 1988).

Interestingly, crop seeds were often deliberately contaminated with seeds of other species to boost sales profits falsely by increasing harvest mass or declaring a higher-rated origin for the goods, e.g. by adding the North American *Ambrosia artemisiifolia* to seeds of other provenances (Nobbe 1876). Today, *Ambrosia*, which had been also introduced as a true American seed import, is a noxious species in Europe due to its allergenic effects (Chauvel et al. 2006).

The distribution of garden plants or forest crops also integrates deliberate and accidental pathways of species transfer, as weeds often co-occur with crops in the soil (Prach et al. 1995; Hodkinson and Thompson 1997). The highly invasive *Chondrilla juncea* was initially introduced to Australia attached to imported vine-stocks (Mack and Lonsdale 2001).

3.2.3 Invasions at the Infra-Specific Level Through Deliberate Introductions

Reintroductions of native species frequently occur in grasses and leguminoses as well as in woody species because their seeds can often be produced at a lower cost abroad. As introduced provenances of native species, especially cultivars, differ genetically from regional populations, their transfer probably provides highly effective, albeit clandestine pathways for invasions at the infra-specific level. In Spain, introduced provenances of *Dactylis glomerata* spread from sown to natural grassland and affected endemic taxa of the same species by competition and hybridisation (Lumaret 1990). Many tree nurseries prefer to use introduced seeds to produce native species. Seeds of the hazelnut (*Corylus avellana*), for example, are mainly imported from Italy and Turkey to Germany (Spethmann 1995). Such introductions are believed to affect regional genetic diversity, and they are economically relevant due to possible decreases in frost tolerance, resistance against pathogens, and rates of establishment and growth (Jones et al. 2001).

3.3 Deliberate Secondary Releases Within the New Range

Human-mediated introductions to a new area are a prerequisite for but no guarantee of the subsequent establishment and spread of species at regional and landscape scales. Inter- or infra-continental transport routes between donor and recipient areas mostly have botanical gardens, tree nurseries or seed companies as introduction foci. From here, invasions rarely occur directly, in contrast to that of *Matricaria discoidea* which spread from the botanical garden in Berlin (Sukopp 1972).

Evidently, most invasions by deliberately introduced species spread from invasion foci which were set by secondary releases during decades and centuries subsequent to the initial introduction of a species to a new range (Kowarik 2003). Horticulture and seed companies function mainly as interfaces between continental and regional scales by determining species availability on the market. By this, they indirectly enhance the further regional functions of plantations as invasion foci.

Socioeconomic factors, such as varying market routes or prices through time, influence customers' choice of a species for long periods after an initial introduction to a region. For example, in a sample of 534 ornamentals, species which had escaped cultivation had been for sale more frequently both in the 19th century and today than was the case for non-escaping species (Dehnen-Schmutz et al. 2006). The economically motivated use of plant species may change conspicuously over time, as the fate of the North American black cherry (*Prunus serotina*) in Europe for the last 350 years exemplifies. A switch from a rare ornamental to a forest crop in the 19th century, and broad usage for other reasons during the following century resulted in manifold invasions and finally in control (Starfinger et al. 2003). Similarly, the increasing popularity of *Rhododendron ponticum* in the British Isles facilitated its subsequent spreading (Dehnen-Schmutz and Williamson 2006).

Forestry is a good example of the changing importance of pathways to invasion in time and space. At the end of the 18th century, hundreds of species were introduced to Europe, tree nurseries and experimental forest plantations being the main targets of intercontinental transport routes. About 500 non-native species were listed in 1787 for a single plantation near Berlin (Kowarik 1992). Of these, only those which were planted afterwards in large quantities at the landscape scale (*Robinia pseudoacacia*, *Quercus rubra*, *Pinus strobus*, *Pseudotsuga menziesii*; Kowarik 2003) became important invaders. Through forestry, in recent decades at least 100 million ha of plantations with non-native species (84 % conifers) have been established in the southern hemisphere (Zobel et al. 1987). Although relatively few in terms of species numbers, invasions by conifers have led to far-reaching ecological and economic consequences (Richardson and Rejmánek 2004).

3.3.1 Cultivation as a Driver in Plant Invasions

Cultivation provides a powerful pathway for subsequent plant invasions at regional and landscape scales by enhancing the establishment of founder populations (Mack 2000) and, even beyond the threshold of naturalisation, species range expansion by bridging adequate but spatially isolated sites (Kowarik 2003).

The maintenance of cultivated individuals by humans may function in protecting non-native plant populations from detrimental environmental effects which may otherwise prevent establishment and further spread (Mack 2000). The vector strength of cultivation is illustrated by the fact that 25 % of 328 cultivated non-native woody species emerged spontaneously in Hamburg's residential areas, as reported by Kowarik (2005). Cultivating species in large quantities provides a high propagule pressure, known to be a decisive driver in plant invasions (Williamson 1996). Several studies demonstrate a close correlation between the quantity of cultivation and subsequent invasion events, for example, for *Eucalyptus* species introduced to South Africa (Rejmánek et al. 2005).

Through regional transfer, deliberate secondary releases may create myriads of potential invasion foci and thereby bridge, often repeatedly, spatial or environmental barriers. Except for agricultural weeds, all problematic invaders in Germany have been frequently dispersed on purpose by a broad array of pathways of secondary releases, which often provide a shift from urban to semi-natural and natural habitats (Kowarik 2003). Deliberate releases also facilitate invasions by amplifying propagule exposure to natural, or other human-mediated, dispersal vectors at the landscape scale. Vertebrates, for example, disperse 50 % of naturalised plant species in Australia (Rejmánek et al. 2005).

In northern Germany, the vector strength of secondary releases was assessed at the landscape scale by analysing the origin of problematic plant populations. The invasion foci of 63–76 % of more than 100 populations could be directly traced back to deliberate releases. Virtually all populations of *Prunus serotina* descended from local forestry plantations. In *Fallopia* species, plantings (20 %), and deposition of garden waste (29 %) and soil (20 %) led to the establishment of over two-thirds of all problematic populations. In *Heracleum mantegazzianum*, plantings (9 %), sowings by beekeepers (20 %), and deposition of propagules with garden waste (18 %) or soil (4 %) provided key pathways for spreading (Schepker 1998; Kowarik 2003).

Dumping waste into rivers may effectively induce further spread of aquarium plants. Anthropogenically increased water temperature may even enhance tropical species in temperate regions (Hussner and Lösch 2005). For both water and terrestrial plants, deliberate releases to "enrich" nature have been quite effective. Invasion of *Elodea canadensis* started soon after 1859 when the species was released into some lakes near Berlin and, for the small region of Frankonia, 75 species are known to have been planted even at natural sites by amateur botanists (Kowarik 2003).

3.3.2 From Clumped to Linear Patterns

As Pauchard and Shea (2006) state, propagule movement tends, at a regional scale, to follow landscape corridors such as rivers or roads. Secondary releases may overlay the resulting linear patterns, as they lead mostly to clumped releases of species. These result initially in clumped populations adjacent to the site of release, which may persist for decades and centuries as indicators of earlier horticulture (Kowarik 2005). Since the end of the 16th century, *Tulipa sylvestris* has been used as an ornamental north of the Alps. In northwest Germany, about 50 % of all populations and 72 % of populations with more than 10,000 individuals are confined to historical gardens and similar sites of early horticulture. When exposed to rivers, a shift to linear dispersal occurred only rarely (Kowarik and Wohlgemuth 2006). Such shifts to long-distance dispersal frequently occur in many successful invaders and may efficiently overlay former release patterns, for example, in *Heracleum mantegazzianum*, *Impatiens glandulifera* and *Fallopia* species in Europe (Pyšek and Prach 1993). In the United States, more than 370,000 ha of *Tamarix* stands result from plantations along rivers and subsequent dispersal by moving water (Pauchard and Shea 2006).

3.4 Accidental Transfer of Non-Target Species

Among the huge diversity of human-mediated modes of accidental transport of species (Thellung 1912; Ridley 1930; Bonn and Poschlod 1998), two principal ways can be distinguished: (1) the transport by direct association of propagules to a conveyer, such as the attachment of seeds to cars by mud, and (2) the transport of propagules associated with goods which are moved by one or more conveyers. Deliberate and accidental transfer of species may coincide, as illustrated by the pathway of crop seed transfer described above.

Where and how quickly vehicles and ships – or people and animals, as living conveyers, move depend on the method and route of transport. Transport efficiency expressed in terms of number, velocity and spatial reach of moved propagules (Carlton and Ruiz 2005) is not necessarily equivalent to vector efficiency. Transport vectors differ noticeably in the way propagules are released during or after transport. We thus emphasise the role of release processes, which can clearly determine the efficiency of a dispersal pathway and resulting invasion patterns.

3.4.1 Transfer by Goods: Spatial-Temporal Separation of Propagule Transport and Release

In deliberate transfers of target species, the processes of transporting and releasing propagules are usually separated in time and space. Release through cultivation regularly occurs after transport and leads initially to clumped patterns of resulting offspring. As an exception to this general rule, transport losses of target species may provide continuous propagule release during transport, resulting in linear patterns of emerging populations. This is most evident in seed crops which are accidentally dispersed by spilling from loading areas of trains or trucks and subsequently emerge along transport corridors (Suominen 1979). Oilseed rape (*Brassica napus*), for example, can establish large populations alongside roads which may persist over long periods and are associated with major transport routes to oilseed processing plants (Crawley and Brown 2004).

Accidental transfer of species associated with goods leads preferentially to discontinuous patterns of release and clumped populations of the emerging offspring. This may occur both during and after transport. Harbours and train stations are hotspots of non-native diversity which have long been recognised (Thellung 1915; Brandes 1993); this is mostly due to the release of propagules during the switch from one conveyer to another. Associated cleaning procedures often enhance clumped propagule release. *Citrus* fruits, for example, were once often protected by hay on their way from southern to northern European regions. At the end of the train journeys, the packing material was usually discarded at the stations. Jauch (1938) reports 814 species which were moved this way, most of them originating from southern Europe. The release of solid ballast material in the vicinity of ports is another example of a highly efficient transport vector associated with clumped patterns of propagule release at the endpoint of transport routes (Thellung 1915; Ridley 1930). Analogous patterns occur when using water as weighting material, but marine release sites facilitate subsequent long-distance dispersal more effectively than do terrestrial ones (Chap. 4).

As another pathway for accidental introductions, wool imports show a clear separation between propagule transport and release, and also illustrate how changes in technology may affect vector efficiency through curbing propagule release. More than 1,600 species, mostly from the southern hemisphere, have arrived in Europe associated with wool (Probst 1949). Only a few (e.g. *Xanthium spinosum*, *Senecio inaequidens*; Thellung 1915; Ernst 1998) crossed the threshold of naturalisation (Table 3.2). The confinement of most wool adventives to the vicinity of wool industry sites reflects how propagules were released after transport. Formerly, waste from wool processing contained high amounts of viable seeds and was dispersed, often as organic manure, adjacent to the factories (Salisbury 1964). Today, wool transports still provide a worldwide transfer of propagules, but the number of wool adventives in the field has decreased conspicuously due to changes in the way waste from wool processing is treated and released (Bonn and Poschlod 1998).

3.4.2 Direct Association with Vehicles: Coincidence of Transport and Release

Traffic routes were associated with plant invasions even before modern road construction started. The early colonisation of North America by European settlers induced multiple range expansions of non-native species, often following trails and primitive roads (Crosby 1986). The inventions of railways and cars in the 19th century substantially increased the role of traffic as a dispersal vector. Apart from an association with transported goods, propagules can be transported directly through adhesive dispersal on the surface of vehicles.

In contrast to most vectors which promote the intercontinental introduction of species to a new range, and also in contrast to transfers by means of transported goods, the direct attachment of propagules to vehicles provides a vector where the processes of reception, transport and release of propagules largely overlap in time and space. Road vehicles, for example, continuously disperse seeds during their journey and this may coincide with the attachment of further diaspores to the vehicle from the roadside flora. Along the route of transport, propagules released by one car can be attached to a following one. Vectors which provide such interlinked chains of reception, transport and release of diaspores lead to linear distribution patterns along transport networks. These retrace the transport route, which acts concurrently as corridor of seed release. This continuous release of seeds during transport also enhances the probability of exposing propagules to other, both natural and human-mediated dispersal vectors.

3.4.2.1 Adhesion to Vehicles

The adhesive association of propagules to the surface of vehicles can occur accidentally along the whole route of a vehicle's journey. The attachment of propagules to cars has been confirmed through analyses of seed samples from the surface of cars (Hodkinson and Thompson 1997). The attachment of propagules to cars as transport vectors is mediated by mud (Clifford 1959; Hodkinson and Thompson 1997). The efficiency of propagule reception by vehicles depends on exposure to potential seed sources. Thus, cars which were driven in rural surroundings and on unpaved roads had much higher seed contents than cars from urban areas and paved roads (Hodkinson and Thompson 1997). Although not tested experimentally, it can be assumed that the attachment potential of propagules to trains is lower than that for cars, as direct contact to seed sources is limited in the former case.

Various parts of vehicles, such as the tires, wheel arches, tire wells, hood and trunk grooves and window washer grooves, can support the accumulation of mud and seeds on the surface of vehicles (Hodkinson and Thompson 1997). Occasional events of long-distance dispersal may also be facilitated by unintended internal transport, e.g. in engine blocks, passenger spaces or trunks. Corresponding to the varying duration of seed adhesion on different parts of vehicles, the transport distances cover a broad spatial scale associated with short- to longer-distance dispersal. As an example, initial roadside populations of coastal species in the United Kingdom were found at great distances from coastal seed sources, probably due to long-distance dispersal by vehicles (Scott and Davison 1985). At the local scale, dispersal from these initial populations was enhanced in the direction of traffic flow, demonstrating also short-distance effects of vehicles on seed dispersal.

3.4.2.2 Transport Routes: from Patterns to Processes

The relevance of traffic as a pathway in plant invasions is usually determined from observed distribution patterns of non-native species along traffic routes. As these depend both on dispersal processes and on characteristic site conditions of roadside and railway corridors, the vector, i.e. the underlying processes causing these distributional patterns, can be retraced only indirectly. This necessitates a differentiation of site- and vector-dependent mechanisms, both of which may enhance plant migration along transportation systems: (1) seed dispersal by vehicles (Hodkinson and Thompson 1997) and (2) high disturbance and altered site conditions which provide safe sites for the establishment of numerous non-native species and form suitable migration corridors (Hansen and Clevenger 2005). Both processes are mutually dependent, as site conditions along transport routes can favour the establishment of populations of non-native species which are dispersed by vehicles and which, in turn, can act as seed sources for subsequent adhesion to vehicles.

It is thus difficult to assess the vector strength of adhesive dispersal by vehicles as such because plant migration patterns along transportation corridors usually reflect the confounding effects of site characteristics and the agency of traffic as well as of other dispersal vectors. Nonetheless, several studies provide evidence that dispersal by vehicles does indeed enhance the range expansion of non-native plant species. First, time series of roadside invasions reveal a linear and sequential range expansion along roadsides (Kopecky 1988; Ernst 1998), indicating that spreading is due to dispersal *within* the corridor, rather than to adjacent seed sources. Second, along transport systems, isolated founder populations of invasive species can be found (Scott and Davison 1985; Ernst 1998) which are likely to result from long-distance dispersal by vehicles. Furthermore, non-native species may contribute a large share of the seeds found in mud samples from vehicles, for example, about 70 % in the sludge of a car wash in Canberra, Australia (Wace 1977).

At the local scale, road verges are among the habitat types with the highest proportion of non-native species and usually comprise more non-native species than does the adjacent landscape (Gelbard and Belnap 2003; Hansen and Clevenger 2005). Roadside populations of non-native plant species can act as focal points for invasion into the adjacent landscape. In grassland habitats, an increase in non-native species richness can be observed up to 100 m from the edge of transportation corridors (Tyser and Worley 1992; Gelbard and Belnap 2003). Roads also enhance plant transfers into protected areas (Tyser and Worley 1992).

Similarly to road traffic, linear distribution of non-native species has been observed along rail tracks (Hansen and Clevenger 2005). Long-distance dispersal along the railway system has been acknowledged as a major driver in the rapid spread of *Senecio inaequidens* (Ernst 1998). In addition to continuous dispersal during train travel, discontinuous release of propagules occurs at stations. However, similarly to road vehicles, it is difficult to unequivocally identify the functioning of railroad tracks as migration or dispersal corridors.

Effects of transport corridors on plant invasions at larger scales are indicated by a positive correlation between the density of the transportation system and the density and richness of non-native plants of an area (Vilà and Pujadas 2001). At the regional scale, migration along transport corridors can cause altitudinal shifts in species distribution (Kopecky 1988).

3.4.3 Role of Living Conveyers

Since the earliest migrations, people and their domesticated animals (Chap. 2) are known to have transferred plant species accidentally at local to continental scales. Propagules may be externally moved through attachment to footwear, clothing, fur or hoofs, and internally through the digestive tract.

Studies on dispersal by livestock illustrate a huge potential for long-distance dispersal. A flock of 400 sheep can move about 8 million diaspores during a single vegetation period, with a retention period of up to 100 days in the fleece of a sheep (Poschlod and Bonn 1998). Wandering herds can cover distances of hundreds of kilometres and lead to a continuous release of propagules during transport. This provided, for example, an efficient spread of *Xanthium spinosum*, which was eventually called shepherd's plague (Thellung 1915). Although transhumance is today less widespread in Europe, long-distance dispersal through domesticated animals still occurs as livestock are moved by trains or other vehicles at least at regional to infra-continental scales, often integrating straw and fodder as additional vectors for plant dispersal. In North America, invasions by *Bromus tectorum* were facilitated in this way (Mack 1981). During the period of colonisation in the southern hemisphere, livestock most likely led to an efficient intercontinental transfer of plant species through the establishment of pasture regimes in the southern hemisphere or even by releasing domesticated animals as a living larder for early voyagers on isolated oceanic islands (Crosby 1986).

High numbers of propagules associated with the dung of cattle, sheep and horses indicate pathways of dispersal via endozoochory (Poschlod and Bonn 1998). Viable seeds of numerous exotic species have been found in horse dung along recreational riding trails (Campbell and Gibson 2001). Although in this case only one species moved into the adjacent forest, the large number of exotic species in horse dung reflects a potential of horseback activities for inducing invasions. Within German biosphere reserves, sheep transfer the invasive *Lupinus polyphyllus* by endozoochory (Otte et al. 2002). Sheep also provide a potential pathway for crop dispersal, as they have been shown to be capable of excreting viable canola seeds for up to 5 days after consumption (Stanton et al. 2003). This could lead to seed transfer from grazed stubble paddocks to habitats outside of cultivated fields, facilitating the establishment of feral crop populations.

Human population size is a good predictor for non-native plant species richness (McKinney 2002), but the understanding of underlying processes is still limited. Despite the enormous increase in human mobility in the last decades, only a few studies have directly analysed the role of humans in moving species by attachment to footwear or clothing (Clifford 1956; Falinski 1972). The anecdotal spread of *Plantago major* in North America as "Englishmen's foot" (Crosby 1986) clearly indicates the measurable strength of this vector. Association with footwear of travelling botanists, tourists and sportsmen may provide an infra- and intercontinental transfer of species, as shown by Clifford (1956) and Powell (1968). Sportsmen have also introduced seeds of invasive species to Hawai'i on their shoes (Higashino et al. 1983). Still little is known about the role of humans in dispersing propagules after consumption. The frequent germination of tomatoes on river banks in Germany suggests an efficient seed transfer from the human digestive tract via sewage to rivers as natural corridors of transport (Schmitz 2004).

3.5 Conclusions

Humans promote plant dispersal at local, regional and continental scales. In processes leading to the introduction of a species to a new range, human agency is a prerequisite for subsequent invasions. Even at more local to regional scales, however, human-mediated dispersal can be crucial for invasion success but seems to be still underestimated. Both accidental and deliberate vectors of plant invasions are engaged in the further spread of non-native species following their initial introduction and sometimes, as in horticulture, the vector of initial introduction can be the same as the one which fosters range expansion in the new area. As a consequence, prevention and management of invasions should address processes which lead to the initial introduction of a species to a new range as well as those enhancing subsequent invasion success within the new range through secondary releases or accidental transfer of propagules.

Distinguishing between the processes of propagule transport and release is useful for a better understanding of vector efficiency and of the spatial patterns of invasions resulting from the release of propagules to the environment. As an additional point, we stress here the role of human intervention as a key driver in both processes (Table 3.3). Often, transport and release of propagules occur both intentionally and accidentally. Additionally, deliberately transported target species may also be released accidentally and, vice versa, unintentionally moved non-target species may be released purposefully into the environment. The classification of vectors according to the underlying human motivation could be a viable approach for the development of policy and management strategies dealing with biological invasions, as it helps to group those vectors which can be governed by similar control measures.

Despite the long history in studying pathways in plant dispersal, our knowledge of the functioning and efficiency of vectors is still limited. In this regard, it is useful to further analyse the strength of different vectors in terms of efficiency of both propagule transport and release and the transition to subsequent stages of invasion success. As has been broadly acknowledged, the focal challenge in assessing the present and future role of vectors is their changing nature over time and space.

Table 3.3 Presence or absence of human motivation as underlying driver in the transport and release of propagules. Both processes can be linked to deliberate or accidental human agency

	Deliberate release of propagules	Accidental release of propagules
Deliberately moved target species	Introduction and subsequent propagation or cultivation of ornamentals	Transport losses in crop seeds by spilling from vehicles or trains
	Deposition of ornamentals as garden waste	
	Planting of introduced species to enrich nature	
Accidentally moved non-target species	Release of propagules with solid or liquid ballast materials	Transport losses of weed seeds associated with crop seeds
	Deposition of propagule-containing waste from wool factories as manure	Deposition of propagules attached to vehicles, animals or to human footwear
	Discharge of propagules which were associated with goods by cleaning procedures (e.g. trains)	
Resulting spatial patterns	Mostly clumped invasion foci by discontinuous release of propagules	Mostly linear invasion foci by continuous release of propagules

References

Bonn S, Poschlod P (1998) Ausbreitungsbiologie der Pflanzen Mitteleuropas: Grundlagen und kulturhistorische Aspekte. Quelle und Meyer, Wiesbaden

Brandes D (1993) Eisenbahnanlagen als Untersuchungsgegenstand der Geobotanik. Tuexenia 13:415–444

Campbell JE, Gibson DJ (2001) The effect of seeds of exotic species transported via horse dung on vegetation along trail corridors. Plant Ecol 157:23–35

Carlton JT, Ruiz GM (2005) Vector science and integrated vector management in bioinvasion ecology: conceptual frameworks. In: Mooney HA, Mack RN, McNeely JA, Neville LE, Schei PJ, Waage JK (eds) Invasive alien species. Island Press, Washington, DC, pp 36–58

Chauvel B, Dessaint F, Cardinal-Legrand C, Bretagnolle F (2006) The historical spread of *Ambrosia artemisiifolia* L. in France from herbarium records. J Biogeogr 33:665–673

Clifford HT (1956) Seed dispersal on footwear. Proc Bot Soc Br Isles 2:129–131

Clifford HT (1959) Seed dispersal by motor vehicles. J Ecol 47:311–315

Conedera M, Krebs P, Tinner W, Pradella M, Torriani D (2004) The cultivation of *Castanea sativa* (Mill.) in Europe, from its origin to its diffusion on a continental scale. Veg Hist Archaeobot 13:161–179

Crawley MJ, Brown SL (2004) Spatially structured population dynamics in feral oilseed rape. Proc R Soc Lond B Biol Sci 271:1909–1916

Crosby AW (1986) Ecological imperialism: the biological expansion of Europe, 900–1900. Cambridge University Press, Cambridge

Dehnen-Schmutz K, Williamson M (2006) *Rhododendron ponticum* in Britain and Ireland: social, economic and ecological factors in its successful invasion. Environ Hist (in press)

Dehnen-Schmutz K, Touza J, Perrings C, Williamson M (2006) The horticultural trade and ornamental plant invasions in Britain. Conserv Biol (in press)

Elton CS (1958) The ecology of invasions by plants and animals. Methuen, London

Ernst WHO (1998) Invasion, dispersal and ecology of the South African neophyte *Senecio inaequidens* in The Netherlands: from wool alien to railway and road alien. Acta Bot Neerl 47:131–151

Falinski JB (1972) Anthropochory in xerothermic grasslands in the light of experimental data. Acta Soc Bot Polon 41:357–368

Forman J (2003) The introduction of American plant species into Europe: issues and consequences. In: Child LE, Brock JH, Brundu G, Prach K, Pyšek K, Wade PM, Williamson M (eds) Plant invasions: ecological threats and management solutions. Backhuys, Leiden, pp 17–39

Franz G (1984) Geschichte des deutschen Gartenbaues. Ulmer, Stuttgart

Gelbard JL, Belnap J (2003) Roads as conduits for exotic plant invasions in a semiarid landscape. Conserv Biol 17:420–432

Gressel J (ed) (2005) Crop ferality and volunteerism. CRC Press, Boca Raton

Hansen MJ, Clevenger AP (2005) The influence of disturbance and habitat on the presence of non-native plant species along transport corridors. Biol Conserv 125:249–259

Higashino PK, Guyer W, Stone CP (1983) The Kilauea wilderness marathon and crater rim runs: sole searching experiences. Newslett Hawaiian Bot Soc 22:25–28

Hodkinson DJ, Thompson K (1997) Plant dispersal: the role of man. J Appl Ecol 34:1484–1496

Hussner A, Lösch R (2005) Alien aquatic plants in a thermally abnormal river and their assembly to neophyte-dominated macrophyte stands (River Erft, North Rhine-Westphalia). Limnologica 35:18–30

Jauch F (1938) Fremdpflanzen auf den Karlsruher Güterbahnhöfen. Beitr Naturkde Forsch Südwestdeutschland 3:76–147

Jones AT, Hayes MJ, Hamilton NRS (2001) The effect of provenance on the performance of *Crataegus monogyna* in hedges. J Appl Ecol 38:952–962

Kopecky K (1988) Einfluß der Straßen auf die Synanthropisierung der Flora und Vegetation nach Beobachtungen in der Tschechoslowakei. Folia Geobot Phytotax 23:145–171

Kornas J (1988) Speirochore Ackerwildkräuter: von ökologischer Spezialisierung zum Aussterben. Flora 180:83–91

Kowarik I (1992) Einführung und Ausbreitung nichteinheimischer Gehölzarten in Berlin und Brandenburg. Verh Bot Ver Berlin Brandenburg Beiheft 3:1–188

Kowarik I (1995) Time lags in biological invasions with regard to the success and failure of alien species. In: Pyšek P, Prach K, Rejmánek M, Wade M (eds) Plant invasions – general aspects and special problems. SPB Academic, Amsterdam, pp 15–38

Kowarik I (2003) Human agency in biological invasions: secondary releases foster naturalisation and population expansion of alien plant species. Biol Invasions 5:293–312

Kowarik I (2005) Urban ornamentals escaped from cultivation. In: Gressel J (ed) Crop ferality and volunteerism. CRC Press, Boca Raton, pp 97–121

Kowarik I, Wohlgemuth JO (2006) *Tulipa sylvestris* in northwestern Germany: a non-indigenous species as indicator of former horticulture. Polish Bot Stud 22 (in press)

Lumaret R (1990) Invasion of natural pastures by a cultivated grass (*Dactylis glomerata* L.) in Galicia, Spain: process and consequence on plant-cattle interactions. In: di Castri F, Hansen AJ, Debussche M (eds) Biological invasions in Europe and the Mediterranean Basin. Kluwer, Dordrecht, pp 392–397
Mack RN (1981) Invasion of *Bromus tectorum* L. into western North America: an ecological chronicle. Agro-Ecosystems 7:145–165
Mack RN (1991) The commercial seed trade: an early disperser of weeds in the United States. Econ Bot 45:257–273
Mack RN (2000) Cultivation fosters plant naturalization by reducing environmental stochasticity. Biol Invasions 2:111–122
Mack RN (2001) Motivations and consequences of the human dispersal of plants. In: McNeely JA (ed) The great reshuffling: human dimensions in invasive alien species. IUCN, Gland, pp 23–34
Mack RN (2003) Global plant dispersal, naturalization, and Invasion: pathways, modes, and circumstances. In: Ruiz GM, Carlton JT (eds) Invasive species: vectors and management strategies. Island Press, Washington, DC, pp 3–30
Mack RN, Lonsdale WM (2001) Humans as global plant dispersers: getting more than we bargained for. Bioscience 51:95–102
Mack RN, Simberloff D, Lonsdale WM, Evans H, Clout M, Bazzaz FA (2000) Biotic invasions: causes, epidemiology, global consequences, and control. Ecol Appl 10:689–710
McKinney ML (2002) Do human activities raise species richness? Contrasting patterns in United States plants and fishes. Global Ecol Biogeogr 11:343–348
Nobbe F (1876) Handbuch der Samenkunde. Physiologisch statistische Untersuchungen über den Gebrauchswert der land- und forstwirtschaftlichen sowie gärtnerischen Saatwaren. Wiegand, Hempel & Parey, Berlin
Otte A, Obert S, Volz H, Weigand E (2002) Effekte von Beweidung auf *Lupinus polyphyllus* Lindl. in Bergwiesen des Biosphärenreservates Rhön. Neobiota 1:101–133
Pauchard A, Shea K (2006) Integrating the study of non-native plant invasions across spatial scales. Biol Invasions 8:399–413
Poschlod P, Bonn S (1998) Changing dispersal processes in the central European landscape since the last Ice Age: an explanation for the actual decrease of plant species richness in different habitats? Acta Bot Neerl 47:27–44
Powell RH (1968) Harmful plant species entering New Zealand 1963–1967. N Z J Bot 6:395–401
Prach K, Hadinec J, Michalek J, Pyšek P (1995) Forest planting as a way of species dispersal. Forest Ecol Manage 76:191–195
Probst R (1949) Wolladventivflora Mitteleuropas. Voigt-Schild, Solothurn
Pyšek P, Hulme PE (2005) Spatio-temporal dynamics of plant invasions: linking pattern to process. Ecoscience 12:302–315
Pyšek P, Prach K (1993) Plant invasions and the role of riparian habitats – a comparison of 4 species alien to Central Europe. J Biogeogr 20:413–420
Pyšek P, Sádlo J, Mandák B (2002) Catalogue of alien plants of the Czech Republic. Preslia 74:97–186
Reichard SH, White P (2001) Horticulture as a pathway of invasive plant introductions in the United States. Bioscience 51:103–113
Rejmánek M, Richardson DM, Higgins SI, Pitcairn MJ, Grotkopp E (2005) Ecology of invasive plants: state of the art. In: Mooney HA, Mack RN, McNeely JA, Neville LE, Schei PJ, Waage JK (eds) Invasive alien species – a new synthesis. Scope 63. Island Press, Washington, DC, pp 104–161
Richardson DM, Rejmánek M (2004) Conifers as invasive aliens: a global survey and predictive framework. Diversity Distrib 10:321–331

Richardson DM, Pyšek P, Rejmánek M, Barbour MG, Panetta FD, West CJ (2000) Naturalization and invasion of alien plants: concepts and definitions. Diversity Distrib 6:93–107
Ridley HN (1930) The dispersal of plants throughout the world. Reeve, Ashford, Kent
Salisbury E (1953) A changing flora as shown in the study of weeds of arable land and waste places. In: Lousley JE (ed) The changing flora of Britain. Botanical Society of the British Isles, Oxford, pp 130–139
Salisbury E (1964) Weeds and aliens. Collins, London
Schepker H (1998) Wahrnehmung, Ausbreitung und Bewertung von Neophyten. Eine Analyse der problematischen nichteinheimischen Pflanzenarten in Niedersachsen. Ibidem, Stuttgart
Schmitz U (2004) Frost resistance of tomato seeds and the degree of naturalisation of *Lycopersicon esculentum* Mill. in Central Europe. Flora 199:476–480
Scott NE, Davison AW (1985) The distribution and ecology of coastal species on roadsides. Vegetatio 62:433–440
Spethmann W (1995) In-situ/ex-situ-Erhaltung von heimischen Straucharten. In: Kleinschmit J, Begemann F, Hammer K (eds) Erhaltung pflanzengenetischer Ressourcen in der Land- und Forstwirtschaft. ZADI, Bonn, pp 68–87
Stanton R, Pratley J, Hudson D (2003) Sheep are potential vectors for the spread of canola (*Brassica napus*) seed. Austral J Exp Agric 43:535–538
Starfinger U, Kowarik I, Rode M, Schepker H (2003) From desirable ornamental plant to pest to accepted addition to the flora? The perception of an alien plant species through the centuries. Biol Invasions 5:323–335
Sukopp H (1972) Wandel von Flora und Vegetation in Mitteleuropa unter dem Einfluß des Menschen. Ber Landwirtsch 50:112–139
Suominen J (1979) The grain immigrant flora of Finland. Acta Bot Fenn 111:1–108
Thellung A (1912) La flore adventice de Montpellier. Mém Soc Sci Nat Cherbourg 38:622–647
Thellung A (1915) Pflanzenwanderungen unter dem Einfluß des Menschen. Beibl Bot Jahrb 3/5:37–66
Tyser RW, Worley CA (1992) Alien flora in grasslands adjacent to road and trail corridors in Glacier National Park, Montana (USA). Conserv Biol 6:253–262
Vilà M, Pujadas J (2001) Land-use and socio-economic correlates of plant invasions in European and North African countries. Biol Conserv 100:397–401
Wace N (1977) Assessment of dispersal of plant species – the car-borne flora in Canberra. Proc Ecol Soc Austral 10:167–186
Willerding U (1986) Zur Geschichte der Unkräuter Mitteleuropas. Wachholtz, Neumünster
Williams DG, Baruch Z (2000) African grass invasion in the Americas: ecosystem consequences and the role of ecophysiology. Biol Invasions 2:123–140
Williamson M (1996) Biological invasions. Chapman & Hall, London
Zobel BJ, Wyk GV, Stahl P (1987) Growing exotic forests. Wiley, New York

4 Is Ballast Water a Major Dispersal Mechanism for Marine Organisms?

STEPHAN GOLLASCH

4.1 Introduction

More than 1,000 non-indigenous aquatic species, ranging from unicellular algae to vertebrates, have been found in European coastal waters, including navigational inland waterways for ocean-going vessels and adjacent water bodies. Approximately half of all non-indigenous species recorded to date have established self-sustaining populations (Gollasch 2006). These introductions are of high concern, as many cases have caused major economical or ecological problems (Chaps. 13–19).

Species are introduced unintentionally (e.g. with ships) or intentionally (e.g. for aquaculture purposes and re-stocking efforts). In shipping, the prime vectors for species transportation are ballast water and in the hull fouling of vessels. Further, a considerable number of exotic species migrates through man-made canals. Examples are the inner-European waterways connecting the Ponto-Caspian region and the Baltic Sea. Also, the Suez Canal "opened the door" for Red Sea species migrations into the Mediterranean Sea and vice versa (Gollasch et al. 2006; Chap. 5).

For the purpose of this contribution, the following inventories of introduced species in coastal waters were considered: North Sea (Gollasch 1996; Reise et al. 1999; Nehring 2002), Baltic Sea (Leppäkoski 1994; Gollasch and Mecke 1996; Leppäkoski and Olenin 2000; Olenin et al. 2005), British Isles (Eno 1996; Eno et al. 1997), Ireland (Minchin and Eno 2002), Mediterranean Sea (Galil and Zenetos 2002; CIESM 2005). Other key publications are a book dealing with aquatic invaders in Europe (Leppäkoski et al. 2002) and a review of marine introduced species in Europe (Streftaris et al. 2005). To broaden the scope, additional datasets from outside Europe were also included: from Australia, Japan, New Zealand and North America, including Hawaii, as well as the inventories of introduced species prepared during the Global Ballast Water Management Programme (GloBallast), i.e. from Brazil, China, India, Iran, South Africa and Ukraine.

Ecological Studies, Vol. 193
W. Nentwig (Ed.)
Biological Invasions

4.2 Vectors

Introduction vectors are defined as the physical means by which species are transported from one geographic region to another (Carlton 2001). Vectors include various natural mechanisms of spread and also anthropogenically aided species dispersal. Since man started to sail the oceans, species have been either intentionally or unintentionally in transport. The Vikings were wide-reach seafarers and may have been responsible for the introduction of the infaunal bivalve *Mya arenaria* to Europe (Petersen et al. 1992) – this actually may be the first ship-mediated species introduction into Europe. When returning from North America, Vikings may have kept live *M. arenaria* aboard as food supply. Alternatively, they may have imported the clam in the solid ballast used on their vessels. It is assumed that Viking ships arrived in Europe from muddy estuaries in North America, and these estuaries would have been highly populated with *M. arenaria*. However, it is also possible that there was a gradual re-expansion of *M. arenaria* into Europe after the last glaciation period. Wolff (2005) states that the transfer of *M. arenaria* by the Vikings may have occurred only on an occasional vessel because there was no direct transport route between North America and Europe in Viking times (Marcus 1980). Greenlanders travelled to North America more frequently, and also between Greenland and Norway, but it is unclear whether these trips were undertaken by the same vessels. As a result, *M. arenaria* was probably first introduced from North America to Greenland and thereafter from Greenland to Europe (Ockelmann 1958; Höpner Petersen 1978, 1999).

Historically, aquaculture and stock transfers of aquatic species resulted in a significant amount of taxa being transported worldwide. Dry and semi-dry ballast is no longer in use with merchant shipping but, during former times, this vector is claimed to have introduced a certain number of species worldwide, e.g. *Littorina littorea* to North America (Carlton 1992), *Chara connivens* to coastal areas of the Baltic Sea and several seashore plants into Europe (Wallentinus 2002). Estimations reveal that more than 480,000 annual ship movements occur worldwide with the potential for transporting organisms. Various calculations have been made on the amount of ballast water carried with the world's fleet of merchant ships – it has been estimated that 2–12 billion t of ballast water are transported annually. In ballast tanks and also other ship vectors including hulls, anchor chains and sea chests, ships may carry 4,000 to 7,000 taxa each day (Gollasch 1996; Carlton, personal communication). One reason for this great diversity of organisms in transit arises from the three different "habitats" inside ballast water tanks: (1) tank walls, (2) ballast water, and (3) the sediment (Taylor et al. 2002).

When calculating the number of new invaders per time unit, every 9 weeks a new species is found. It should be noted that this value is regionally very different and, in Europe, there are indications that a new species was found every

3 weeks over the time period 1998–2000 (ICES WGITMO 2004). However, not all species recorded form self-sustaining populations (Minchin and Gollasch 2002). In many cases, it is impossible to clearly identify the introduction vector. In bivalves, for example, introductions may be attributed to larval transport in ballast water releases, adults in hull fouling of ships or imports as (non-)target species for aquaculture activities.

4.3 Relative Vector Importance

Previous studies have shown that the most prominent invasion vectors are shipping and aquaculture activities (Streftaris et al. 2005; Gollasch 2006). An exception from this general trend is the Mediterranean Sea where the dominating invasion "vector" is the opening of the Suez Canal, enabling Red Sea species to migrate into the south-eastern Mediterranean Sea and vice versa. This phenomenon is also known as Lessepsian migration (cf. Ferdinand de Lesseps planned the Suez Canal which opened in 1869). However, this is not considered to be a vector but rather a removal of a migration barrier, and has been included here only for reasons of comparison (Chap. 5).

The relative vector importance is regionally very different. Assessing all available data as outlined above, on a global scale species introduced with hull fouling seem to slightly dominate those associated with ballast water and

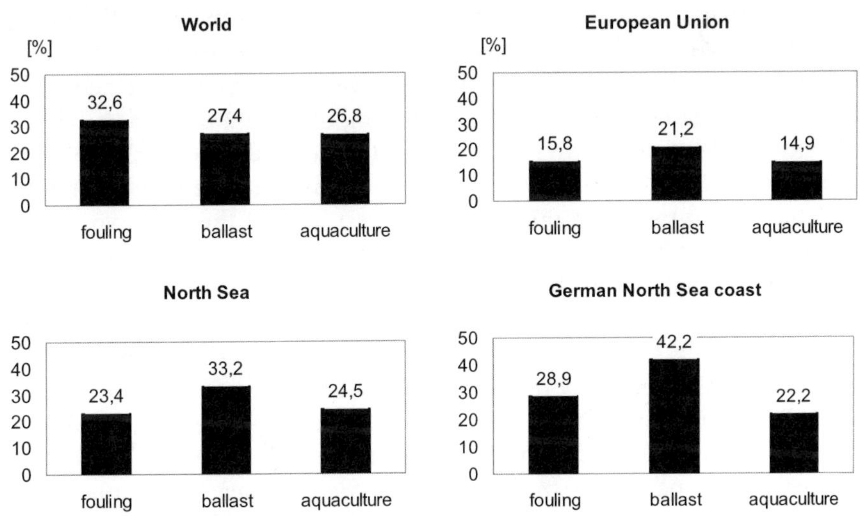

Fig. 4.1 Relative importance of the invasion vectors hull fouling, ballast water and aquaculture efforts per region

aquaculture. In waters of the European Union, ballast water-mediated species introductions prevail and this trend becomes even clearer when extracting the data for the North Sea (Fig. 4.1).

4.4 Ballast Water

Ballast water is in use in shipping to, for example, strengthen structures and to submerge the propeller when no cargo is carried. Ballast water has long been suspected as major vector for species introductions. One of the first assumptions that a species was introduced by ballast water to regions outside its native range was made by Ostenfeld (1908) after a mass occurrence of the Asian phytoplankton algae *Odontella* (*Biddulphia*) *sinensis.* The species was first recorded in the North Sea in 1903. The first study to sample ships' ballast water was carried out 70 years later by Medcof (1975), followed by many others. In Europe, 14 ship sampling studies have been undertaken. More than 1,500 ballast tank samples were taken, of which approx. 80 % represent samples from ballast water and nearly 20 % from ballast tank sediments. Almost 600 vessels have been sampled since 1992. The total number of taxa identified overall was more than 1,000. The diversity of species found included bacteria, fungi, protozoans, algae, invertebrates of different life stages including resting stages, and fishes with a body length up to 15 cm. The most frequently collected taxa were diatoms, copepods, rotifers, and larvae of Gastropoda, Bivalvia and Polychaeta (Gollasch et al. 2002).

The above-mentioned shipping studies have clearly shown that an enormous number of taxa can be found in ballast tanks at the end of ship voyages. However, en-route studies based on daily sampling frequencies showed that organisms in ballast water die out over time. The most significant decrease in organism densities occurs during the first 3 days of the voyage, and after 10 days most individuals were found dead (e.g. Gollasch et al. 2000; Olenin et al. 2000). However, exceptions from this general trend occurred. In one study, most taxa died out during the first days of the ship's voyage but harpacticoid copepods increased in numbers towards the end of the voyage, documenting that certain species reproduce in ballast water tanks (Gollasch et al. 2000).

Species with a high potential to cause unwanted impacts in the receiving environment are frequently transported in ballast water. This refers mainly to phytoplankton species which may produce toxins – these species, when introduced to areas in close proximity to aquaculture farms, are of great concern. Further, human pathogens including *Vibrio cholerae* have been found in ballast water.

One invader well known for its negative impact is the Ponto-Caspian zebra mussel *Dreissena polymorpha*. When very abundant, the mussel may clog the water intakes of power plants and municipal waterworks, one notable example being the North American Great Lakes. Although this species causes fouling problems, it is unlikely that it was introduced to North America in the hull fouling of ships. The oceanic voyage is too long for these freshwater organisms to survive. As a result, certain species, including those causing fouling problems as adults, are introduced as larvae in ballast water.

Ship design continuously improves, resulting in ever faster and larger vessels. Consequently, ship arrivals are more frequent, ballast tanks increase in size and the time an organism needs to survive in a ballast water tank is reduced. As a result, the volume and frequency of ballast water discharges increase, which also enhances the likelihood of species surviving in the new habitat after ballast water discharge. By implication, each new generation of ships has the potential to increase the risk of invasion.

4.5 Risk-Reducing Measures

The vectors shipping, aquaculture and stocking may play a different role in the future, as regulatory instruments are either in place or developing with the aim to minimize the number of new species introductions. In aquaculture and stocking, the International Council for the Exploration of the Sea (ICES) has updated its Code of Practice on the Introductions and Transfers of Marine Organisms (ICES 2005). This instrument provides (voluntary) guidelines to avoid unwanted effects of moved species and unintentional introductions of non-target taxa. ICES member countries planning new marine (including brackish) species introductions are requested to present to the ICES Council a detailed prospectus on the rationale, the contents of the prospectus being detailed in the Code. Having received the proposal, the Council may then request its Working Group on Introductions and Transfers of Marine Organisms (WGITMO) to evaluate the prospectus. WGITMO may request more information before commenting on a proposal. If the decision is taken to proceed with the introduction, then only progeny of the introduced species may be transplanted into the natural environment, provided that a risk assessment indicates that the likelihood of negative genetic and environmental impacts is minimal, that no disease agents, parasites or other non-target species become evident in the progeny to be transplanted, and that no unacceptable economic impact is to be expected. A monitoring programme of the introduced species in its new environment should be undertaken, and annual progress reports should be submitted to ICES for review at WGITMO meetings, until the review process is considered complete.

The International Maritime Organization (IMO), the United Nations body which deals, e.g. with minimizing pollution from ships, has developed two conventions relevant to biological invasions.

1. The International Convention on the Control of Harmful Anti-Fouling Systems on Ships: this Convention was developed as a consequence of the unwanted impact from poisonous antifouling paints based on tri-butyl-tin (TBT) in the aquatic environment. Eventually, the use of TBT was banned and TBT-free antifouling paints are currently being developed and tested. However, it is feared that alternative ship coatings may not be as effective, possibly resulting in more species arriving in new habitats with ship hull fouling.
2. The International Convention for the Control and Management of Ships' Ballast Water and Sediments: although being of limited effectiveness, ballast water exchange in open seas is recommended as a partial solution to reduce the number of species in transit. In the future, ballast water treatment will eventually be required. This Convention was adopted in 2004 and is now open for signature by IMO member states. IMO is currently developing 15 guidelines to address certain key issues in the Convention in greater detail.

4.6 Ballast Water Management Options

Ballast water exchange has been suggested as a management tool for vessels on transoceanic voyages. Ballast water taken onboard ships in ports or coastal areas would be exchanged for deep oceanic water, the background assumption being that, in the open sea, fewer organisms will be present and also plankton species are unlikely to survive in coastal areas when the ballast water is discharged in the next port of call. This water exchange approach is also recommended for cases of vessels travelling between two freshwater ports, as the salinity increase would likely kill any freshwater organisms pumped onboard in the freshwater ports.

The water replacement efficiency during ballast water exchange depends on, e.g. the ballast tank design. Trials have shown that three times volumetric exchange of ballast water results in approx. 95 % removal of phytoplankton cells and approx. 60 % removal of zooplankton organisms. However, the 5 % of phytoplankton surviving may amount to millions of specimens (Taylor et al. 2002).

Since the mid-1990s, roughly 20 initiatives on ballast water treatment have been completed or are still ongoing. The treatment options considered to date include filtration, use of hydrocyclones, heat treatment, coagulation/flocculation, pH adjustment and chemical treatment, including electrolytical generation of agents from seawater and UV.

4.7 Conclusions

The relative importance of invasion vectors is difficult to assess because not all introduced species can clearly be attributed to any one vector. However, shipping seems to be the prime invasion vector today. In shipping, key vectors are ballast water and hull fouling. Their relative importance is regionally very different, being strongly influenced by local economies and shipping patterns. However, as shown above, any ship design improvement which results in larger and faster ships will favour ballast water-mediated introductions. Consequently, the relative vector importance should be revisited over time. Further, trade scenarios and shipping patterns may change over time. When planning mitigation measures aimed at reducing the number of new species introductions, the prime invasion vectors should be addressed first; the findings of such assessments may change over time.

Exotic species will definitely continue to spread, although the timely implementation of the above-mentioned measures may significantly reduce the invasion rate. As newly found species are usually reported with a certain time lag due to publishing procedures in scientific journals, the number of first records in the current decade will likely increase in the future. To solve this problem, a new European journal of applied research on biological invasions in aquatic ecosystems has been launched and will be published frequently to announce new findings of biological invaders (http://www.aquaticinvasions.ru/). Timely publication of new introduced species is not only of academic interest. It may also result in an early warning instrument with the aim to develop eradication programmes of certain species. The success of rapid response measures to eliminate newly introduced species is dependent on early detection. Successful efforts are known from Australia and North America. In Europe, early detection and rapid response scenarios are currently developing.

The management of already established species requires more effective international cooperation of neighbouring states. Following the precautionary principle, emphasis should be placed on the prevention of species introductions because, once established, their secondary spread is difficult or impossible to control. This approach comes particularly into focus noting the impacts certain invasions have caused, including implications for native species, fisheries, aquaculture and human health. It is therefore hoped that the above-mentioned regulatory instruments enter into force soon and will timely be implemented with the aim to reduce future species introductions and their potential negative impacts, resulting in an improved protection of the world oceans. It seems logical to address the most prominent invasion vectors first – as shown above, these are likely ballast water and hull fouling of ships.

Acknowledgements. The manuscript preparation was supported by the European Union FP 6 specific targeted research project DAISIE (SSPI-CT-2003-511202).

References

Carlton JT (1992) Introduced marine and estuarine molluscs of North America: an end-of-the-20th-century perspective. J Shellfish Res 11:489–505

Carlton JT (2001) Introduced species in U.S. coastal waters: environmental impacts and management priorities. Pew Oceans Commission, Arlington, VA

CIESM (2005) Atlas of exotic species in the Mediterranean Sea. www.ciesm.org/atlas

Eno NC (1996) Non-native species in British waters: effects and controls. Aquat Conserv Mar Freshwater Ecosyst 6:215–228

Eno NC, Clark RA, Sanderson WG (1997) Non-native species in British waters: a review and directory. Joint Nature Conservation Committee, Peterborough

Galil B, Zenetos A (2002) A sea change – exotics in the eastern Mediterranean Sea. In: Leppäkoski E, Gollasch S, Olenin S (eds) Invasive aquatic species of Europe: distribution, impacts and management. Kluwer, Dordrecht, pp 325–336

Gollasch S (1996) Untersuchungen des Arteintrages durch den internationalen Schiffsverkehr unter besonderer Berücksichtigung nichtheimischer Arten. Dissertation, Universität Hamburg, Verlag Dr. Kovac, Hamburg, 314 pp

Gollasch S (2006) Overview on introduced aquatic species in European navigational and adjacent waters. Helgoland Mar Res 60:84–89

Gollasch S, Mecke R (1996) Eingeschleppte Organismen. In: Lozan JL, Lampe R, Matthäus W, Rachor E, Rumohr H, v Westernhagen H (eds) Warnsignale aus der Ostsee. Parey, Berlin, pp 146–150

Gollasch S, Lenz J, Dammer M, Andres HG (2000) Survival of tropical ballast water organisms during a cruise from the Indian Ocean to the North Sea. J Plankton Res 22:923–937

Gollasch S, Macdonald E, Belson S, Botnen H, Christensen J, Hamer J, Houvenaghel G, Jelmert A, Lucas I, Masson D, McCollin T, Olenin S, Persson A, Wallentinus I, Wetsteyn B, Wittling T (2002) Life in ballast tanks. In: Leppäkoski E, Gollasch S, Olenin S (eds) Invasive aquatic species of Europe: distribution, impacts and management. Kluwer, Dordrecht, pp 217–231

Gollasch S, Galil BS, Cohen A (eds) (2006) Bridging divides – maritime canals as invasion corridors. Springer, Dordrecht

Höpner Petersen G (1978) Life cycles and population dynamics of marine benthic bivalves from the Disko Bugt area of West Greenland. Ophelia 17:95–120

Höpner Petersen G (1999) Five recent *Mya* species, including three new species and their fossil connections. Polar Biol 22:322–328

ICES (2005) ICES code of practice on the introductions and transfers of marine organisms 2005. ICES, Copenhagen

ICES WGITMO (2004) Report of the ICES working group on introductions and transfers of marine organisms (WGITMO). WGITMO, Cesenatico

Leppäkoski E (1994) Non-indigenous species in the Baltic Sea. In: Boudouresque CF, Briand F, Nolan C (eds) Introduced species in European coastal waters. CIESM, Monaco, Eur Comm Ecosyst Res Rep 8:67–75

Lepp‰koski E, Olenin S (2000) Non-native species and rates of spread: lessons from the brackish Baltic Sea. Biol Invasions 2:151–163

Leppäkoski E, Gollasch S, Olenin S (eds) (2002) Invasive aquatic species of Europe: distribution, impacts and management. Kluwer, Dordrecht

Marcus GJ (1980) The conquest of the North Atlantic. Boydell Press, Woodbridge

Medcof JC (1975) Living marine animals in a ships' ballast water. Proc Natl Shellfish Assoc 65:54–55

Minchin D, Eno NC (2002) Exotics of coastal and inland waters of Ireland and Britain. In: Leppäkoski E, Gollasch S, Olenin S (eds) Invasive aquatic species of Europe: distribution, impacts and management. Kluwer, Dordrecht, pp 267–275

Minchin D, Gollasch S (2002) Vectors – how exotics get around. In: Leppäkoski E, Gollasch S, Olenin S (eds) Invasive aquatic species of Europe: distribution, impacts and management. Kluwer, Dordrecht, pp 183–192

Nehring S (2002) Biological invasions into German waters: an evaluation of the importance of different human-mediated vectors for nonindigenous macrozoobenthic species. In: Leppäkoski E, Gollasch S, Olenin S (eds) Invasive aquatic species of Europe: distribution, impacts and management. Kluwer, Dordrecht, pp 373–383

Ockelmann WK (1958) The zoology of East Greenland. Marine Lamellibranchiata. Medd Grønland 122(4):1–256

Olenin S, Gollasch S, Jonusas S, Rimkute I (2000) En-route investigations of plankton in ballast water on a ship's voyage from the Baltic to the open Atlantic coast of Europe. Int Rev Hydrobiol 85:577–596

Olenin S, Daunys D, Dauniene E (eds) (2005) Baltic sea alien species database. http://www.ku.lt/nemo/mainnemo.html

Ostenfeld CH (1908) On the immigration of *Biddulphia sinensis* Grev. and its occurrence in the North Sea during 1903–1907. Medd Komm Havundersogelser Plankton 1(6):1–25

Petersen KS, Rasmussen KL, Heinemeler J, Rud N (1992) Clams before Columbus? Nature 359:679

Reise K, Gollasch S, Wolff WJ (1999) Introduced marine species of the North Sea coasts. Helgoländer Meeresunters 52:219–234

Streftaris N, Zenetos A, Papathanassiou E (2005) Globalisation in marine ecosystems: the story of non-indigenous marine species across European seas. Oceanogr Mar Biol Annu Rev 43:419–453

Taylor A, Rigby G, Gollasch S, Voigt M, Hallegraeff G, McCollin T, Jelmert A (2002) Preventive treatment and control techniques for ballast water. In: Leppäkoski E, Gollasch S, Olenin S (eds) Invasive aquatic species of Europe: distribution, impacts and management. Kluwer, Dordrecht, pp 484–507

Wallentinus I (2002) Introduced marine algae and vascular plants in European aquatic environments. In: Leppäkoski E, Gollasch S, Olenin S (eds) Invasive aquatic species of Europe: distribution, impacts and management. Kluwer, Dordrecht, pp 27–54

Wolff WJ (2005) Non-indigenous marine and estuarine species in The Netherlands. Zool Med Leiden 79:1–116

5 Waterways as Invasion Highways – Impact of Climate Change and Globalization

Bella S. Galil, Stefan Nehring, and Vadim Panov

5.1 Introduction

The earliest civilizations flourished on the banks of navigable rivers. Indeed, their first monumental hydrological construction projects were concerned with irrigation and transport: around 2200 B.C., the first navigable canal, the Shatt-el-hai, linking the Tigris and Euphrates rivers in Mesopotamia, was excavated; in the 6th century B.C., a canal was built which joined the Nile with the northern Red Sea and, in the 4th century B.C., the Grand Canal in China connected Peking to Hangzhou, a distance of almost 1,000 km. The technological innovations of the 18th century led to an expansion of the network of navigable inland waterways, followed in the 19th century and the early part of the 20th century by the excavation of two interoceanic canals: the Suez Canal, which opened a direct route from the Mediterranean Sea to the Indo-Pacific Ocean, and the Panama Canal, which afforded passage between the Atlantic and Eastern Pacific oceans.

Canals connecting rivers over watersheds or seas across narrow land bridges "dissolve" natural barriers to the dispersal of aquatic organisms, thereby furnishing these with many opportunities for natural dispersal as well as for shipping-mediated transport. The introduction of alien aquatic species has proven to be one of the most profound and damaging anthropogenic deeds – involving both ecological and economic costs. Globalization and climate change are projected to increase aquatic bioinvasions and reduce environmental resistance to invasion of thermophilic biota. Navigable waterways serving as major invasion corridors offer a unique opportunity to study the impact of these processes.

Ecological Studies, Vol. 193
W. Nentwig (Ed.)
Biological Invasions

5.2 The Watery Web – Inland Waterways of Europe

The complex European network of inland waterways was created over a period of more than 200 years (see Ketelaars 2004 for review; Fig. 5.1). The network comprises over 28,000 km of navigable rivers and canals, and extends from the Atlantic Ocean to the Ural Mountains, connecting 37 countries in Europe and beyond. This immense aquatic web connects previously isolated watersheds and has facilitated "all-water" transport from the Caspian and Black seas to the Baltic and North seas and beyond.

Four invasion corridors have been traced between the southern and northern European seas (Jazdzewski 1980; Panov et al. 1999; Nehring 2002; Bij de

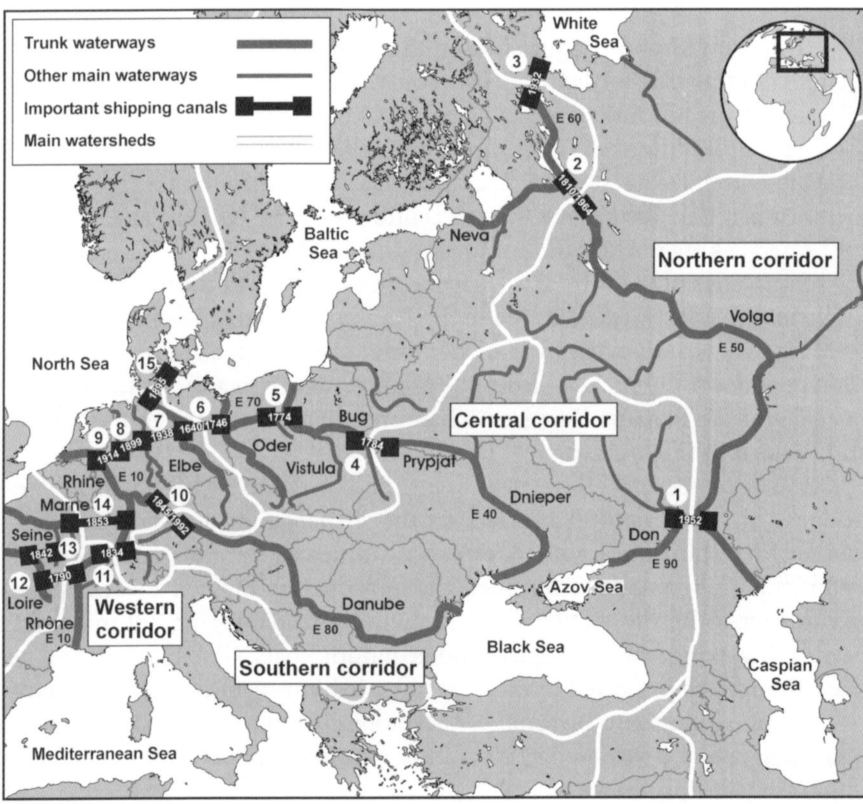

Fig. 5.1 Important European waterways and invasion corridors of aquatic species. Canal number: *1* Volga-Don Canal, *2* Volga-Baltic Canal, *3* White Sea–Baltic Sea Canal, *4* Bug-Prypjat Canal, *5* Vistula-Oder Canal, *6* Havel-Oder Canal, *7* Mittelland Canal, *8* Dortmund-Ems Canal, *9* Rhine-Herne Canal, *10* Ludwig Canal and Main-Danube Canal, *11* Rhine-Rhône Canal, *12* Canal du Centre, *13* Canal de Briar, *14* Rhine-Marne Canal, *15* Kiel Canal (dates of the openings are indicated; for further explanations, see text)

Vaate et al. 2002; Slyn'ko et al. 2002; Van der Velde et al. 2002; Ketelaars 2004). The largest, comprising 6,500 km of main waterways and 21 inland ports of international importance, is the "northern corridor" linking the Black and Azov seas with the Caspian Sea via the Azov-Caspian waterway (E90, including the Volga-Don Canal, opened in 1952, no. 1 in Fig. 5.1), with the Baltic and White seas via the Volga-Baltic waterway (E50, the Volga-Baltic Canal, first opened in 1810, reopened after major reconstruction in 1964, no. 2 in Fig. 5.1), and the White Sea–Baltic Sea waterway (E60, White Sea–Baltic Sea Canal, opened in 1932, no. 3 in Fig. 5.1). The "central corridor" connects the Black Sea with the Baltic Sea region via Dnieper (E40) and the Bug-Prypjat Canal (opened in 1784, no. 4 in Fig. 5.1). The Vistula-Oder Canal (opened in 1774, no. 5 in Fig. 5.1) and the Havel-Oder Canal (first opened in 1640, reopened after major reconstruction in 1746, no. 6 in Fig. 5.1) connect the "central corridor" with the Elbe River and the North Sea. Since 1938, the Elbe is directly connected with the Rhine via the Mittelland Canal (no. 7 in Fig. 5.1), Dortmund-Ems Canal (no. 8 in Fig. 5.1), and the Rhine-Herne Canal (no. 9 in Fig. 5.1; Jazdzewski 1980; Nehring 2002). The "southern corridor" owes its origins to Charlemagne who, in the 8th century, began digging the *Fossa Carolina* between the Rezat, a tributary of the Rhine, and the Altmühl river flowing into the Danube. Heavy rainfall thwarted his plan and over 1,000 years passed before another emperor, Louis I, constructed the Ludwig Canal with 101 locks which connects the Danube with the Main River, a tributary of the Rhine (opened in 1845, destroyed in World War II, no. 10 in Fig. 5.1). Reconstruction of the Main-Danube Canal began in 1959 and, in 1992, the Danube (E80) and the Rhine (E10) were finally connected (Nehring 2002; Bij de Vaate et al. 2002; Van der Velde et al. 2002; WSV 2005). The "western corridor" links the Mediterranean with the North Sea via the Rhône (E10) and the Rhine-Rhône Canal (opened in 1834, no. 11 in Fig. 5.1). Although of little commercial import today, numerous old navigable canals in France and the Benelux countries, including the Canal du Centre (opened in 1790, no. 12 in Fig. 5.1), the Canal de Briar (opened in 1842, no. 13 in Fig. 5.1) and the Rhine-Marne Canal (opened in 1853, no. 14 in Fig. 5.1), connected major river basins and may have served as early dispersal routes for alien species from the Mediterranean to the North Sea basin.

5.3 Aquatic Highways for Invasion

The precise number of aquatic species, primarily of Ponto-Caspian origin, which benefited from the extensive network described above and extended their ranges far and wide is as yet unknown but we estimate that 65 species may have spread through European waterways. Some Ponto-Caspian species are considered as pests: *Dreissena polymorpha*, which spread across Western

Europe in the 19th century, exerts a significant impact upon community structure and functions, by modifying spatial and food chain resources. Although heavy water pollution reduced *Dreissena* populations by the mid-20th century, the improvement of water quality since the 1980s promoted their recovery (Chap. 15). *Dreissena* populations nowadays have again attained densities of up to 40,000 individuals m^{-2} in German waterways (Nehring 2005). The amphipod *Chelicorophium curvispinum*, which spread via the central and southern corridors in the 20th century, has radically altered the communities by covering hard substrates with a layer of muddy tubes up to 4 cm thick (Van der Velde et al. 2002). Since 1996, its population in the Rhine has been reduced from more than 10,000 to 500 individuals m^{-2} because of heavy predation pressure exerted by another Ponto-Caspian amphipod, *Dikerogammarus villosus* (Haas et al. 2002). After its initial introduction in 1995 via the southern corridor into the Main River, *D. villosus* has achieved wide dispersal via the Rhine (Chap. 15) and several canals in northern Germany (Fig. 5.1), and this in record time – by 2000, it was observed more than 1,000 km away in the Oder (Nehring 2005). The phenomenally successful invasive amphipod has become a major component of the macrobenthic assemblages in German waterways, and significantly impacts their ecosystem (Haas et al. 2002).

At least five Mediterranean macroinvertebrate species invaded the Rhine and neighbouring basins through the "western corridor". One of these alien species, the euryhaline isopod *Proasellus coxalis*, has established populations in German inland waters as well in North Sea estuaries (Nehring 2002). The other invasion corridors (northern, central and southern) have served as important routes for Ponto-Caspian species to disperse to the North Sea and Baltic Sea basins. At least six Ponto-Caspian macroinvertebrate species invaded Western Europe waters using the central corridor and one species, *Dreissena polymorpha*, probably dispersed also along the northern corridor (Nehring 2002; Bij de Vaate et al. 2002; Van der Velde et al. 2002). Following the opening of the new Main-Danube Canal in 1992, however, the "southern corridor" has proven to be the most important dispersal route into Western Europe for Ponto-Caspian species.

To date, 14 macroinvertebrate and fish species originating in the Danube have established populations in the Main and Rhine river systems, and some have spread further through the Mittelland Canal into the Ems, Weser and Elbe rivers and up to the Oder (Nehring 2005). Four species from the Rhine have recently been recorded from the Danube (e.g. the clam *Corbicula fluminalis*; Tittizer and Taxacher 1997). Twenty-five of the 44 established alien macrozoobenthic species recorded from inland waters of Germany are considered to have arrived through navigable waterways (Aet Umweltplanung 2006). In the Rhine, more than 20 % of the species and more than 90 % of the biomass are represented by alien species – the Rhine is an "international waterway" in the full sense of the word.

Organisms spread through waterways mainly through larval and postlarval drifting, active dispersal, and transport on ships' hulls. An examination of the hulls and cooling water filters of vessels plying the Danube-Main waterways revealed the presence of six alien species, underscoring the importance of that vector (Reinhold and Tittizer 1999). By contrast, we may owe the high influx of Ponto-Caspian biota through the "southern corridor" to an engineering development: to compensate for the drain deficit of the Main River, more than 100 million m^3 of Danube water are transferred annually through the Main-Danube Canal into the Main (Nehring 2002). It stands to reason that additional organisms originating in the Danube ecosystem and the Ponto-Caspian region will spread via the Main-Danube Canal, especially mobile species which have already been observed in the upper and middle Danube, such as the amphipods *Chelicorophium sowinskyi*, *Dikerogammarus bispinosus* and *Obesogammarus obsesus* (Nehring 2002; Bij de Vaate et al. 2002). Recently, this was proven true – *O. obsesus* was first found in the Rhine in 2004, and further records in 2005 establish this Ponto-Caspian species as new member of the Rhine biocoenosis (Aet Umweltplanung 2006).

Recent invasions of three Ponto-Caspian onychopod crustaceans – *Cercopagis pengoi*, *Evadne anonyx*, *Cornigerius maeoticus* – into the eastern Gulf of Finland (Baltic Sea) may indicate the increasing significance of the Volga-Baltic inland waterway in shipping-mediated long-distance intracontinental transfer of invasive species in a south–north direction (Rodionova and Panov 2005; Rodionova et al. 2005). Southward dispersal via inland waterways has been shown for the northern corridor (Slyn'ko et al. 2002). Among the first invasive alien species established in the basins of the North and Baltic seas, the Chinese mitten crab, *Eriocheir sinensis*, made use of inland waterways to spread to the basins of the Black Sea (Zaitsev and Ozturk 2001), Volga River (Slyn'ko et al. 2002) and Caspian Sea (Robbins et al. 2006). Eastward spread of alien species via the Azov-Caspian waterway (southern part of the northern corridor) is likely the most intensive among the European waterways: since the opening of the Volga-Don Canal in 1952, at least 17 alien species have been introduced to the previously isolated Caspian Sea (Grigorovich et al. 2003), including the invasive Atlantic ctenophore *Mnemiopsis leidyi*, significantly affecting commercial fisheries and the whole Caspian ecosystem (Shiganova et al. 2004).

Estuaries are subjected to a two-sided invasion pressure – both through inland waterways and through coastal activities such as aquaculture – and thus represent hot spots for the occurrence of aquatic alien species (Nehring 2006). Estuarine ports servicing both inland waterways and oceanic shipping are prone to inoculations of trans-oceanic biota and may occasionally promote secondary spread of alien biota upstream. In 1985, the brackish water polychaete *Marenzelleria neglecta* (=*M.* cf. *viridis*) was introduced in ballast water to the German Baltic Sea coast. Within years, soft-bottom community structure was totally changed by this invasive species (Zettler 1997). Since

1996, it has become increasingly abundant in German North Sea estuaries (Nehring and Leuchs, unpublished data). Its spread is attributed to the Kiel Canal (opened 1895, no. 15 in Fig. 5.1), which connects the brackish Baltic Sea (Kiel Bight) with the brackish waters of the Elbe estuary at the North Sea coast.

5.4 Hot and Hotter – the Role of Temperature in European Waterways Invasions

Increasing water temperature – in groundwater, surface runoff, streams or rivers – has a significant impact on the spread of alien species. The Asiatic clam *Corbicula fluminea* was first found in Europe in 1989 at the confluence of the Rhine and Meuse rivers near the port of Rotterdam, by 1990 it was recorded in the Rhine, in 1997 in the Danube, in 1998 in the Elbe, in 2000 in the drainage basin of the Seine River, and in 2003 in the Saône and Rhône rivers and in the Canal du Midi, clearly dispersing along the web of navigational waterways (Vincent and Brancotte 2002). It has been suggested that the successful dispersal of the Asiatic clam in European waters is correlated with winter water minima exceeding 2 °C (Schöll 2000). Seeing that winter inland water temperatures in Germany are frequently below this value, *C. fluminea* should seldom be seen. Yet, man-induced discharge of warmer waters – industrial and residential – into the waterways raises their temperature above 2 °C and promotes the establishment of this species (Fig. 5.2). In fact, downstream from cooling water outlets of power stations, populations of *C. fluminea* reach densities of more than 3,000 individuals m^{-2} (Haas et al. 2002).

The past two decades have seen a dramatic increase in invasion rates of Ponto-Caspian species in the eastern Gulf of Finland (the northern end of the Volga-Baltic waterway): at least 50 % of established alien species were first recorded after 1986. Moreover, the number of Ponto-Caspian aliens established in the eastern Gulf of Finland in the past half century is five times as high as the number of alien species originating elsewhere. The invasions of *Cercopagis pengoi*, *Evadne anonyx* and *Cornigerius maeoticus*, introduced to the gulf by vessels using the "northern corridor", occurred after 1990. This period was characterized by significant declines in shipping activity via the Volga Baltic waterway due to the economic crisis in Russia; yet, environmental changes in the gulf increased its invasibility to warm-temperate Ponto-Caspian species (Panov et al. 2006). Most likely, slight changes in the temperature regime of the eastern Baltic resulted also in the recent range extension of *Dreissena polymorpha*. Distribution of this temperate species in the Baltic Sea in the century and half since its first introduction was limited to latitudes below 60°N, despite available pathways (inland waterways from the Ponto-Caspian to northwest of Russia) and suitable mechanisms of introduction

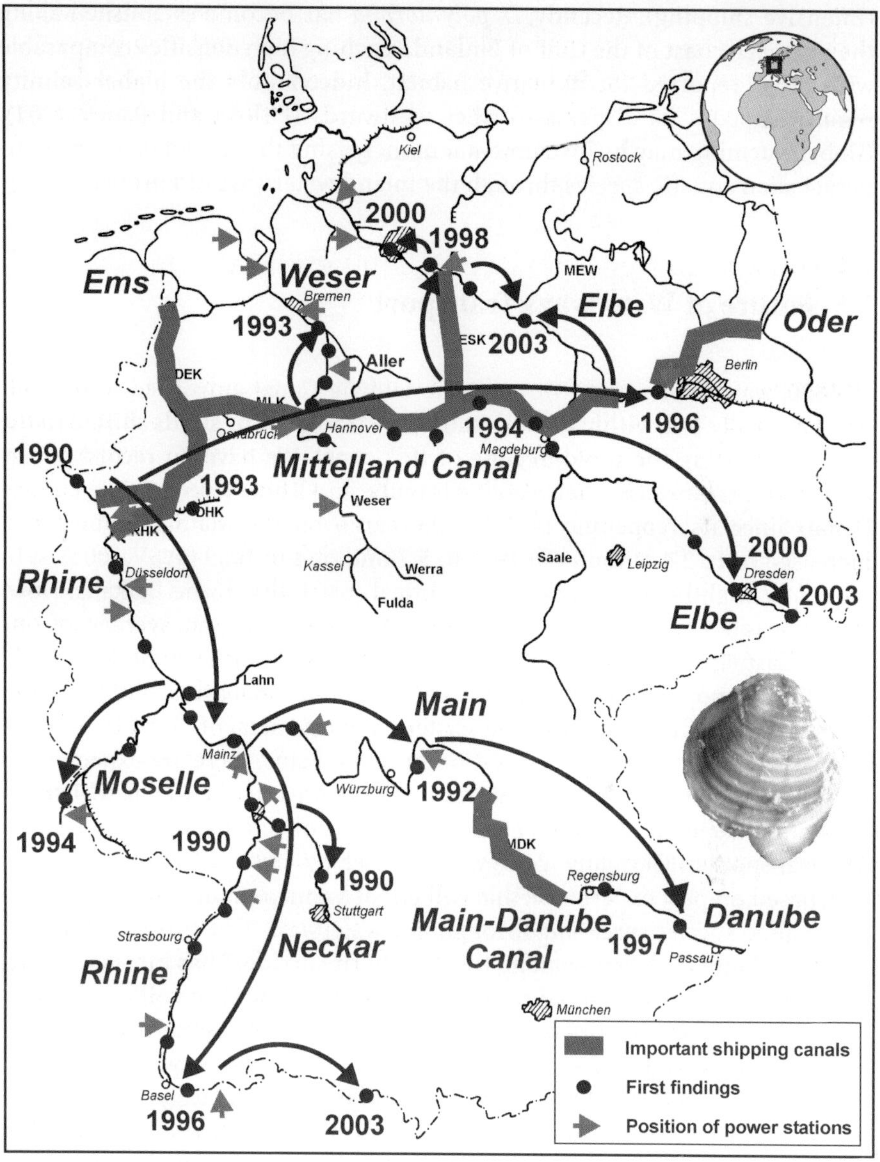

Fig. 5.2 Spread of the invasive Asiatic clam *Corbicula fluminea* in German waterways: first findings, spreading direction, and locations of power stations over 2.5 million MWh (modified after Schöll 2000)

(intensive shipping). Recently, *D. polymorpha* has become established along the northern coast of the Gulf of Finland, reaching high densities comparable with those reported for its native habitat. Indeed, only the higher salinity regime impedes its dispersal further westwards (Orlova and Panov 2004). Global warming may be instrumental in increasing the spread of warm-temperate alien aquatic species through the inland waterways of Europe.

5.5 Future of Waterways Transport

In 1850, waterborne cargo through the Ludwig Canal amounted to 0.2 million t annually but, within a few decades, the transport of goods shifted to the railways, causing the rapid decline of this canal. We have no record of any alien species transferred between the Danube and Rhine rivers via the Ludwig Canal. Since its reopening in 1992, the cargo on the Main-Danube Canal increased from 2.4 million t in 1993 to 5.2 million t in 1999 (WSV 2005), subsequently slightly declining due to political instability in the Balkans. However, the rate of invasion remains constant: since 1992, an average of one Ponto-Caspian species a year arrives through the Main-Danube Canal and establishes a population in the Rhine and neighbouring basins. With the completion of the first round of the eastern enlargement of the EU, and the improvement of the political situation in the Balkans, there are expectations for greater waterborne trade volumes on the Main-Danube Canal, which can accommodate up to 18 million t annually. It is predicted that the inland waterway transport in Germany alone will increase by 43 % by 2015 (WSV 2005) and, based on past experience, this will entail a concomitant rise in the number of alien species spreading through this waterway.

The volume of shipping along the "northern corridor" has increased to its pre-1990 level (20 million t of cargo annually, including crude oil), and is currently limited by poor upkeep of the system, including some derelict locks and waterways, and two strategic bottlenecks: the Azov-Caspian and Volga-Baltic waterways (UN 2005). Russia plans to integrate its waterway network into the European one, focusing on the Volga–Don–Danube corridor. The Russian network is due to open to international shipping by 2010. The expected increase in waterborne transport will doubtlessly be followed by a rise in the number of alien species in this cross-continental system of rivers, canals, lakes and inland seas. Policy and management should be aware that this increase in waterborne transport will facilitate the transfer of invasive species through the European web of inland waterways. Control and reduction of the dispersal of alien species may entail the installation of barriers such as deterrent electrical systems as well as chloride- or pH-altered locks (Clarkson 2004; Nehring 2005).

5.6 Suez and Panama – the Interoceanic Canals

The seawater-fed Suez Canal serves as a nearly unidirectional conduit for Red Sea and Indo-Pacific biota into the Mediterranean. Despite impediments such as the canal's long length and shallowness, and strong variations in turbidity, temperature and salinity, more than 500 Red Sea species have been recorded from the Mediterranean Sea and many have become established along the Levantine coast, with some extending their range westwards to Tunis, Sicily and the Adriatic Sea (Galil 2000). The Suez Canal has provided access for over 80 % of all alien fish, decapod crustaceans and molluscs in the Mediterranean Sea (www.ciesm.org/atlas). Red Sea aliens now dominate the community structure and function of the Levantine littoral and infralittoral zones, having replaced some local populations of native species. Some alien species are considered as pests or cause nuisance whereas other invaders are of commercial value – Red Sea prawns and fish presently constitute nearly half of the trawl catches along the Levantine coast (Goren and Galil 2005).

By contrast, the triple-locked Panama Canal is a freshwater corridor between the Atlantic and Pacific oceans. Although the fresh waters of Lake Gatun connect the Rio Chagres on the Caribbean slope and the Rio Grande on the Pacific slope, facilitating the intermingling of their formerly isolated faunas (Smith et al. 2004), the lake forms an effective barrier to the dispersal of marine biota, which generally cannot tolerate hyposaline conditions. Only seven Atlantic decapod crustacean species have been collected from the Pacific drainage and a single Pacific crab from the Atlantic drainage, but none are known to have established populations outside the canal (Abele and Kim 1989). Apart from the euryhaline Atlantic tarpon, *Megalops atlanticus*, regularly reported near the Pacific terminus of the Panama Canal and around Coiba Island, no fish have established populations along the Pacific coast of Panama beyond the Miraflores lagoon, although several species, predominantly blennies and gobies, have breeding populations in the canal (Hildebrand 1939; McCosker and Dawson 1975; Gunter 1979).

It has been assumed that organisms progress through canals as a result of "natural" dispersal, by autochthonous active or passive larval or adult movements, unaided either directly or indirectly by human activity (other than the construction of the canal as such). Indeed, a temporal succession of directional ("stepping stones") records from the Red Sea, the Suez Canal, and along the coasts of the Levant confirms a species status as a naturally dispersing Red Sea alien. However, dispersal could also result from anthropogenic translocation – already Fox (1926) wrote "It is, of course, well known that ships have in more than one instance dispersed marine organisms from one part of the world to another. This must apply equally to transport through the Suez Canal". Shipping has been implicated in the dispersal of numerous neritic organisms, from protists and macrophytes to fish (Carlton 1985). The transport on the hulls of ships of boring, fouling, crevicolous or adherent species is certainly the most ancient vector of aquatic species introduction. Slower-moving and frequently moored vessels, such as tugs and barges permanently employed in canal operations and maintenance, may have a larger share than other vessels in transport from one end of the canal to the other. Fouling generally concerns small-sized sedentary, burrow-dwelling or clinging species, although larger species characterized by life histories which include a life stage appropriate for such dispersal may be disseminated as well (Zibrowius 1979). Ballast water is usually taken into dedicated ballast tanks or into empty cargo holds when offloading cargo, and discharged when loading cargo or bunkering (fuelling). Water and sediment carried in ballast tanks, even after voyages of several weeks' duration, have been found to contain many viable organisms. Since the volume of ballast water may be as much as a third of the vessel's deadweight tonnage, it engenders considerable concern as a key vector of introduction. However, it is seldom possible to ascertain precise means of transmission, as some organisms may conceivably be transported by several modes (Chap. 4).

In addition to serving as corridors for autochthonous or shipping-based invasions of alien species, canals facilitate aquatic invasions globally by increasing the overall volume of ship-borne trade and changing the patterns of maritime transport. The opening of the Suez Canal in 1869, and the Panama Canal in 1914, had an immediate effect on shipping and trade, markedly altering global shipping routes. The Suez and Panama canals are the world's greatest shortcuts and its densest shipping lanes: about 6 and 3.4 % of total world seaborne cargo passes through these respectively (The Economist, 23 July 2005). What possible effects could climate change and globalization have on marine invasions through these canals?

The Suez Canal has a tropical sea at one end and a subtropical sea at the other, the annual temperature range on the Mediterranean side (15–30.5 °C) being greater than that in the Gulf of Suez (23.5–28.5 °C). Red Sea aliens, originating in tropical waters, require "temperatures high enough for the reproductive processes and development of eggs, and minimum winter temperatures above their lethal limits" to establish populations in the Mediterranean (Ben Tuvia 1966). For some of the most successful Red Sea invasive species, the initiation of explosive population growth coincided with a rise in winter water temperatures. The abrupt rise in the populations of the Red Sea lizard fish *Saurida undosquamis*, the Red Sea goldband goatfish, *Upeneus moluccensis*, and other fish and penaeid prawns was attributed to a rise of 1–1.5 °C in the Levantine surface seawater temperature during the winter months of 1955–1956 (Ben Yami 1955; Chervinsky 1959; Ben Yami and Glaser 1974). The appearance of six Red Sea fish species, in addition to a proliferation of previously rare thermophilic Mediterranean species, in the Adriatic Sea since the mid-1980s was correlated with a rise in eastern Adriatic surface temperatures in 1985–1987 and 1990–1995 (Dulcic and Grbec 2000; Dulcic and Lipej 2002).

Similarly, a considerable increase in the number of Red Sea fish, decapods and molluscs along the south-western Anatolian coast and in the southern Aegean Sea has been attributed to a more extensive inflow of the Asia Minor Current, resulting in a westward flow of warm, salty water from the Levantine Sea (Galil and Kevrekidis 2002; Bilecenoglu et al. 2002; Corsini et al. 2002; Kumulu et al. 2002; Yokes and Galil 2004; Yokes and Rudman 2004; Katagan et al. 2004). Global warming would likely have a significant influence on the establishment and distribution of Red Sea species entering through the Suez Canal. Rising seawater temperature may change the pool of species which could establish themselves in the Mediterranean, enable temperature-limited Red Sea species to expand beyond their present distributions in the Mediterranean, and impact on a suite of population characteristics (reproduction, survival) determining interspecific interactions and, therefore, the dominance and prevalence patterns of alien species, providing Red Sea aliens with a distinct advantage over native Mediterranean biota.

The Panama Canal has a tropical sea at either end but, although the annual temperature range on the Pacific side is greater than that on the Atlantic side, due to seasonal upwelling and episodic El Niño events, the "rigorous physical perturbations" on the Caribbean side mean that the former is "characterized by the presence of rich stenothermal biotic communities" (Glynn 1972). A shift in weather patterns may have incalculable biotic consequences across the isthmus.

5.7 Globalization and Shipping – "Size Matters"

Expanding global trade engenders greater volumes of shipping, and economic development of new markets brings about changes to shipping routes. The world's seaborne trade amounted to more than 6.7 billion t in 2004. Almost 40 % of the cargo originated in Asia, and much of it was destined for Europe and North America (The Economist, 26 November 2005).

The Suez Canal benefited from the development of Middle Eastern oilfields, being closely associated with the oil trade from the Gulf – oil shipments constituted over 70 % of total traffic volume in 1966 (Quéguiner 1978). The closure of the canal in 1967–1975 launched a rapid increase in tanker sizes and the emergence of VLCC (very large crude carriers, with capacities of 150,000–300,000 t) and ULCC (ultra-large crude carriers, with capacities exceeding 300,000 t) vessels specifically designed for long-haul routes. Although recent tanker traffic has been competing with the SUMED pipeline for the transmission of oil from the Gulf of Suez to the Mediterranean and the alternate route around the Cape of Good Hope, thousands of laden and partly laden oil tankers transit the canal annually, transporting about 1.3 million barrels day^{-1} (=174,200 t day^{-1}; www.eia.doe.gov/emeu/cabs). At present, the

Suez Canal accommodates Suezmax class tankers of 200,000 t maximum cargo. In order to attract larger vessels to use the waterway, the Suez Canal Authority has been expanding the channel to accommodate ULCCs with oil cargos of up to 350,000 t by 2010.

Whereas earlier progress through the Suez Canal might have been restricted to euryhaline and generally hardy species, it is now mainly depth-restricted. Formerly, most Red Sea aliens occupied the Mediterranean littoral and infralittoral to a depth of 60 m and, with few exceptions, were not found in deeper waters (Galil 1989; Golani 1996; Bilecenoglu and Taskavak 1999). However, recent records of the typically deepwater Red Sea molluscs *Ergalatex contracta* Huart 1996 and *Mactrinula tryphera* Melvill 1899 off the Levantine coast conceivably indicate the general entry of deepwater invaders (Mienis 2004). The increase in canal depth to accommodate larger vessels not only facilitates the invasion of species showing upper depth ranges (as adults or larvae) which otherwise would not permit passage but, in addition, the enlargement of the canal increases current velocities (Soliman 1995). Implications of faster current on the transport of biota through the canal are clear: "With gradually improving chances for planktonic larvae to pass the Canal a steeply increasing invasion of Red Sea animals into the Mediterranean can be expected – an immigration which in a not too far future might radically change the whole faunal composition of its eastern basin" (Thorson 1971: p. 846). The profound changes wrought on the eastern Mediterranean biota commenced with the opening of the Suez Canal. The influx of Red Sea biota is rooted in the continuous expansion of the canal, which has altered its hydrography and hydrology and enhanced its potential as a "corridor" facilitating the passage of greater numbers of organisms.

Half of the cargo transiting the Panama Canal originates in or is destined for US ports, and China and Japan are the next-biggest users. The share of the world's trade transiting through the Panama Canal was reduced from 5.6 % in 1970 to 3.4 % in 2004 because the 80,000 dead-weight-ton Panamax, the largest vessels able to traverse the canal, are only half as large as newly built container ships. Vessels too big for the waterway use other routes to travel between Asia and the US coast. Alternatively, goods are either unloaded at US West Coast ports and transported by road and rail or they are trans-shipped between Panama's Atlantic and Pacific ports. To maintain its market share, the Panama Canal Authority plans to construct a third lane and new locks to accommodate ships twice as big as Panamax vessels (The Economist, 23 July 2005).

For decades, the permissible draft of the Suez Canal and the dimensions of the locks in the Panama Canal determined ship sizes, and delayed and limited the construction of more stable, larger vessels with smaller ratios of ballast water and hull surface to cargo, thus impacting on patterns of shipping-transported biota worldwide. The Suez Canal and the Panama Canal have induced profound economic, political and social changes, facilitating globalization by reducing costs of ship-borne cargo. By increasing maritime trade, contracting and altering shipping routes, and influencing vessel size, both canals have had profound impacts on ship-borne bioinvasions. Whereas the Suez Canal has been serving as a veritable gateway for Red Sea species into the Mediterranean Sea, the waters of Lake Gatun and the upper locks of the Panama Canal have reduced its suitability for marine biota.

Acknowledgements. This review has been supported by the European Commission with the FP 6 integrated project ALARM (contract GOCE-CT-2003-506675, VP) and the specific targeted research project DAISIE (contract SSPI-CT-2003-511202, BG and VP).

References

Abele LG, Kim W (1989) The decapod crustaceans of the Panama Canal. Smithsonian Contrib Zool 482:1–50

Aet Umweltplanung (2006) Aquatic aliens in German inland and coastal waters. http://www.aquatic-aliens.de

Ben Tuvia A (1966) Red Sea fishes recently found in the Mediterranean. Copeia 2:254–275

Ben Yami M (1955) 1954-1955: over-fishing or bad season (in Hebrew)? Fish Bull Haifa 6:10–14

Ben Yami M, Glaser T (1974) The invasion of *Saurida undosquamis* (Richardson) into the Levant Basin – an example of biological effect of interoceanic canals. Fish Bull Haifa 72:359–373

Bij de Vaate A, Jazdzewski K, Ketelaars HAM, Gollasch S, Van der Velde G (2002) Geographical patterns in range extension of Ponto-Caspian macroinvertebrate species in Europe. Can J Fish Aquat Sci 59:1159–1174

Bilecenoglu M, Taskavak E (1999) General characteristics of the Turkish marine ichthyofauna. Zool Middle East 18:41–56

Bilecenoglu M, Taskavak E, Bogac Kunt K (2002) Range extension of three Lessepsian migrant fish (*Fistularia commersoni*, *Sphyraena flavicauda*, *Lagocephalus suezensis*) in the Mediterranean Sea. J Mar Biol Assoc UK 82:525–526

Carlton JT (1985) Transoceanic and interoceanic dispersal of coastal marine organisms: the biology of ballast water. Oceanogr Mar Biol Annu Rev 23:313–371

Chervinsky J (1959) A systematic and biological comparison between the lizardfish (*Saurida grandisquamis*) from the Mediterranean and the Red Sea (with English abstract). Fish Bull Haifa 19:10–14

Clarkson RW (2004) Effectiveness of electrical fish barriers associated with the Central Arizona Project. N Am J Fish Manage 24:94–105

Corsini M, Kondilatos G, Economidis PS (2002) Lessepsian migrant *Fistularia commersonii* from the Rhodes marine area. J Fish Biol 61:1061–1062

Dulcic J, Grbec B (2002) Climatic change and Adriatic ichthyofauna. Fish Ocean 9:187–191

Dulcic J, Lipej L (2002) Rare and little-known fishes in the Eastern Adriatic during last two decades (1980-2001). Periodicum Biol 104:185–194

Fox HM (1926) I. General part. Zoological results Cambridge expedition to the Suez Canal, 1924. Trans Zool Soc Lond 22:1–64

Galil BS (1989) Bathymetric distribution and habitat preferences of Lessepsian migrant Decapoda along the Mediterranean coast of Israel, or: do decapods have cold feet. In: Spanier E, Steinberger Y, Luria M (eds) Environmental quality and ecosystem stability. ISEEQS, Jerusalem, vol IV-B, pp 147–153

Galil BS (2000) A sea under siege – alien species in the Mediterranean. Biol Invasions 2:177–186

Galil BS, Kevrekidis K (2002) Exotic decapods and a stomatopod off Rhodes Island (Greece) and the Eastern Mediterranean Transient. Crustaceana 75:925–930

Glynn PW (1972) Observations on the ecology of the Caribbean and Pacific coasts of Panama. Bull Biol Soc Washington 2:13–30

Golani D (1996) The marine ichthyofauna of the eastern Levant – history, inventory and characterization. Israel J Zool 42:15–55

Goren M, Galil BS (2005) A review of changes in the fish assemblages of Levantine inland and marine ecosystems following the introduction of non-native fishes. J Appl Ichthyol 21:1–7

Grigorovich IA, Therriault TW, MacIsaac HJ (2003) History of aquatic invertebrate invasions in the Caspian Sea. Biol Invasions 5:103–115

Gunter G (1979) Marine fishes of Panama as related to the Canal. Gulf Res Rep 6:267–273

Haas G, Brunke M, Streit B (2002) Fast turnover in dominance of exotic species in the Rhine river determines biodiversity and ecosystem function: an affair between amphipods and mussels. In: Leppäkoski E, Olenin S, Gollasch S (eds) Alien species in European waters. Kluwer, Dordrecht, pp 426–432

Hildebrand SF (1939) The Panama Canal as a passageway for fishes, with lists and remarks on the fishes and invertebrates observed. Zoologica 24:15–45

Jazdzewski K (1980) Range extensions of some gammaridean species in European inland waters caused by human activity. Crustaceana suppl 6:84–106

Jazdzewski K, Konopacka A, Grabowski M (2002) Four Ponto-Caspian and one American gammarid species (Crustacea, Amphipoda) recently invading Polish waters. Contrib Zool 71:115–122

Katagan T, Kocatas A, Zengin M, Ates AS (2004) An indo-pasific Stomatopod from the sea of Marmara: *Erugosquilla massavensis* (Kossmann, 1880). Crustaceana 77:381–383

Ketelaars HAM (2004) Range extensions of Ponto-Caspian aquatic invertebrates in continental Europe. In: Dumont HJ, Shiganova TA, Niermann U (eds) Aquatic invasions in the Black, Caspian, and Mediterranean seas. Kluwer, Dordrecht, pp 209–236

Kumulu M, Eroldogan OT, Aktas M, Gocer M (2002) A new shrimp record for the Turkish Seas: *Melicertus Hathor* (Burkenroad, 1959) (Penaeidae: Crustacea). Israel J Zool 48:246–247

Leppäkoski E, Gollasch S, Gruszka P, Ojaveer H, Olenin S, Panov V (2002) The Baltic – a sea of invaders. Can J Fish Aquat Sci 59:1175–1188

McCosker JE, Dawson CE (1975) Biotic passage through the Panama Canal, with particular reference to fishes. Mar Biol 30:343–351

Mienis H (2004) New data concerning the presence of Lessepsian and other Indo-Pacific migrants among the molluscs in the Mediterranean Sea with emphasize on the situation in Israel. Turkish J Aquat Life 2:117–131

Nehring S (2002) Biological invasions into German waters: an evaluation of the importance of different human-mediated vectors for nonindigenous macrozoobenthic species. In: Leppäkoski E, Olenin S, Gollasch S (eds) Alien species in European waters. Kluwer, Dordrecht, pp 373–383

Nehring S (2005) International shipping – a risk for aquatic biodiversity in Germany. In: Nentwig W, Bacher S, Cock MJW, Dietz H, Gigon A, Wittenberg R (eds) Biological invasions – from ecology to control. Neobiota 6:125–143

Nehring S (2006) Four arguments why so many alien species settle into estuaries, with special reference to the German river Elbe. Helgoland Mar Res 60:127–134

Orlova MI, Panov VE (2004) Establishment of the zebra mussel, *Dreissena polymorpha* Pallas, in the Neva Estuary (Gulf of Finland, Baltic Sea): distribution, population structure and possible impact on local unionid bivalves. Hydrobiologia 514:207–217

Panov VE, Krylov PI, Telesh IV (1999) The St. Petersburg harbour profile. In: Gollasch S, Leppäkoski E (eds) Initial risk assessment of alien species in Nordic coastal waters. Nordic Council of Ministers, Copenhagen, pp 225–244

Panov VE, Bychenkov DE, Berezina NA, Maximov AA (2003) Alien species introductions in the eastern Gulf of Finland: current state and possible management options. Proc Est Acad Sci Biol Ecol 52:254–267

Panov VE, Rodionova NV, Bolshagin PV, Bychek EA (2006) Invasion biology of Ponto-Caspian onychopod cladocerans (Crustacea: Cladocera: Onychopoda). Hydrobiologia (in press)

Quéguiner J (1978) The Mediterranean as a maritime trade route. Ocean Manage 3:179–189

Reinhold M, Tittizer T (1999) The transport of macroinvertebrates by a river vessel's cooling water filter (in German). Wasser Boden 51:61–66

Robbins RS, Sakari M, Baluchi N, Clark PF (2006) The occurrence of *Eriocheir sinensis* H. Milne Edwards, 1853 (Crustacea: Brachyura: Varunidae) from the Caspian Sea region. Iran Aquat Invasions 1:31–32

Rodionova N, Panov V (2005) Establishment of the Ponto-Caspian predatory cladoceran *Evadne anonyx* in the eastern Gulf of Finland, Baltic Sea. Aquat Invasions 1:7–12

Rodionova NV, Krylov PI, Panov VE (2005) Invasion of the Ponto-Caspian predatory cladoceran *Cornigerius maeoticus maeoticus* (Pengo, 1879) into the Baltic Sea. Oceanology 45:66–68

Schöll F (2000) Temperature as a regulative factor for the dispersal of *Corbicula fluminea* (O.F. Müller 1774) (in German). Hydrol Wasserbewirtsch 44:318–321

Shiganova TA, Dumont HJ, Sokolsky AF, Kamakin AM, Tinenkova D, Kurasheva EK (2004) Population dynamics of *Mnemiopsis leidyi* in the Caspian Sea, and effects on the Caspian ecosystem. In: Dumont HJ, Shiganova TA, Niermann U (eds) Aquatic invasions in the Black, Caspian, and Mediterranean seas. Kluwer, Dordrecht, pp 71–111

Slyn'ko YV, Korneva LG, Rivier IK, Shcherbina KH, Papchenkov VG, Orlova MI, Therriault TW (2002) Caspian-Volga-Baltic invasion corridor. In: Leppäkoski E, Olenin S, Gollasch S (eds) Alien species in European waters. Kluwer, Dordrecht, pp 339–411

Smith SA, Bell G, Bermingham E (2004) Cross-cordillera exchange mediated by the Panama Canal increased the species richness of local freshwater fish assemblages. Proc R Soc Lond Biol Sci 271:1889–1896

Soliman GF (1995) The influence of the last widening and deepening of the Suez Canal on its tidal motion. Bull Natl Inst Oceanogr Fish 21:35–66

Thorson G (1971) Animal migrations through the Suez Canal in the past, recent years and the future (a preliminary report). Vie Milieu suppl 22:841–846

Tittizer T, Taxacher M (1997) First record of *Corbicula fluminea/fluminalis* (Müller 1774) (Corbiculidae, Mollusca) in the Danube (in German). Lauterbornia 31:103–107

UN (2005) Inventory of most important bottlenecks and missing links in the European waterway network. Resolution no 49. United Nations Organization, New York

Van der Velde G, Nagelkerken I, Rajagopal S, Bij de Vaate A (2002) Invasions by alien species in inland freshwater bodies in Western Europe: the Rhine delta. In: Leppäkoski E, Olenin S, Gollasch S (eds) Alien species in European waters. Kluwer, Dordrecht, pp 360–372

Vincent T, Brancotte V (2002) Present distribution and spreading modes of *Corbicula* spp. in France. Bull Soc Zool Fr 127:241–252

WSV (2005) Wasser- und Schifffahrtsverwaltung des Bundes. http://www.wsv.de

Yokes B, Galil BS (2004) New record of alien decapods from the southwestern coast of Turkey. Rapp Réunions Comm Int Explor Sci Mer Méditerr 37:556

Yokes B, Rudman WB (2004) Lessepsian opisthobranchs from southwestern coast of Turkey; five new records for Mediterranean. Rapp Réunions Comm Int Explor Sci Mer Méditerr 37:557

Zaitsev YU, Ozturk B (eds) (2001) Exotic species in the Aegean, Marmara, Black, Azov and Caspian seas. Turkish Marine Research Foundation, Istanbul

Zettler M (1997) Populations dynamics, growth and production of the neozoon *Marenzelleria* cf. *viridis* (Verrill, 1873) (Polychaeta: Spionidae) in a coastal water of the southern Baltic Sea. Aquat Ecol 31:177–186

Zibrowius H (1979) Serpulidae (Annelida Polychaeta) de l'océan Indien arrivés sur des coques de bateaux à Toulon (France). Rapp Réunions Comm Int Explor Sci Mer Méditerr 25/26(4):133–134

Section II

Traits of a Good Invader

Short Introduction

Wolfgang Nentwig

One of the striking features of successful invasive species is that they are a minority among alien species. Aliens, again, are only a small subgroup of all species. Consequently, the process of becoming a successful invasive alien species can be understood as a strong selection sequence. An assemblage of special characteristics, also called traits, is obviously important to belong to the spearhead of aliens.

Determining the traits of successful invaders ought to be a very rewarding task because it would enable us to determine the characteristics of the most harmful invaders. This would facilitate early diagnosis of potential invaders and would also offer a promising approach to avoid alien invasive species. It is obvious that scientists have been focussing on the nature of these traits for a long time but, do they really exist? At least, we do not get any simple answers for different groups of invasive alien species which have been investigated in this respect (Chaps. 7 and 8).

How do invasive species fit into the framework of ecological theories? Is it true, as some have suggested, that aliens are inherently superior to native species because they are so successful? And, what is the basis of this superiority? It has also been suggested that invasives possess novel weapons which might explain their success. This super-weapon theory may sound strange but was, among others, responsible for the resurrection of allelopathy in plants and intensified the discussion of gradual differentiation in chemical defence among alien populations. If successful aliens principally conquer only empty niches or develop a loss of defence against natural enemies, as some theories suggest, this would bear good prospects for our battle. Or are such universal theoretical explanations not applicable to the reality of a strange mixture of invasive alien species (Chap. 6)?

Short Introduction

6 Integrating Ecological and Evolutionary Theory of Biological Invasions

Ruth A. Hufbauer and Mark E. Torchin

6.1 Introduction

While research on biological invasions is becoming more predictive (e.g., Mack 1996; Kolar and Lodge 2001; Peterson 2003; Arim et al. 2006; Mitchell et al. 2006), significant challenges lie ahead. Indeed, it is still not clear what leads some introduced species to remain benign while others become aggressive invaders. Here, we review some principal ecological and evolutionary hypotheses employed to explain biological invasions. We present an overview of these hypotheses, and suggest approaches to integrate them into a more comprehensive framework that will allow potential interactions among them to be examined.

Biological invasions are spatially and temporally continuous processes, encompassing transport, establishment and spread phases (Sakai et al. 2001). We focus here on the spread, or demographic expansion, of non-native species that are established, since this stage will ultimately determine an invader's impact in a novel environment. Demographic expansions of introduced species can encompass changes within individuals, such as increase in size or fecundity, and within populations, such as increase in geographic spread and density. We refer to demographic expansions of introduced species as invasion success.

Generally, a species is considered invasive if it has significant ecological, environmental or economic impacts in its novel range. Measuring such impacts is not trivial, and thus categorizing introduced species as invasive is often vague and inconsistent. Three issues make measuring impact difficult. First, not all populations of an invasive species will exhibit the same demographic patterns, with some expanding rapidly or attaining high densities and others remaining small. Second, invasive species may not have parallel ecological, environmental and economic impacts where they invade. While these different types of impacts are generally linked, in some cases introduced

Ecological Studies, Vol. 193
W. Nentwig (Ed.)
Biological Invasions

species can have significant negative ecological and environmental effects while having positive economic effects (e.g., cows). Third, it is important to have a baseline against which changes associated with invasion can be judged. We argue that comparing the ecology of an introduced species to that of populations in its native range will set the most relevant baseline by which to measure changes in ecology resulting from translocation. If we first understand the causes of demographic variation among populations within the native range, then we can compare introduced populations to this gradient (Torchin et al. 2001).

For some species, population demographics may be similar between the native and novel ranges, and mechanisms of their success may be no different. Such species may be quite benign when introduced, or if they cause economic problems in their native range, then often they will in their introduced range as well (e.g., insect pests such as Western corn root worm, *Diabrotica virgifera*, which is a native pest in North America, and an invasive pest in Europe; Chap. 2). If a species is more abundant, dense, or widespread in the novel range than in the native range, then – by definition – something has fundamentally changed in its ecology or perhaps evolution. Because invasiveness is a combined function of the invaded community and the invader, the changes leading to greater success in the new range can be extrinsic changes in the environment that favor the invading species, or they can be intrinsic to the invading species. We refer to species that experience significant positive demographic changes that contribute to invasion success as strong invaders, and those with no change or significant negative changes as weak invaders. As Hierro et al. (2005) point out, there are remarkably few comparative data between the native and novel ranges of species documenting whether the demography of invaders changes (but see Torchin et al. 2001).

Introduced species that are weak invaders can be important economically and environmentally, and also give rise to interesting and urgent ecological questions. However, weak invaders are less useful in helping us address questions regarding fundamental ecological and evolutionary changes that appear to underlie the most damaging invasions (e.g., tamarisk and zebra mussel in North America). To further an understanding of the causes of biological invasion, it is vital to know whether or not most species considered to be "invasive" are strong invaders that have experienced dramatic demographic changes relative to their native range.

Herein, we propose a metric to quantify the continuum from weak invaders to strong invaders. Response ratios are used to compare the means of experimental treatments (X_E) and controls (X_C), where $R=X_E/X_C$ (Hedges et al. 1999). If $R>1$, then the experimental treatment is larger than the control, and if $R<1$, then the experimental treatment is smaller than the control. For comparative research on biological invasions, the response ratio we propose is the ratio of a measure of performance in the introduced range (P_I) to that in the native range (P_N). Thus, this metric measure of invasion success represents a relative measure of average individual or population performance in the introduced versus native environments. The response ratio, R, is simply the ratio of performance in the introduced to the native environments ($R=P_I/P_N$), and provides a quantitative measure of demographic change between the introduced and native environment. A response ratio of 1 indicates no change between the introduced and native range. A response ratio greater than 1 shows positive change, and a response ratio less than 1 shows a negative change relative to the native environment. If P_I and P_N are normally distributed, and P_N is unlikely to be negative, then the log of the response ratio (L) is approximately normally distributed (Hedges et al. 1999), making it a statistically tractable metric. Replicate measures of either individual or population performance (e.g., average body size, fecundity, seed set) are required to evaluate whether R differs significantly from 1 (or L from zero) for a given species.

Strong invasion ($R>1$) can have one or more of several underlying causes, such as important changes in the biotic environment (e.g., reduced competition or reduced suppression by natural enemies), or intrinsic evolutionary changes associated with the introduction. Quantifying whether strong invasion occurs is simply a first fundamental step (that has often been skipped) in coming to a more general understanding of the causes of biological invasions.

The response ratio (R) that we have defined for demographic change in introduced species is linked to whether or not a species is labeled as "invasive" (Fig. 6.1). Nevertheless, as labels of invasiveness typically are based on subjective assessments of ecological and economic impacts, R will not perfectly predict invasiveness. The outliers, however, are likely to be particularly interesting cases.

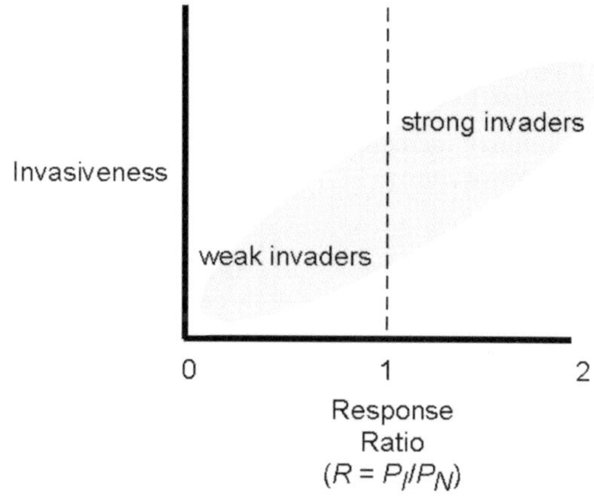

Fig. 6.1 Relationship between the response ratio and invasiveness. When $R=1$, the populations in the introduced range have not experienced demographic change (see text). Strong invaders are those with $R>1$. Weak invaders are those with $R=1$. While some populations or species may be considered quite invasive and have $R<1$, we predict that few species will lie in the *upper left quadrant*. While some species with $R>1$ may not be considered ecologically and economically invasive, we predict few species will lie in the *lower right quadrant*

6.2 Hypotheses to Explain Biological Invasion

Demographic expansions of invasive species are influenced by multiple biotic and abiotic factors. Our aim is to review ecological and evolutionary hypotheses of invasion success, and to explain relationships among biotic factors, including characteristics of both the invaded communities and the invaders themselves (Table 6.1). We evaluate whether the hypotheses can help explain strong invasion, or not. We first outline multiple hypotheses, and then focus on four: two ecological (enemy release, biotic resistance) and two evolutionary (evolution of increased competitive ability, hybridization). We further explore links between these four hypotheses. Most existing work integrating across hypotheses focuses on ecological hypotheses only (though see Facon et al. 2006). We offer ideas connecting ecological and evolutionary hypotheses, and suggest ways in which a more clear and integrated framework may help explain seemingly contradictory results.

Table 6.1 Main ecological and evolutionary hypotheses to explain invader success following introduction, and whether they can help explain strong invasion (R>1)

Hypotheses	Selected early and recent references	R>1
Ecological hypotheses		
Preadaptation/disturbance	Baker and Stebbins (1965), Sax and Brown (2000)	No
Inherent superiority	Elton (1958), Sax and Brown (2000)	No
Novel weapons	Callaway and Ridenour (2004)	Yes
Empty niche	Elton (1958), Hierro et al. (2005)	No
Mutualist facilitation/ invasional meltdown	Simberloff and Von Holle (1999), Richardson et al. (2000)	No
Biotic resistance	Elton (1958), Levine et al. (2004)	Yes
Enemy release	Darwin (1859), Torchin and Mitchell (2004)	Yes
Evolutionary hypotheses		
Hybridization	Ellstrand and Schierenbeck (2000)	Yes
Evolution of increased competitive ability	Blossey and Nötzold (1995)	Yes
Founder events	Mayr (1954)	Yes
Genetic architecture	Lee (2002)	No

6.2.1 Ecological Hypotheses

There are multiple ecological hypotheses to explain invasions (Table 6.1). In addition to this review, the reader is referred to Sax and Brown (2000), Hierro et al. (2005) and Mitchell et al. (2006). When a species establishes in a new range, it may be successful if it is *preadapted* to facets of the new environment (Baker and Stebbins 1965; Sax and Brown 2000). Such a species may have a long association with humans and ecosystems modified by humans that native species do not have. For example, Eurasian species that have long been adapted to agricultural settings may be quick to invade regions where agriculture is relatively new. The native species in those regions are not adapted to the agricultural setting, and are at a disadvantage when preadapted species from an agricultural area are introduced. While this hypothesis may explain the success of introduced species relative to native species, it does not explain how introduced species might be more successful in the new range relative to the native range, becoming strong invaders. Such species might lie in the upper left quadrant of Fig. 6.1, being relatively invasive, yet having low *R*.

Several hypotheses suggest that invasive species are in some way simply inherently superior to the native species in the communities that they invade (e.g., Elton 1958; Sax and Brown 2000). For example, invaders may be superior competitors (Sax and Brown 2000), or they may be superior predators (e.g., brown tree snakes; Wiles et al. 2003). A related idea, the *novel weapons hypothesis*, proposes that invasive species possess biochemical weapons that are novel to the native inhabitants of the invaded community (Callaway and Ridenour 2004; Vivanco et al. 2004), but are less effective in the native range where the competing species have evolved in the presence of those weapons. Much of the evidence for this intuitively appealing hypothesis comes from plants in the genus *Centaurea* that are thought to be allelopathic (Callaway and Ridenour 2004; Vivanco et al. 2004). However, there is some debate about the evidence for allelopathy in these systems (Blair et al. 2005).

The *empty niche hypothesis* posits that invasive species are able to use resources not used by native species, or use them more efficiently and thereby create a new ecological niche in a community (Elton 1958). This hypothesis may help explain invasion, but not strong invasion, unless the niche used in the novel range is in some way more favorable or broader than that used in the native range. This hypothesis is linked to the idea that species-rich communities are more difficult to invade than species-poor communities, which is discussed below under the biotic resistance hypothesis.

In several respects, preadaptation, inherent superiority, and the empty niche hypotheses are related. Each proposes that the invaded environment is suitable for invasion from the outset, and that similarly, the invasive species has the capability of invading that environment without any intrinsic ecological or evolutionary changes being required. Thus, when preadaptation, inherent superiority, or an empty niche lead to strong invasion, the external environment of the new range must be more favorable to the invasive species than its native range.

A positive feedback between introduced species may enable them to become invasive, a process termed *invasional meltdown* (Simberloff and Von Holle 1999; Richardson et al. 2000). Often, mutualists are important in facilitating invasional meltdown. Similarly to the enemy release hypothesis described below, mutualists are likely to be lost during the process of invasion. Some species rely upon mutualists for successful passage through crucial stages of their life cycles. For example, obligatorily outcrossing plants that rely upon pollination can not reproduce sexually without a pollinator. Other species engage in mutualisms that are not obligate but that increase their fitness in many situations, such as plants harboring mycorrhizal fungi. For introduced species engaged in mutualisms to become invasive, their mutualists must be replaced by mutualists from the novel range, or their mutualists from the native range must be introduced to the novel range as well. Without their mutualists, they may remain latent for years. For example, Pierce's disease (*Xylella fastidiosa*), a bacterial disease of many woody plants, did not become invasive in North America until an efficient vector (glassy-winged sharpshooter, *Homalodisca coagulata*) was introduced (Redak et al. 2004). While the hypothesis that mutualisms facilitate invasions is well-supported, it may not explain strong invasion unless other factors and interactions (in the case of Pierce's disease, the availability of many susceptible hosts) also come into play.

The invasion process may "filter out" parasites (including specialized herbivores), pathogens, or other natural enemies that occur in an invading host's native range through several mechanisms (Keane and Crawley 2002; Torchin et al. 2003). The *enemy release hypothesis* (ERH) posits that this filtering process releases introduced species from the top-down population regulation exerted in the native range, enabling strong invasion. A key prediction of the ERH is that introduced populations harbor fewer natural enemies compared to populations within their original range (Williams 1954; Elton 1958). Additionally, it is often postulated that there should be a corresponding shift to a higher proportion of generalist enemies relative to the native range, since generalists are more likely to shift to novel species. Another prediction of the ERH is that introduced species may gain a competitive edge because they are less likely to be affected by natural enemies than are native competitors (Elton 1958; Keane and Crawley 2002). These key predictions of the ERH are not mutually exclusive, but the mechanisms may not occur simultaneously. Growing evidence indicates that introduced species have fewer enemies than where they are native (reviewed in Torchin and Mitchell 2004). Whether this loss of enemies drives the unusual demographic expansion of some introduced species remains equivocal (e.g., Lampo and Bayliss 1996; Beckstead and Parker 2003; Reinhart et al. 2003). Additionally, the extent to which introduced species suffer less from enemies relative to native competitors (and how this affects competitive interactions) remains uncertain (e.g., Settle and Wilson 1990; Calvo-Ugarteburu and McQuaid 1998; Blaney and Kotanen 2001).

The *biotic resistance hypothesis* (BRH) focuses on the communities into which species are introduced, and aims to understand how they differ in their ability to resist invasion (Elton 1958; Tilman 1997; Levine and D'Antonio 1999; Maron and Vilà 2001), and to suppress population growth of invaders (Levine et al. 2004). Most studies of biotic resistance have focused on competitors and the role of species diversity in shaping competitive regimes. However, there is growing appreciation that the importance of this process varies strongly with the setting in which competitive interactions occur. For example, the relative abilities of species to compete successfully is influenced by their interactions with higher trophic levels (e.g., Schierenbeck et al. 1994; Courchamp et al. 2000; Chase et al. 2002). Invading species also experience resistance directly from enemies that are able to use the novel organisms as prey or hosts. Introduced species invading communities with close relatives are more likely to accumulate natural enemies and experience stronger competition than is the case in the absence of relatives (Strong et al. 1984; Mack 1996; Torchin and Mitchell 2004). Hence, the enemy release and the biotic resistance hypotheses are fundamentally linked.

6.2.2 Evolutionary Hypotheses

Most research on the mechanisms underlying biological invasions has focused on the ecological explanations outlined above, despite the pioneering symposium of Baker and Stebbins (1965) addressing the potentially critical role of evolution in the success of colonizing species. For evolution to contribute to the success of introduced species, it must increase relative fitness of individuals in the population, and thereby increase population growth rates. A challenge is understanding whether and how evolution may lead to more rapid increases in fitness in introduced populations than in native populations, leading to strong invasions.

The *evolution of increased competitive ability* (EICA) hypothesis is an evolutionary corollary to the enemy release hypothesis that proposes that escape from natural enemies alters the selection regime, such that costly defenses against enemies no longer enhance fitness. The evolutionary loss of defenses enables resources to be directed toward growth and reproduction, or other traits influencing performance (Blossey and Nötzold 1995). Thus, EICA addresses a specific scenario in which the selective regime changes for introduced species but not for native species, and thus can explain strong invasion. Research on this hypothesis has principally focused on plants, but the basic premise of EICA should be valid for other taxa, too. The EICA hypothesis leads to two key predictions. First, introduced populations should exhibit a loss of defenses against natural enemies when compared to native populations. Second, when grown or reared in the novel environment, individuals from introduced populations should be more fit than individuals from native populations (or, in a reciprocal transplant, introduced individuals should be locally adapted to their new environment). While some studies examining EICA have found a loss of defense in introduced species, only a subset have demonstrated altered resource allocation facilitating demographic expansion of the invader (Bossdorf et al. 2005).

The EICA hypothesis proposes a specific mechanism for adaptive evolution. However, the genetic variation available for adaptive evolution might be expected to be limited in introduced species due to bottlenecks in population size that can be associated with the introduction process (Nei et al. 1975). In fact, reduced genetic variation in the invaded region compared to the native region is often considered evidence for bottlenecks (e.g., Tsutsui et al. 2000; Hassan et al. 2003). Population bottlenecks may lead to inbreeding depression and may limit adaptive evolution, particularly when populations remain small for multiple generations (e.g., Lande 1980). This may not always be the case, however, as founder effects that can be associated with bottlenecks may actually convert epistatic variance into additive genetic variance and thereby enhance the potential for a response to selection (see below; Goodnight 1987; Bryant and Meffert 1995). In addition, multiple introductions from different native origins, even if each imposes a strong bottleneck in population size, may enhance variation, particularly if there is significant genetic structure among populations in the native range. When individuals from those populations cross in the new range, they can generate introduced populations that can harbor greater genetic variation than is found in any single population in the native range (Kolbe et al. 2004). Variation can also be increased if introduced species cross with related species (native or introduced; see Chap. 16).

Abbott (1992) and Ellstrand and Schierenbeck (2000) highlighted the role of *hybridization* between species and gene flow among distinct genotypes in invasions, and proposed that it increases the invasiveness of exotic species by generating genetic variation, evolutionary novelty or hybrid vigor (e.g., Vilà and D'Antonio 1998). Hybridization may initially reduce fitness (Arnold et al. 2001), but a combination of selection and backcrossing may result in individuals with higher fitness than is the case for the hybridizing parents (Arnold and Hodges 1995; Arnold et al. 2001). Whether hybridization influences the demographic success of introduced species is still under debate: many invasive taxa are of hybrid origin (Ellstrand and Schierenbeck 2000), but few data connect hybridization or outcrossing directly to changes that would increase invasion success. Two aspects of introductions make hybridization a good candidate for explaining strong invasions. First, when backcross or other hybrid individuals have higher fitness than the parents, that effect often is restricted to novel habitats (Lexer et al. 2003), and the introduced range is by definition a novel habitat. Thus, introduced taxa of hybrid origin may be able to invade areas that are unavailable to the parental species. Second, we propose that hybridization might be expected to be more frequent in the novel range, because the introduction process unites distinct genotypes of a species or distinct species whose distributions do not overlap in the native range.

In addition to leading to bottlenecks in population size, and potentially to increased hybridization, introductions may be viewed as *founder events*. The genetic consequences of founder events have long been a proposed mechanism for rapid evolution, and even speciation (e.g., Mayr 1954; Barton and Charlesworth 1984; Gavrilets and Hastings 1996; Regan et al. 2003). Founder events may lead to "genetic revolutions" through breaking up what have been called co-adapted gene complexes (Mayr 1954), and can increase additive genetic variation in phenotypic traits (Goodnight 1987; Bryant and Meffert 1995). Such genetic revolutions poise the population for rapid response to a new selective regime. Thus, founder events may be implicated in strong invasion, as they lead to changes that do not occur in the native populations of an invasive species. Lee (2002) proposed that the genetic architecture and the amount of available additive genetic variance contribute to whether an introduced species will become invasive, or not. While this is likely quite true, and an important area of research that is much neglected, genetic architecture alone may not be adequate to explain strong invasion. Rather, founder effects, or other changes associated with introductions such as release from enemies (e.g., EICA), must also be invoked for genetic architecture to contribute to strong invasion.

6.3 Proposed Refinements to Hypotheses, Predictions and Tests

There likely is room for improvements to all of the above hypotheses. Here, we offer several suggestions for refining the enemy release and evolution of increased competitive ability hypotheses, then move on to discuss synergies among these hypotheses.

6.3.1 Refining the Enemy Release Hypothesis

Given the multiple scenarios for interactions between enemies, introduced species and native species, it is important to evaluate critically our expectations for the outcomes of such interactions. This should include differences in predictions for generalist and specialist enemies (Müller-Schärer et al. 2004; Joshi and Vrieling 2005). An analytical framework combining comparisons of

(1) differential parasitism in populations of a single species, both in its native and introduced region (within-species, cross-regional comparison), (2) native versus introduced species in the same region (between-species comparison of ecologically analogous competitors), (3) introduced species that are not invasive with introduced species that are invasive (between introduced species), and (4) population growth rates and enemy abundance among populations within a range (within species, within region) will help fully evaluate enemy release and the role of enemies in biological invasions. When used in combination, such comparisons will clarify the extent to which natural enemies keep their host populations in check, and the consequences of release from these natural enemies on population growth. To our knowledge, this joint approach has not yet been employed.

6.3.2 Refining the Evolution of Increased Competitive Ability Hypothesis

Evidence for adaptive evolution in invaders supporting the EICA hypothesis, particularly local adaptation to the new environment, is growing (e.g., Leger and Rice 2003; Lee et al. 2003; Blair and Wolfe 2004; Maron et al. 2004), but is not found in all cases (e.g., Willis et al. 1999; van Kleunen and Schmid 2003). Similarly to cases dealing with enemy release, much of the conflict in the data may be due to the specific comparisons that are made. For example, most of the data come from common garden experiments with plants, largely comprised of specimens that simply happened to be available for study, without particular knowledge of the origin of an invasion (but see Maron et al. 2004). Often, these are not comprehensive samples taken from across the native and introduced ranges (but see Blair and Wolfe 2004). If common gardens include inappropriate comparisons, then the data may not reflect the role of evolution in invasions.

6.4 Recent Syntheses and Synergies Between Hypotheses

Efforts have been made to synthesize across hypotheses. Shea and Chesson (2002) view biological invasions from the perspective of community ecology theory. They discuss invasions in terms of niche opportunities that are determined largely by resources, natural enemies, and the physical environment. Top-down population regulation is exerted by both specialist and generalist enemies, and for invaders, specialists may be relatively rare in the new range. The availability of resources that are present in a habitat is governed by other members of the community using those resources. Thus, Shea and Chesson (2002) bring together aspects of the enemy release, biotic resistance, and empty niche hypotheses.

Facon et al. (2006) propose a framework that combines ecological and evolutionary perspectives on biological invasions. Additionally, similarly to Shea and Chesson (2002), they bring together work focusing on the invasibility of communities (e.g., BRH) and the properties of invaders (e.g., ERH, inherent superiority). They argue that for invasion to occur, there must be a match between the invaded ecosystem and the invader. If that match exists from the outset, then all that is required for invasion is migration of a potential invader into the ecosystem. If that match does not exist, then in addition to migration, either the ecosystem must be altered, such as through changes in land use, or the invader must be altered, such as through evolutionary changes associated with the founder event, or with a response to natural selection following migration. Changes in both the ecosystem and the invader are possible, and may be necessary prior to some invasions.

Blumenthal (2005) proposes a conceptual link between enemy release and resource availability (the *resource-enemy release hypothesis*, or RERH) that differs from simply combining the enemy release and the biotic resistance hypotheses. Rather, he notes first that species that are able to take advantage of rich resources typically are fast growing, and not well defended against enemies. Because of this, enemy release may provide the greatest benefit to species that are adapted to high-resource conditions. This creates an interaction between resource availability and enemy release, amplifying the potential for invasion (Fig. 6.2). This also has implications for management, as it suggests that successful control of invaders may require both increasing enemy load (e.g., through biological control) and reducing resource availability.

Another conceptual integration of hypotheses is offered in Mitchell et al. (2006). They take a comparative perspective on invaders in their native and

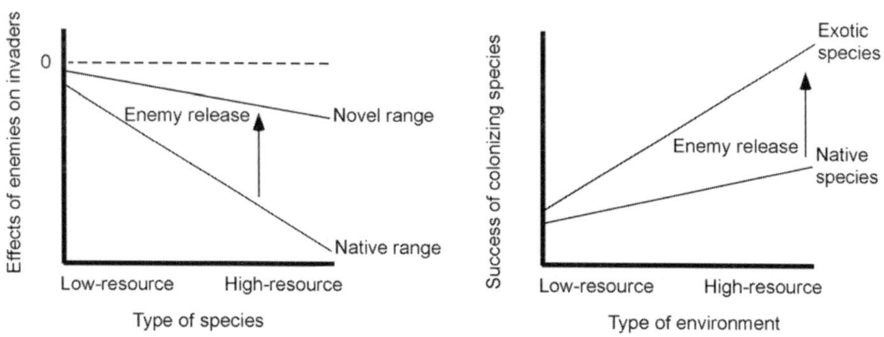

Fig. 6.2a, b Illustration of how resources and enemy release may interact to cause invasion. **a** Species adapted to environments with high resources are inhibited by enemies in their native range, and therefore have great potential to benefit from escaping their natural enemies. **b** Although all high-resource species will benefit from high resource availability, resource availability will increase the advantage of introduced species relative to native species due to enemy release (modified from Blumenthal 2005)

introduced ranges, examining how their population dynamics are influenced by several types of interspecific interactions and the abiotic environment. The biotic interactions they examined comprise predation, parasitism, mutualism, and competition, making this the first theoretical integration of these key groups.

Furthermore, Mitchell et al. (2006) highlight that phylogenetic relationships between an invader and the members of the invaded community may play a critical role in the outcome of an introduction. Introduced species invading communities with close relatives should be more susceptible to enemies in that community, and thus accumulate a broader suite of enemies compared to invasions of communities in which relatives are absent (Strong et al. 1984; Mack 1996). However, they may also be more likely to gain mutualists from their relatives. The comparative importance of enemies and mutualists, and their relative differences in host specificity will determine whether invading a community containing close relatives is a disadvantage for invaders. The relatedness of members of the community may also be an indicator of suitability of the abiotic environment, when related species have similar environmental requirements.

Additional connections and synergies between hypotheses are possible, and for the field to advance, those links should be clarified and formalized in a predictive framework. We offer an initial attempt at such a framework (Table 6.2), focusing on interactions among enemy release, biotic resistance, and evolutionary change (including both evolution of increased competitive ability, and hybridization). By formalizing the connections among hypotheses to explain invasion, we can generate specific, testable predictions. These predictions can guide research efforts, and resultant data can feed back into the predictive framework.

For example, we suggest that genetic variation and changes associated with introductions may interact directly with enemy release and biotic resistance in at least two key ways. First, with a severe bottleneck in population size upon introduction, a species will lose genetic variation, but is also more likely to lose natural enemies (Torchin et al. 2002). Thus, adaptive evolution may be most limited by lack of genetic variation when enemy release is likely to be greatest, setting up a useful comparison between two potentially opposing pathways for invasion success (Drake 2003; Table 6.2, hypothesis 4a). Additionally, reductions in variation will define the context within which new interactions with enemies develop, and may affect the ability of the invader to defend against enemies.

A second way that genetic variation may interact with enemy release and biotic resistance is through changes associated with hybridization. Often, it is only in novel environments that hybrid and backcross offspring are fitter than their parents. One key novel aspect of a new range may be the lack of enemies. Thus, ways in which hybridization alters interactions with natural enemies could be particularly important. One mechanism may be simply that

Table 6.2 Initial conceptualization of a predictive framework including three main hypotheses with their predictions, and examples of how these may be integrated to generate refined hypotheses and predictions

Hypotheses	Predictions
1. Introduced populations experience a release from enemies (ERH)	a. Introduced populations have fewer enemies compared to populations in the native range b. Reduction in enemy diversity leads to less damage compared to populations in the native range, which facilitates demographic expansion
2. Introduced species experience biotic resistance in new regions (BRH)	a. Introduced species compete with native species b. Introduced species accumulate novel enemies over time
3. Introduced species diverge evolutionarily from native populations	a. Introduced populations experience a reduction in genetic diversity compared to native populations (bottlenecks) b. Multiple independent introductions to one region increases genetic diversity (hybridizing and outcrossing with native species or other invaders) c. Introduced species locally adapt to new biotic environments, and members of the invaded community adapt to the presence of the invader (adaptive evolution)
4. (1a+3a) Introductions with few enemies have experienced a bottleneck in population size	a. Adaptive evolution may be limited by lack of genetic variation when enemy release is greatest
5. (1a+2a+3 c) Reduction of parasitism gives introduced species an advantage when competing with heavily parasitized native species	a. Resources diverted from supporting parasites to support host growth and reproduction b. EICA, costly defenses selected against, freeing resources for growth and reproduction
6. (2b+3 c) Introduced species will experience an increase in enemies over time as enemy relationships in the new range evolve	a. Historical invasions will have a subset of enemies from the range of origin and a subset from the current range b. Historical invasions will be parasitized by a greater proportion of local parasite species compared to contemporary invasions c. Historical invasions provide a glimpse into the evolutionary future of contemporary invasions
7. (1a+2a+2b+1b) Introduced species invading communities with phylogenetically similar species will experience a smaller demographic advantage from lack of parasites	a. Parasites will shift from native species to use phylogenetically similar introduced species b. Introduced species invading areas with close relatives should be less invasive
8. (1a+3b+3 c) Hybridization and outcrossing can alter relationships with natural enemies, competitors or other novel aspects of the environment, directly facilitating invasion, and providing variation necessary for adaptive evolution	a. Hybridization will promote adaptation to a novel environment with low enemy load (e.g., EICA)

hybridization may increase genetic variation in resource allocation, providing variants that shunt resources away from defense (e.g., Floate et al. 1993; Fritz et al. 2001). These variants may be strongly selected for in a novel environment lacking enemies. Therefore, hybridization may promote the evolutionary response predicted by the evolution of increased competitive ability hypothesis. Similarly, hybridization may provide the variation necessary to respond to selection imposed by changes in the competitive regime associated with the new range (Table 6.2, hypothesis 8a).

Clearly, the variables that transform with introduction to a new range (e.g., enemies, competitors, mutualists, genetic variation) can all change at once, and should be considered together. Figure 6.3 illustrates a simple conceptual model for how invasiveness may vary with three of these variables. We predict that in many cases, invasiveness will increase with release from enemies and competitors, and with an increase in genetic variation. The speed and direction of adaptive evolution will be influenced by extrinsic factors (e.g., enemies and competitors) and intrinsic factors (e.g., genetic diversity). Thus, the change depicted in invasiveness may be due to either ecological or evolutionary processes, or a combination of both. Of course, exceptions to the general trend shown in Fig. 6.3 are known to exist. For example, the Argentine ant (*Linepithema humile*) has accrued an advantage with loss of genetic variation (Tsutsui et al. 2000), although the change in natural enemies has yet to be evaluated in this case.

Figure 6.3 can be used in two ways: knowing how enemies, competitors and genetic diversity differ between the native and introduced range, one can make initial predictions of relative invasiveness of species (e.g., whether R is greater or less than 1). Alternatively, using a measure of invasiveness based on the response ratio, we can make initial predictions of how these three variables might have changed between the native and introduced range with respect to the direction of change along the axes, and in some cases, the relative magnitude of change. Testing such predictions should provide new

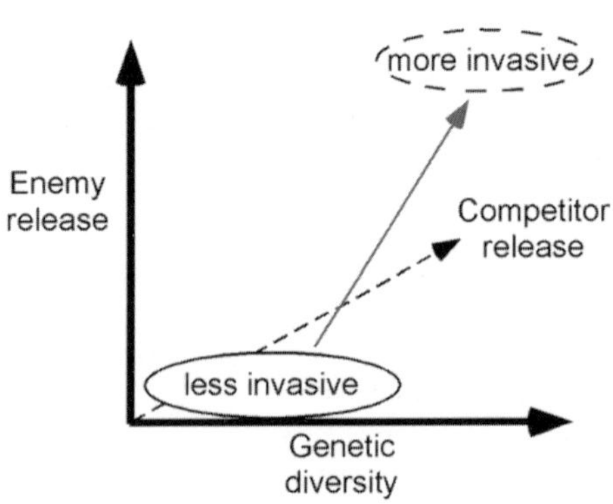

Fig. 6.3 Prediction of relative invasiveness when combining three of the variables likely to change with introductions (see text for further information)

insights into the relative importance of these factors for different taxa in setting the stage for introduced species to become invasive.

6.5 Conclusions

Research on biological invasions is at an exciting junction. The vast amount of empirical research has enabled deductions that move beyond individual case studies, and the theoretical stage is set for advancement. With this review of ecological and evolutionary hypotheses to explain invasion, we offer a quantitative measure for a key aspect of invasiveness – that of ecological change between the native and introduced range – and an example of how synthesizing across hypotheses to form a predictive framework can guide future empirical and theoretical work.

References

Abbott RJ (1992) Plant invasions, interspecific hybridization and the evolution of new plant taxa. Trends Ecol Evol 7:401–405

Arim M, Abades SR, Neill PE, Lima M, Marquet PA (2006) Spread dynamics of invasive species. Proc Natl Acad Sci USA 103:374–378

Arnold ML, Hodges SA (1995) Are natural hybrids fit or unfit relative to their parents? Trends Ecol Evol 10:67–71

Arnold ML, Kentner EK, Johnston JA, Cornman S, Bouck AC (2001) Natural hybridisation and fitness. Taxon 50:93–104

Baker HG, Stebbins GL (1965) The genetics of colonizing species. Academic Press, New York

Barton NH, Charlesworth B (1984) Genetic revolutions, founder effects, and speciation. Annu Rev Ecol Syst 15:133–164

Beckstead J, Parker IM (2003) Invasiveness of *Ammophila arenaria*: release from soil-borne pathogens? Ecology 84:2824–2831

Blair A, Wolfe LM (2004) The evolution of an invasive plant: an experimental study with *Silene latifolia*. Ecology 85:3035–3042

Blair AC, Hanson BD, Brunk GR, Marrs RA, Westra P, Nissen SC, Hufbauer RA (2005) New techniques and findings in the study of a candidate allelochemical implicated in invasion success. Ecol Lett 8:1039–1047

Blaney CS, Kotanen PM (2001) Effects of fungal pathogens on seeds of native and exotic plants: a test using congeneric pairs. J Appl Ecol 38:1104–1113

Blossey B, Nötzold R (1995) Evolution of increased competitive ability in invasive non-indigenous plants – a hypothesis. J Ecol 83:887–889

Blumenthal D (2005) Interrelated causes of plant invasion. Science 310:243–244

Bossdorf O, Auge H, Lafuma L, Rogers WE, Siemann E, Prati D (2005) Phenotypic and genetic differentiation between native and introduced plant populations. Oecologia 144:1–11

Bryant EH, Meffert LM (1995) An analysis of selectional response in relation to a population bottleneck. Evolution 49:626–634

Callaway RM, Ridenour WM (2004) Novel weapons: invasive success and the evolution of increased competitive ability. Frontiers Ecol Environ 2:436–443

Calvo-Ugarteburu G, McQuaid CD (1998) Parasitism and introduced species: epidemiology of trematodes in the intertidal mussels *Perna perna* and *Mytilus galloprovincialis*. J Exp Mar Biol Ecol 220:47–65

Chase JM, Abrams PA, Grover JP, Diehl S, Chesson P, Holt RD, Richards SA, Nisbet RM, Case TJ (2002) The interaction between predation and competition: a review and synthesis. Ecol Lett 5:302–315

Courchamp F, Langlais M, Sugihara G (2000) Rabbits killing birds: modelling the hyperpredation process. J Anim Ecol 69:154–164

Darwin C (1859) On the origin of species by means of natural selection, 1st edn. Harvard University Press, Cambridge, MA

Drake JM (2003) The paradox of the parasites: implications for biological invasion. Proc R Soc Lond Ser B Biol Sci 270:S133–S135

Ellstrand NC, Schierenbeck KA (2000) Hybridization as a stimulus for the evolution of invasiveness in plants? Proc Natl Acad Sci USA 97:7043–7050

Elton CS (1958) The ecology of invasions by animals and plants. Methuen, London

Facon B, Genton BJ, Shykoff JA, Jarne P, Estoup A, David P (2006) A general eco-evolutionary framework for understanding bioinvasions. Trends Ecol Evol 21:130–135

Floate KD, Kearsley MJC, Whitham TG (1993) Elevated herbivory in plant hybrid zones: *Chrysomela confluens*, *Populus* and phenological sinks. Ecology 74:2056–2065

Fritz RS, Hochwender CG, Lewkiewicz DA, Bothwell S, Orians CM (2001) Seedling herbivory by slugs in a willow hybrid system: developmental changes in damage, chemical defense, and plant performance. Oecologia 129:87–97

Gavrilets S, Hastings A (1996) Founder effect speciation: a theoretical reassessment. Am Nat 147:466–491

Goodnight CJ (1987) On the effect of founder events on epistatic genetic variance. Evolution 41:80–91

Hassan M, Harmelin-Vivien M, Bonhomme F (2003) Lessepsian invasion without bottleneck: example of two rabbitfish species (*Siganus rivulatus* and *Siganus luridus*). J Exp Mar Biol Ecol 291:219–232

Hedges LV, Gurevitch J, Curtis PS (1999) The meta-analysis of response ratios in experimental ecology. Ecology 80:1150–1156

Hierro JL, Maron JL, Callaway RM (2005) A biogeographical approach to plant invasions: the importance of studying exotics in their introduced and native range. J Ecol 93:5–15

Joshi J, Vrieling K (2005) The enemy release and EICA hypothesis revisited: incorporating the fundamental difference between specialist and generalist herbivores. Ecol Lett 8:704–714

Keane RM, Crawley MJ (2002) Exotic plant invasions and the enemy release hypothesis. Trends Ecol Evol 17:164–170

Kolar CS, Lodge DM (2001) Progress in invasion biology: predicting invaders. Trends Ecol Evol 16:199–204

Kolbe JJ, Glor RE, Schettino LRG, Lara AC, Larson A, Losos JB (2004) Genetic variation increases during biological invasion by a Cuban lizard. Nature 431:177–181

Lampo M, Bayliss P (1996) Density estimates of cane toads from native populations based on mark-recapture data. Wildlife Res 23:305–315

Lande R (1980) Genetic variation and phenotypic evolution during allopatric speciation. Am Nat 116:463–479

Lee CE (2002) Evolutionary genetics of invasive species. Trends Ecol Evol 17:386–391

Lee CE, Remfert JL, Gelembiuk GW (2003) Evolution of physiological tolerance and performance during freshwater invasions. Integrated Comp Biol 43:439–449
Leger EA, Rice KJ (2003) Invasive California poppies (*Eschscholzia californica* Cham.) grow larger than native individuals under reduced competition. Ecol Lett 6:257–264
Levine JM, D'Antonio CM (1999) Elton revisited: a review of evidence linking diversity and invasibility. Oikos 87:15–26
Levine JM, Adler PB, Yelenik SG (2004) A meta-analysis of biotic resistance to exotic plant invasions. Ecol Lett 7:975–989
Lexer C, Welch ME, Durphy JL, Rieseberg LH (2003) Natural selection for salt tolerance quantitative trait loci (QTLs) in wild sunflower hybrids: implications for the origin of *Helianthus paradoxus*, a diploid hybrid species. Mol Ecol 12:1225–1235
Mack RN (1996) Predicting the identity and fate of plant invaders: emergent and emerging approaches. Biol Conserv 78:107–121
Maron JL, Vilà M (2001) When do herbivores affect plant invasion? Evidence for the natural enemies and biotic resistance hypotheses. Oikos 95:361–373
Maron JL, Vilà M, Bommarco R, Elmendorf S, Beardsley P (2004) Rapid evolution of an invasive plant. Ecol Monogr 74:261–280
Mayr E (1954) Change of genetic environment and evolution. In: Huxley J, Hardey AC, Ford EB (eds) Evolution as a process. Allen and Unwin, London, pp 157–180
Mitchell CE, Agrawal AA, Bever JD, Gilbert GS, Hufbauer RA, Klironomos JN, Maron JL, Morris WF, Parker IM, Power AG, Seabloom EW, Torchin ME, Vázques DP (2006) Biotic interactions and plant invasions. Ecol Lett 9:726–740
Müller-Schärer H, Schaffner U, Steinger T (2004) Evolution in invasive plants: implications for biological control. Trends Ecol Evol 19:417–422
Nei M, Maruyama T, Chakraborty R (1975) The bottleneck effect and genetic variability in populations. Evolution 29:1–10
Peterson AT (2003) Predicting the geography of species' invasions via ecological niche modeling. Q Rev Biol 78:419–433
Redak RA, Purcell AH, Lopes JRS, Blua MJ, Mizell RF, Andersen PC (2004) The biology of xylem fluid-feeding insect vectors of *Xylella fastidiosa* and their relation to disease epidemiology. Annu Rev Entomol:243–270
Regan JL, Meffert LM, Bryant EH (2003) A direct experimental test of founder-flush effects on the evolutionary potential for assortative mating. J Evol Biol 16:302–312
Reinhart KO, Packer A, van der Putten WH, Clay K (2003) Plant-soil biota interactions and spatial distribution of black cherry in its native and invasive ranges. Ecol Lett 6:1046–1050
Richardson DM, Allsopp N, D'Antonio CM, Milton SJ, Rejmanek M (2000) Plant invasions – the role of mutualisms. Biol Rev Cambridge Philos Soc 75:65–93
Sakai AK, Allendorf FW, Holt JS, Lodge DM, Molofsky, With KA, Baughman S, Cabin RJ, Cohen JE, Ellstrand NC, McCauley DE, O'Neil P, Parker IM, Thompson JN, Weller SG (2001) The population biology of invasive species. Annu Rev Ecol Syst 32:305–332
Sax DF, Brown JH (2000) The paradox of invasion. Global Ecol Biogeogr 9:363–371
Schierenbeck KA, Mack RN, Sharitz RR (1994) Effects of herbivory on growth and biomass allocation in native and introduced species of *Lonicera*. Ecology 75:1661–1672
Settle WH, Wilson LT (1990) Invasion by the variegated leafhopper and biotic interactions: parasitism, competition, and apparent competition. Ecology 71:1461–1470
Shea K, Chesson P (2002) Community ecology theory as a framework for biological invasions. Trends Ecol Evol 17:170–176
Simberloff D, Von Holle B (1999) Positive interactions of nonindigenous species: invasional meltdown? Biol Invasions 1:21–32
Strong DR, Lawton JH, Southwood SR (1984) Insects on plants; community patterns and mechanisms. Harvard University Press, Cambridge, MA

Tilman D (1997) Community invasibility, recruitment limitation, and grassland biodiversity. Ecology 78:81–92

Torchin ME, Mitchell CE (2004) Parasites, pathogens, and invasions by plants and animals. Frontiers Ecol Environ 2:183–190

Torchin ME, Lafferty KD, Kuris AM (2001) Release from parasites as natural enemies: increased performance of a globally introduced marine crab. Biol Invasions 3:333–345

Torchin ME, Lafferty KD, Kuris AM (2002) Parasites and marine invasions. Parasitology 124:S137–S151

Torchin ME, Lafferty KD, Dobson AP, McKenzie VJ, Kuris AM (2003) Introduced species and their missing parasites. Nature 421:628–630

Tsutsui ND, Suarez AV, Holway DA, Case TJ (2000) Reduced genetic variation and the success of an invasive species. Proc Natl Acad Sci USA 97:5948–5953

van Kleunen M, Schmid B (2003) No evidence for an evolutionary increased competitive ability in an invasive plant. Ecology 84:2816–2823

Vilà M, D'Antonio CM (1998) Hybrid vigor for clonal growth in *Carpobrotus* (Aizoaceae) in coastal California. Ecol Appl 8:1196–1205

Vivanco JM, Bais HP, Stermitz FR, Thelen GC, Callaway RM (2004) Biogeographical variation in community response to root allelochemistry: novel weapons and exotic invasion. Ecol Lett 7:285–292

Wiles GJ, Bart J, Beck RE, Aguon CF (2003) Impacts of the brown tree snake: patterns of decline and species persistence in Guam's avifauna. Conserv Biol 17:1350–1360

Williams JR (1954) The biological control of weeds. In: Rep 6th Commonwealth Entomology Congr, London, pp 95–98

Willis AJ, Thomas MB, Lawton JH (1999) Is the increased vigour of invasive weeds explained by a trade-off between growth and herbivore resistance? Oecologia 120:632–640

7 Traits Associated with Invasiveness in Alien Plants: Where Do we Stand?

PETR PYŠEK and DAVID M. RICHARDSON

7.1 History of the Search for Traits and Shifts in Research Focus

Any organism must be equipped for life in a given environment, otherwise it will die. The fundamental question is *how well* does an organism need to be "equipped", or what syndrome of traits must it possess to survive and flourish at a given locality. In the current human-mediated biodiversity crisis, where alien species play an important role, we need to know whether some species are inherently better equipped to become invasive when moved to new areas by humans. If so, we can identify such species and consider management options to prevent, or at least reduce the damaging effects of biological invasions.

Despite the importance of chance and timing in the establishment and spread of alien plants (Crawley 1989), invasions are clearly not entirely random events (Crawley et al. 1996). Much of the early work on invasions was directed at collating traits associated with invasiveness (Booth et al. 2003). The question of whether is it possible to determine a set of traits that predispose a species to be invasive has been a central theme since the emergence of invasion ecology as a discrete field of study (Richardson and Pyšek 2006).

Many studies have attempted to profile successful invaders, starting with Herbert Baker's attempt to identify the traits of an "ideal weed" (Baker 1965), an idea now considered simplistic (Perrins et al. 1993). Baker defined as a weed a plant growing "entirely or predominantly in situations markedly disturbed by man (without, of course, being deliberately cultivated plants)". To him, weeds included plants that encroached onto agricultural land (agrestals), and those occurring in waste places (ruderals; Baker 1965). There was no explicit reference to the status of the species as being native or alien. Perhaps it was the two species pairs he used to document different traits of "weedy"

Ecological Studies, Vol. 193
W. Nentwig (Ed.)
Biological Invasions

and "non-weedy" plants (alien and native congeners in the genera *Eupatorium* and *Ageratum*) that made followers consider Baker's "ideal weed" to be synonymous with "ideal invader" (i.e., an alien plant spreading from sites of introduction). Nevertheless, Williamson (1993) concluded that there is no consistency of life history and reproductive behavior across all weeds, and the same holds for invading plants (Williamson and Fitter 1996).

Work undertaken in the post-Baker era has shown that identifying traits consistently associated with invasiveness is difficult (Alpert et al. 2000); this resulted in a widespread pessimism in the mid-1990s (Crawley 1987; Roy 1990). However, Rejmánek (1996) showed that such traits are a crucial ingredient for explaining (and therefore predicting) invasions. Rejmánek's paper probably stimulated attempts to find correlates of invasiveness across vascular plants, because this is when comparative studies based on large species sets started to appear. Studies comparing species pairs or a few congeners started to be published some 15 years earlier (Fig. 7.1), possibly because data needed for comparative multispecies studies have only recently become available. Classification of whole floras with respect to alien or native status of their members, with reliable information based on objective criteria, is still far from being standard, even two decades after the SCOPE project on biological invasions (Richardson et al. 2000a; Pyšek et al. 2004a).

After a period of stagnation in late 1990s, the relative contribution of comparative multispecies studies has been increasing recently (Fig. 7.1). This is obviously due to improved data availability, the advent of online databases,

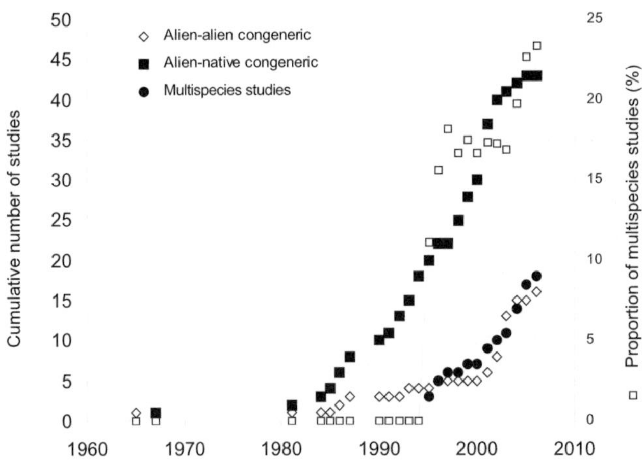

Fig. 7.1 *Left axis* Increase in the cumulative number of studies using different approaches to the analysis of the effect of species traits on invasiveness: comparison of invasive aliens with their native congeners or related genera within a family, comparison of alien congeners with different levels of invasiveness, and comparative analyses of large species sets and whole floras (multispecies studies). *Right axis* Increase in number of multispecies studies, expressed as a proportion of the cumulative total number of studies

and better communication among researchers. Powerful computing facilities and statistical techniques also contributed to fairly promising results in some studies, which in turn probably stimulated further work.

Kolar and Lodge (2001) analyzed 16 invasion studies each containing at least 20 plant or animal species, and concluded that sound generalizations have emerged. Gilpin (1990) pointed out that there is pattern in the available data on invasions, and suggested that further efforts in the study of invasions should be self-consciously statistical. If ecologists are unable to predict outcomes of individual cases, then they should focus on uncovering broader, general ecological patterns (Cadotte et al. 2006b).

7.2 Comparative Analyses of Multispecies Datasets: Every Picture Tells a Story

7.2.1 Methodological Approaches: what is Being Compared?

Can plant invasions be explained and predicted based on the traits of invading species? Has there been real progress, or are we floundering as much as we were 20 years ago? The aim of the present paper is to reassess the potential of studies seeking for plant traits that determine species invasiveness, and to identify such plant characteristics. To achieve this, we need to deal with research approaches first. Just as invasions are notoriously idiosyncratic, so too are the approach methodologies that have been applied to study them.

Multispecies comparative studies need to be classified according to the approach, type of comparison, scale and data character, including measures of occurrence of the species present (Table 7.1). The 18 studies summarized in Table 7.2 use some measure of the occurrence of alien species in the invaded territory, or the presence of aliens and native species as response variables, and explain these by using plant traits. In our analysis, we concentrate on species biological, ecological and physiological traits. We excluded distributional characteristics of alien plants such as the size of native ranges, although such variables are clearly among the best predictors of invasiveness (Rejmánek 1996, 2000). Size of the occupied range is certainly a convenient measure of ecological versatility (Prinzing et al. 2002), but this characteristic results from the interplay of "primary" biological, ecological and physiological traits, and finding a significant link between range size and invasiveness does not tell us much about what traits a plant needs to become a successful invader.

To adopt the correct approach to this issue (column A in Table 7.1), the question being asked needs to be clearly defined (Hamilton et al. 2005). The

Table 7.1 Classification of research strategies for multispecies comparative studies of species traits

A. Approach	B. Comparison	C. Scale	D. Data character[b]	E. Occurrence measures	F. Analytical methods
1. Target-area	1. Native–alien	1. Local/habitat	1. Inventory (>80 % of species)	1. Presence	1. Simple comparison
2. Source-area	2. Alien–alien[a]	2. Regional	2. Database (20–80 % of species)	2. Abundance/frequency	2. Phylogenetic corrections
		3. Continental	3. Species list (<20 % or 20–50 species)	3. Distribution range	3. Net effects: residence time
				4. Historical dynamics	4. Net effects: distribution

[a] Lists of alien species may include all taxa (including casuals), or be restricted to naturalized/invasive species (sensu Richardson et al. 2000a; Pyšek et al. 2004a)
[b] Classification follows Cadotte et al. (2006b)

target-area approach focuses on a pool of species that are alien to a region, and attributes the variation in their success to differences in their traits. It asks the question "what traits distinguish successful invaders from those aliens that have not invaded successfully?" (Hamilton et al. 2005). Alternatively, aliens in a target region can be compared with natives of that region. The target-area approach has been more commonly applied than the source-area approach (Table 7.2).

The source-area approach (sensu Pyšek et al. 2004b) asks the question "do traits of species that become invasive from a geographic 'source' region differ from those species from the same region that fail to invade?" Such an approach can either identify traits that allow species to pass through early phases (i.e., transport and establishment) of the invasion process (Hamilton et al. 2005), or if factors associated with the chance of being transported are controlled for, provide relatively unbiased estimates of the role of traits associated with naturalization and invasion. The source-area approach is more convenient for identifying net effects of traits because it eliminates or reduces the bias and variation associated with different species origins, and pathways and distance of introduction. In this approach, it is crucial that geographic origin and size of the native range of source-pool species are taken into account, to eliminate these biases as much as possible (Pyšek et al. 2004b); otherwise, the life-history traits of invading species can be confused with environmental circumstances associated with dispersal. Unfortunately, the source area approach has been used in only three studies (Table 7.2). We believe that the main constraint to its wider use is not the lack of information on potential source pools, but rather the lack of knowledge on how these species are performing as invaders elsewhere.

Native–alien comparisons (column B in Table 7.1) explore whether the traits of native species in a target area differ from those of alien species that invaded that area. It asks the question "what traits of the invading species enhance their potential to increase in abundance over native species?" (Hamilton et al. 2005). When interpreting the results of alien–native comparisons, one must bear in mind *which* aliens and *which* natives are being compared. Not all aliens spread (Richardson et al. 2000a), but some natives expand their ranges into human-made habitats, or increase in abundance and/or range following human-induced landscape changes (so-called apophytes, expansive species, see Pyšek et al. 2004a; Alpert et al. 2000). Thompson et al. (1995) compared invasive aliens with expanding natives, and concluded that while invasive species differ significantly from non-invasive species (Thompson 1994), the attributes of invasive aliens are not unique, but most are shared by expanding native species. Comparing these two groups may indicate whether being an alien alone exerts specific effects that are not seen in native expanding taxa (Leishman and Thompson 2005; Hamilton et al. 2005).

Analogically, the alien–alien comparison of two or more invading alien congeners exhibiting different levels of invasiveness asks the question "what

Table 7.2 Overview of comparative studies of species traits and their effect on plant invasiveness, usin large species numbers[a]

Author	Region of invasion/ source region	Research strategy	Species number	Comparison
Thompson et al. (1995)	England, Scotland, Ireland, The Netherlands	A1-B1-C2-D3-E1-F1	211 alien	Target/invasive alien vs. expand ing native
Pyšek et al. (1995)	Czech Republic	A1-B12-C2-D3-E12-F1	132 alien	Target/1. natura ized neophyte v native; 2. withir alien
Andersen (1995)	Denmark	A1-B12-C2-D3-E1-F1	93 alien, 40 native	Target/1. alien with native; 2. alien in ruderal and seminatur habitats
Crawley et al. (1996)	British Isles	A1-B1-C2-D1-E1-F2	1,169 alien	Target/naturalized alien (incl. archaeophyte) native
Williamson and Fitter (1996)	Great Britain	A1-B1-C2-D1-E1-F1	1,777 native +alien	Target/invasive alien vs. native
Pyšek (1997)	Central Europe	A1-B1-C2-D2-E1-F2	2,223 native, 457 alien	Target/alien wi native
Pyšek (1997)	New Zealand: Auckland	A1-B2-C2-D1-E14-F2	615 alien	Target/within alien
Goodwin et al. (1999)	Canada: New Brunswick/Europe	A2-B2-C2-D3-E1-F2	165 species pairs (invading+non-invading)	Source/within alien
Thébaud and Simberloff (2001)	Europe, North America: California, Carolinas/reciprocal	A2-B2-C23-D2-E1-F1	651 alien	Source/within alien

Scale	Relevant traits compared	Results
Reg.	Life strategy, height, lateral spread, flowering time, propagule weight, growth form, clonal growth, seed bank, dispersal mode, canopy structure, habitat preference	Aliens more likely clonal, polycarpic perennials with erect leafy stems than are native species, and having transient seed banks
Reg.	Origin, height, life form, life strategy, dispersal mode, pollen vector, human use, habitat preferences	1. aliens more likely than natives to be C- and CR-strategists, dispersed by humans, with preferences for dry, warm and nutrient-rich habitats; 2. successful aliens in seminatural habitats are tall hemicryptophytes escaped from cultivation; 3. successful aliens in human-made habitat are therophytes or geophytes, introduced spontaneously
Reg.	Dispersal mode	Aliens, especially in seminatural habitats, more often with fleshy fruits and dispersed by wind than are natives
Reg.	Life form, height, seed weight, dispersal mode, seedling relative growth rate, flowering time, pollination mode	Aliens taller, with larger seeds, no or protracted dormancy, flowering earlier or later, with more pronounced r- or K-strategies
Reg.	Life form, max. height, spread (height:width), leaf area, leaf shape, leaf longevity, age at 1st flowering, flowering time, seedling relative growth rate, season of seed dispersal, season of germination, pollen vector, fertilization method, breeding system, compatibility	Invasive species are tall, taller than wide, more often phanerophytes, have large leaves, are insect-pollinated and prefer fertile habitats; natives tend to be more often monoecious
Reg.	Clonal growth	Aliens are more often non-clonal
Reg.	Clonal growth, introduction pathways, habitat preferences	Clonal aliens are more often introduced deliberately, more likely to increase in numbers and less likely to occur in dry habitats than are non-clonal aliens
Reg.	Growth form, height, flowering period	Invading species are taller and have longer flowering period than those that do not invade
Cont.	Height	Species are not taller in their introduced range

Table 7.2 (*Continued*)

Author	Region of invasion/ source region	Research strategy	Species number	Comparison
Cadotte and Lovett-Doust (2001)	Canada: SW Ontario (Essex, Hamilton-Wenthworth)	A1-B1-C2-D1-E1-F1	1,330 native, 484 alien	Target/alien wit native
Prinzing et al. (2002)	Argentina: Buenos Aires and Mendoza provinces/Europe	A2-B2-C2-D2-E1-F2	197 alien	Source/within alien
Pyšek et al. (2003)	Czech Republic	A1-B2-C2-D2-E4-F4	668 alien	Target/within alien (neophyte)
Lake and Leishman (2004)	Australia: Sydney	A1-B12-C1-D3-E1-F1	57 alien	Target/alien–ali (invasive vs. non invasive); alien–native
Sutherland (2004)	USA	A1-B12-C3-D1-E1-F1	19,960 native+alien	Target/1. alien with native; 2. invasive with no invasive alien (i ruderal habitats
Hamilton et al. (2005)	Eastern Australia (regional: Royal National Park+continental)	A1-B2-C23-D3-E2-F23	152 alien	Target/within alien (introduce in last 200 years
Lloret et al. (2005)	Mediterranean islands	A1-B2-C2-D2-E2-F2	354 alien	Target/within na uralized alien
Pyšek and Jarošík (2005)	Czech Republic	A1-B2-C2-D2-E2-F34	668 alien	Target/within alien (neophyte)
Cadotte et al. (2006a)	Canada: SW Ontario	A1-B2-C2-D1-E2-F23	1,153 alien	Target/within alien

[a] See Table 7.1 for codes describing the research strategy adopted by individual studies in terms of approach, type of comparison, scale, data character, occurrence measures, and analytical methods used. Only significant results are presented in the last column. Studies are ranked chronologically. Reg., regional, cont., continental

Scale	Relevant traits compared	Results
Reg.	Growth form, clonal growth, breeding system, pollen vector, flowering period, dispersal mode, fruit size, seed number per fruit	Aliens more likely annuals and biennials, hermaphrodites with longer flowering period and with small fruits, less likely to be dispersed by animals; in seminatural habitats, aliens are also more likely trees with many seeds per fruit
Reg.	Habitat preferences, life strategy, dispersal vectors, use by humans	Species that invade have r-strategy, prefer warm, dry, sunny and nitrogen-rich habitats, and are more often used by humans
Reg.	Origin, introduction pathway, growth form, life strategy, time of flowering, dispersal mode, propagule size	Geographical proximity, early flowering, annual growth form, CSR strategy, and human use contribute to early arrival
Local/ habitat	Specific leaf area, leaf texture and hairiness, seed weight, growth form, dispersal mode, vegetative propagation, flowering duration, canopy height	Invasive species have higher specific leaf area than do alien non-invasive and native species, and have more hairy leaves in some situations; aliens have softer leaves; invasive species in disturbed sites have smaller seeds and flower longer than do natives; invasives dispersed more by wind and vertebrates, less by ants; aliens more propagated vegetatively than are natives
Cont.	Vegetative reproduction, breeding system, compatibility system, pollen vector, shade tolerance, growth form, life form, morphology, toxicity	1. aliens less likely clonal and wetland plants than are natives; 2. invasive aliens more likely than non-invasives to be monoecious, self-incompatible, perennial and woody
Reg. +cont.	Specific leaf area, height, seed weight	Small seeds correlated with success at both scales, high SLA at continental scale – both with and without phylogenetic correction (all three traits correlated with abundance at both scales, if minimum residence time not controlled for)
Reg.	Growth form, vegetative propagation, leaf size, morphology (spinescence, pubescence, succulence), life form, height, breeding system, pollen vector, flowering, fruit type, seed size, dispersal mode	Species that reproduce vegetatively, have large leaves, flower in summer and for longer period, or are dispersed by wind and animals have highest abundance; succulent and fleshy fruits favor ruderal and seminatural habitats, respectively
Reg.	Introduction pathway, human use, origin, growth form, life strategy, height, flowering time, dispersal mode, propagule size	Life strategy, origin and dispersal mode have direct effect, height and growth form interact with minimum residence time; aliens from America and Asia dispersed by water are more frequent
Reg.	Clonal growth, flowering, origin, growth form, breeding system, habitat preferences	Abundant aliens have longer flowering duration, originated from Europe or Eurasia, and grow in variable soil moistures

traits enhance the potential of an invasive species to increase in abundance and/or distribution over less-successful alien species?" The two types of comparison are comparatively frequently represented: within the 18 studies summarized in Table 7.2, there are nine native–alien and 13 alien–alien comparisons (some studies use both approaches, e.g., Pyšek et al. 1995; Sutherland 2004; Lake and Leishman 2004).

7.2.2 Data, Scale and Analysis

We included only studies based on statistically tested data; the analyzed traits of alien species had to be tested either against those of native species (native–alien approach), or against differing invasion success (alien–alien approach). This is why some papers, often cited regarding traits typical of invaders, are not considered here – they do not compare with "control" datasets (Timmins and Williams 1987), or they are theoretical studies building profiles from examples, but without primary data analysis (e.g., Noble 1989; Roy 1990; Richardson and Cowling 1992).

Multispecies comparative studies also differ in the number of species involved in comparison, whether the species compared are characterized on presence/absence only or some quantitative measure of their occurrence (or some other measure of the extent of invasion) is used, and in the way data are analyzed. Here, we follow the scheme recently suggested by Cadotte et al. (2006b). A complete species inventory (D1 in Table 7.1) results from a concerted, usually long-term effort to record all extant taxa within a large region. A database (D2) covers large representative group of species, 20–80 % of the total number in a region, or a complete inventory from a subregion; it is usually rather a complete list of species that occur in a large, representative habitat or ecosystem. A species list (D3) includes <20 % of flora in a region, or 50–200 species in total, and is selected on some a priori criteria, such as a sample from a particular habitat. Available multispecies comparative studies are evenly distributed with respect to the data character, with six studies in D1, D2 and D3 each. Of the 18 studies, 16 are based on more than 100 species, and six on more than 1,000 species (Table 7.2).

The fourth category delimited by Cadotte et al. (2006b), termed species groups, deals with comparisons made on limited numbers of species selected according to some criteria, often congeners or confamilials; these are dealt with below. Congeneric studies have received much focus thus far (Fig. 7.1), given their utility in reducing the influence of phylogenetic effects, and the sense of comparative value (Cadotte et al. 2006b). The major reason, nevertheless, is that data on congeners are easier to get – one can collate them in the field in a specifically designed case study. To produce a reliable flora list for a large region (namely, D1 and D2 types in Table 7.1) is much more difficult and not a matter of simple decision; whether it will be possible to analyze such alien flora is beyond the researcher's control, as she/he must rely on data that have been collated by others.

The size of the region needs to be appropriate for the questions asked, but larger areas (>100,000 km^2, the scale of political regions) are preferable, since species inventories are usually compiled for political regions, are biogeographically arbitrary, and are repeatable (see discussion in Cadotte et al. 2006b). The vast majority of studies (16 in Table 7.2) were conducted at the regional scale, with only a few (Thébaud and Simberloff 2001; Sutherland 2004) addressing the problems at a continental scale. The study by Hamilton et al. (2005) is the only one that compares the effect of studied variables between regional and continental scales. In most studies, cross-species comparisons treating species as independent data points were conducted without explicit consideration of phylogenetic relatedness among species (Cadotte et al. 2006b). However, incorporating phylogenetic information can elucidate the extent to which changes in invasiveness may be correlated with changes in other traits through a particular phylogeny (Harvey and Pagel 1991; Cadotte et al. 2006b). Using phylogenetic corrections may, or may not provide different results (Harvey et al. 1995). However, the same results with and without using phylogenetic corrections indicates that throughout the phylogeny of alien species there have been multiple and independent correlated evolutionary divergences between invasion success and the trait examined (as found, e.g., by Hamilton et al. 2005 for seed mass and specific leaf area). The list of studies that used phylogenetic corrections (coded F2 in Table 7.2) clearly indicates that the frequency of its application has been increasing recently, presumably with the gradual improvement in availability of phylogenetic trees for multispecies assemblages. Alternatively, phylogenetic bias can be reduced by comparing invading and non-invading congeners (Goodwin et al. 1999).

The variety of methods, approaches, scales, and measures used in comparative multispecies studies of species invasiveness makes it dangerous to draw generalizations without taking the character of individual studies into account. What can be thus inferred about species traits and their effects on the invasiveness of plant species? Is the message consistent?

7.2.3 Main Findings of Comparative Multispecies Studies (1995–2005)

Although the multispecies studies test different hypotheses, simply because the inventories and databases contain different information (Cadotte et al. 2006b), some traits have been tested frequently enough for a pattern to emerge (Table 7.2 and see below).

Growth form (usually separating species into annual, biennial, perennial, shrubs, and trees) and *life form* (following Raunkiaer's scheme) are the most frequently analyzed traits – obviously because these data are readily available.

Compared with natives, alien species tend to be longer-lived, i.e., phanerophytes, polycarpic (Thompson et al. 1995; Williamson and Fitter 1996), but also shorter-lived in other studies (Pyšek et al. 1995; Cadotte and Lovett-Doust 2001). This broader context supports the conclusions of Crawley et al. (1996) that aliens need to "try harder" than native species. They suggested that there are two characteristic groups of aliens that find vacant niches at different ends of niche axes: aliens that are more K-strategists (long-lived, tall, and with big seeds) than native K-strategists (woody and thicket-forming species that are capable of displacing native vegetation), and those that are more r-strategists than native r-strategists (small, rapidly maturing, long-flowering species that soon succumb to interspecific competition during secondary succession). Recently, Lloret et al. (2005) provided support for the hypothesis that invasion success may be triggered by functional traits quantitatively different from those occurring in the native flora, in which some life forms may be more saturated than others (Mack 2003).

Within aliens, the role of life form seems to be stage-specific: annuals are promoted in terms of early arrival (Pyšek et al. 2003), but invasiveness seems to be associated with long-lived life forms (Sutherland 2004), and to be habitat-specific: therophytes do better in disturbed, hemicryptophytes in seminatural vegetation (Pyšek et al. 1995). Unlike life histories, Grime's life strategy shows no consistent pattern across studies (Table 7.2).

Not surprisingly, alien species originating on the same continent have tended to arrive earlier in Central Europe (Pyšek et al. 2003), but aliens from more distant regions tend to be more frequent or abundant than those from the same continent. There are only two datasets to support this, but both are very representative and based on large numbers of species from Europe (Pyšek and Jarošík 2005) and North America (Cadotte et al. 2006a).

Plant height is often subjected to testing, for the same reason as growth form. Two studies based on British flora (Williamson and Fitter 1996; Crawley et al. 1996) found aliens to be taller than native species, the latter by using phylogenetic corrections. As far as within-alien comparisons are concerned, although aliens do not seem to be generally taller in their invasive ranges (Thébaud and Simberloff 2001), several studies provided evidence that tallness is associated with invasiveness (Goodwin et al. 1999), and with a higher abundance in some types of habitats (Pyšek et al. 1995), or with increased invasiveness in interaction with other features such as life strategy (Pyšek and Jarošík 2005). We should note, however, that some recent, sophisticated studies that considered height found no relationship between height and invasiveness (Hamilton et al. 2005; Lloret et al. 2005; Cadotte et al. 2006b).

Clonality, along with the ability of *vegetative reproduction* and good lateral growth, is positively associated with invasiveness, but its effect is context-dependent. The results depend on whether a large set of aliens, including casual species (sensu Richardson et al. 2000a), is compared, or the comparison is restricted to naturalized or even invasive species only. In the former case, non-clonal species tend to be overrepresented among aliens (Pyšek 1997; Sutherland 2004), but the situation may be reversed in the latter. For more limited data, such as naturalized aliens (Thompson et al. 1995) or smaller species sets in specific habitats (Lake and Leishman 2004), clonal aliens may appear overrepresented compared to clonal natives, become more abundant than non-clonal aliens (Lloret et al. 2005), or increase their abundance at a faster rate (Pyšek 1997).

Only two studies considered *specific leaf area* (SLA), but both concluded that high SLA promotes invasiveness (Lake and Leishman 2004; Hamilton et al. 2005). This is worth mentioning because congeneric studies strongly indicate that this physiological measure is important (see below, Fig. 7.2). On the contrary, seedling relative growth rate (RGR) was not found significant in two studies; a paper exploring its effect on the distribution of 33 woody species invasive in New Zealand did not find seedlings' RGR nor their survival to be related to invasiveness either (Bellingham et al. 2004).

Breeding system and *sex habit* were evaluated in two studies using large species sets. For Britain (Williamson and Fitter 1996) and Ontario (Cadotte and Lovett-Doust 2001), it was concluded that alien species are less often monoecious and more likely hermaphroditic than natives. This provides some support, in broader context, for predictions about the importance of a sexual partner being present (Baker 1965). However, Sutherland (2004) found no significant difference for the North American flora (and even indicated that invasive species on this continent are more likely to be monoecious than are non-invasives). In the same vein, there is no evidence that self-compatibility is more common among aliens than among natives (Williamson and Fitter 1996; Sutherland 2004); Sutherland (2004) even reports the opposite – that aliens are more likely to be self-incompatible. Since congeneric studies addressing this issue are rare (Table 7.3), the main support for the importance of being able to reproduce sexually in the new region is from case studies, e.g., Nadel et al. (1992) for *Ficus*, and Daehler and Strong (1996) for *Spartina*. In these two genera, sudden events that allowed taxa to reproduce sexually – the formation of an allopolyploid taxon (*Spartina*), and the arrival of a pollinator (*Ficus*) – triggered widespread invasions.

Pollen vector has little value in explaining invasion success. Williamson and Fitter (1996) found British aliens to be more likely insect-pollinated than were native species. Using the flora of the same country, Crawley et al. (1996) came to the same result but only for cross-species comparisons, not with phylogenetic corrections applied. None of the four other studies reported significant effects of pollen vector, neither for aliens compared with natives, nor within themselves (Table 7.2). Again, this result seems to be fairly robust because it is strongly supported by congeneric comparisons (Fig. 7.2, Table 7.4).

Timing of flowering is very important, based on 11 studies, seven of which yielded significant results (Tables 7.2, 7.4). Although several studies comparing native and aliens found no significant differences in flowering phenology (cf. Thompson et al. 1995; Williamson and Fitter 1996), and Crawley et al. (1996) did so only without applying phylogenetic corrections, other studies clearly show that it is advantageous for an alien to flower for a more extended period, compared to a native (Lake and Leishman 2004 for Australia; Cadotte and Lovett-Doust 2001 for Ontario). The pattern becomes even more distinct in within-alien comparisons: early-flowering species had higher chances to be introduced early to Central Europe (Pyšek et al. 2003), European species invading in Canada flowered longer than their non-invading congeners (Goodwin et al. 1999), and alien species on Mediterranean islands that flower in summer and over longer periods are more abundant (Lloret et al. 2005). Aliens with longer flowering periods are also more abundant in Ontario (Cadotte et al. 2006b). Interestingly, the pattern of flowering found for the British flora by Crawley et al. (1996) – aliens flower earlier or later than natives – supports the "aliens try harder" concept suggested by these authors (see above).

When *propagule size* was compared for a number of native and alien species, its effect was found to be non-significant (Thompson et al. 1995; Williamson and Fitter 1996), or the results were ambiguous – seeds of aliens were reported to be bigger (Crawley et al. 1996) or very small (Cadotte and Lovett-Doust 2001), or the probability of aliens having seeds smaller than those of natives was associated with disturbed habitats (Lake and Leishman 2004). Most within-alien studies exploring the correlation between seed size and invasion success also yielded non-significant results (Pyšek et al. 2003; Pyšek and Jarošík 2005; Lloret et al. 2005; Cadotte et al. 2006b), the only exception being for Eastern Australia, where small seeds were found to be associated with invasion success at both regional and continental scales (Hamilton et al. 2005). These mostly ambiguous results (as in studies on congeners, Table 7.4) may be partly explained by there being two contrasting groups of aliens – short-lived herbs and woody species, having on average small and large seeds, respectively – each of them successful in different environments. Another reason may be that having both small and large seeds brings about potential pros and cons for an alien plant. Small seeds are correlated with increased seed output (Henery and Westoby 2001), are easily dispersed by wind, and persist longer in soil than do large seeds (Thompson et al. 1993). Large seeds are better for establishment (Harper 1977), and more attractive to vertebrate dispersers (Richardson et al. 2000b). It is, nevertheless, encouraging that Hamilton et al. (2005) in their excellent study, considering phylogenies, net effects and different scales, found small seeds to be significantly correlated with invasion success.

Studies addressing the effect of *dispersal mode* and efficiency did not arrive at consistent conclusions (Table 7.2). Aliens were reported to be more likely dispersed by humans than were native species (Pyšek et al. 1995; Crawley et al. 1996), and less likely by water, wind (Crawley et al. 1996) and animals (Cadote and Lovett-Doust 2001). These results emerged from analyses of large floras, but datasets based on fewer species indicated the opposite – both Andersen (1995) and Lake and Leishman (2004) found aliens to be more often dispersed by wind and vertebrates, or having more fleshy fruits, which indirectly implies the latter. To relate dispersal syndrome to invasion success within aliens is even more difficult – the few available studies highlighted wind, animals (Lloret et al. 2005), and water (Pyšek and Jarošík 2005) as dispersal vectors leading to higher abundance, whereas other studies did not find significant results (Pyšek et al. 1995, 2003; Prinzing et al. 2002). Comparative multispecies studies are constrained by plants being effectively dispersed by many vectors, each of them most efficient under specific circumstances.

Results of comparative large-scale studies on *habitat preferences* are not easy to interpret, as they reflect variation in habitats present in target areas and the variety of approaches used. Affinity for dry habitats seems to be a feature typical of alien species (Thompson et al. 1995; Prinzing et al. 2002; Sutherland 2004). Surprisingly, only two studies (Williamson and Fitter 1996; Prinzing et al. 2002) out of seven indicated affinity of aliens to fertile soils.

7.2.4 Biases to Bear in Mind: Residence Time, Scale and Stage

There are biases that need to be considered when interpreting the results of comparative multispecies studies. Analyses of several pools of alien species have shown that the more time alien species have spent in their introduced ranges, the more likely they are to have become widespread (Pyšek and Jarošík 2005; Cadotte et al. 2006b). To take this into account when exploring net effects of traits requires knowledge of introduction dates, and such data are notoriously hard to obtain for whole floras (Kolar and Lodge 2001; but see Pyšek et al. 2003; Hamilton et al. 2005). Real residence time can be reliably inferred from the date of first reporting (Pyšek and Jarošík 2005; Hamilton et al. 2005). The potential confounding effect of residence time can be demonstrated by the use by humans, which promotes invasiveness in terms of probability of arrival to the new region (Pyšek et al. 2003), and abundance (Pyšek et al. 1995; Prinzing et al. 2002). However, this effect may be mediated through the residence time – plants introduced for utility reasons arrived significantly earlier in the Czech Republic than those planted as ornamentals, and accidental introductions were the latest (Pyšek et al. 2003). If minimum residence time is included into the model, the effect of human use becomes non-significant (Pyšek and Jarošík 2005; Hamilton et al. 2005).

The scale of study may represent another potential bias. Studies at a single spatial scale are unlikely to discern the drivers of invasion patterns (Collingham et al. 2000; Lloret et al. 2004; Pyšek and Hulme 2005), and the effect of a given trait may differ at various scales (Hamilton et al. 2005; Lloret et al. 2005).

When interpreting multispecies studies, one must be aware of the type of comparison and approaches used by the primary researchers. For example, if aliens are compared with natives, then large seeds were identified as typical of aliens (Crawley et al. 1996), but if the analysis is made within aliens, then another study indicated small seeds as promoting invasion success (Hamilton et al. 2005). Similarly, Prinzing et al. (2002) did not find dispersal by wind and vertebrates important for species that reached Argentina from the European source pool, while Lake and Leishman (2004) did find this important for the invasion success of aliens in different habitats. Dispersal by wind and animals may not play a role in the chance of a species to overcome major oceanic barriers, where humans are the main vector, while more natural dispersal agents become important for spread in the new territory (Rejmánek et al. 2005). These seeming contradictions also indicate that the effects of individual traits depend on the stage of the invasion process. Social and economic factors are crucial at the introduction stage, biogeographical and ecological factors at the stage of naturalization, and ecological and evolutionary principles are crucial mediators of invasiveness (Perrings et al. 2005).

7.2.5 Message from Comparative Multispecies Studies

Cadotte et al. (2006b) generalized that the success of plant invaders was related to a short life cycle, dispersal syndrome, large native range size, non-random taxonomic patterns, presence of clonal organs, occupation of disturbed habitats, and early time since introduction. In this review, we do not deal with range sizes nor habitat affinities, and the effect of introduction time (Rejmánek 2000; Pyšek and Jarošík 2005) and taxonomic patterns are evident (Daehler 1998; Pyšek 1998). Nevertheless, our survey of the 18 studies indicates that the effect of life history is more complicated, and the results reported for dispersal syndrome are far from unambiguous. Presence of clonal organs and ability of vigorous spatial growth certainly promote invasiveness, but these traits are context-specific (Table 7.2). This illustrates that even within the limited number of comparative multispecies studies available to date, different researchers include slightly different datasets (compare Table 7.3 with Table 2 in Cadotte et al. 2006b) and interpret them slightly differently.

Our review suggests that comparative multispecies studies provide strong support only for height, vigorous vegetative growth, early and extended flowering, and attractiveness to humans, as traits universally associated with invasiveness in vascular plants (Table 7.4). Studies reporting these findings are not numerous but fairly robust, as they were tested in different regions of the world and are based on different floras. They have potentially useful implications for screening protocols (Daehler et al. 2004).

There are, however, several fundamental limitations of multispecies comparative studies carried out to date. For one, accurate data on many traits of interest are not available for most plant species, not even for very widespread and abundant species. Good data are available for plant height, growth form, seed mass and very general "dispersal syndrome", but data on growth rates, palatability, seed production, and many other traits that are crucial for invasion success are incomplete or of highly dubious quality. Nevertheless, as researchers continue to collect life-history and population-level data, the information contained in inventories will continue to improve and contribute to elucidating the role species traits play in plant invasions (Cadotte et al. 2006b).

7.3 Studies of Congeners and Confamilials

Although some interesting patterns have emerged from the studies reviewed above, it is clear that uncovering a set of traits associated with invasiveness applicable to all vascular plants, and for all of the world's biomes, is totally unrealistic. A trait or set of traits that potentially confers invasiveness to an African *Acacia* in Australia cannot be expected to do the same for a European grass in North America. Nonetheless, there is value in continuing the search for traits determining invasiveness at a fine taxonomic scale, or for particular life forms or "functional types" (Rejmánek and Richardson 1996).

Much work has thus focused on congeners, confamilials, and otherwise taxonomically and phylogenetically related species (Fig. 7.2). Especially for congeners, such comparisons reduce phylogenetic problems that bedevil interspecific comparisons (Rejmánek and Richardson 1996; Cadotte and Lovett-Doust 2001). This approach involves pairing invasive species with native species or non-invasive congeners – if a consistent difference can be identified between invader and native, then that difference might help to explain invasiveness in some taxa (Daehler 2003).

Following the pioneering studies of Baker (1965) and Harris (1967), congeneric studies of invasiveness started to appear in the 1980s. This interest seems to have been stimulated by the SCOPE program on biological invasions launched in 1982 (Fig. 7.1). Our analysis is based on 46 comparisons of aliens with their native congeners or confamilials, and 18 studies that compare two or more alien congeners differing in their invasiveness (Fig. 7.2). The increase in the number of available studies indicates that alien–native congeneric comparisons have been increasing faster (Fig. 7.1). That there are more studies of this kind simply reflects that there are more such natural experiments available to researchers. Many prominent invaders have native congeners in invaded regions, but the sets of alien congeners invading in the same region, and differing in the degree of invasiveness or status (naturalized vs. casual) are certainly more limited.

7.3.1 Assumptions for Congeneric Studies

In attempting to determine traits of invasive species, ecologists often use species native to the invaded range as control species. Muth and Pigliucci (2006) argue that because many native species themselves are aggressive colonizers, comparisons using this type of control do not necessarily yield relevant information, and suggest that comparing introduced invasives vs. introduced non-invasives is more appropriate. As in the multispecies comparisons discussed above, however, whether native–alien or native–native comparisons are preferred (if we have a choice of such research strategy at all) depends on a concrete situation. If the invader outcompetes the native congener, or they at least coexist in the same type of habitat, then the comparison is relevant and the questions asked are the same as in comparable multispecies studies.

We identified 64 studies comparing one or more pairs of congeners (50 studies), or species from the same family (14 studies). In total, species from 21 families are represented (Fig. 7.2). The criterion for a pair/group of congeneric alien taxa to be included was that they differ in their degree of invasiveness. There was always a notable invasive species, and one or more other alien species that could be considered non-invasive (or at least much less invasive). Therefore, we did not consider studies comparing non-invasive aliens with native species (e.g., Blaney and Kotanen 2001). There is no reason to expect differences in traits explicable by alien status only, i.e., without reference to invasive potential (Crawley et al. 1996). It is also important how non-invasive counterparts of invaders are defined. Probably the best approach in terms of eliminating biases was used by Burns (2004). In this study, non-invasive species were those that were cultivated in the same region as invasive congener, but did not escape from cultivation. Ideally, the "invasive" and "non-invasive" species used in a comparison should have been present in a region for approximately the same period of time, and have experienced the same opportunities for sampling a range of potentially invasible sites (e.g., through human-mediated dissemination). Where this was clearly not the case, comparisons were not included in our analysis.

7.3.2 Searching for Generalities Within Genera

For each study, we recorded all traits that were subjected to statistical testing in primary papers, and those that were reported to differ significantly between aliens and natives, or invasive and non-invasive aliens. In total, there were 27 traits subjected to testing in at least three studies; these were classified into morphological, physiological, reproductive, and "response" traits – the latter comprise the response of species to external factors such as herbivory, burning, or soil conditions (Table 7.4). In terms of structure of the traits considered, alien–native studies focus more on physiology (45 % of trait/congeners comparisons) than do within-alien studies; the corresponding value for the latter is 18 %, accounted for almost exclusively by measurements of seedling RGR. Comparison of alien congeners deal disproportionately more with reproductive characteristics. In total, the studies analyzed comprised 222 between-congener/confamilial comparisons of individual traits (Table 7.3). The list of analyzed studies covers the majority of published data, and can therefore be regarded as a highly representative sample of current research.

Since we, as in the case of comparative multispecies studies, recorded not only the number of significant results for individual traits, but also their proportion among all tested cases for a given trait, the results summarized in Fig. 7.2 and Table 7.4 provide a reasonably robust assessment of important traits associated with invasiveness in plants.

Some traits associated with invasiveness in multispecies comparative studies also emerge as important in congeneric studies (Table 7.4). Surprisingly, neither approach revealed an unambiguous and positive effect of high biomass on invasiveness, but both did so for plant height. As far as biomass is concerned, a substantial proportion of the 15 studies found its effect non-significant or even opposite, i.e., the native congener had higher biomass than the alien one (Schierenbeck et al. 1994; Smith and Knapp 2001).

Growth rate and allocation to growth appear important; closely associated with this is the capacity for vigorous spatial growth (Fig. 7.2, Table 7.4); this seems to contradict the results of Daehler (2003) who reviewed the performance of co-occurring native and alien species, and did not find higher growth rates, competitive ability nor fecundity to be characteristic of the latter. Rather, the relative performance of invaders and co-occurring natives depended on growing conditions (Daehler 2003). That aliens, compared to natives, or invaders compared to less-invasive taxa, exhibited faster growth is a very robust result in our analysis (Fig. 7.2), and the question arises whether

Table 7.3 Frequencies of pairwise species comparisons classified according to groups of traits and approaches

Group of traits	Number of comparisons		Total	Comparisons (%)	
	Alien–native	Within alien		Alien–native	Within alien
Morphological	31	5	36	17.6	10.9
Physiological	79	8	87	44.9	17.4
Reproductive	49	25	74	27.8	54.3
Response	17	8	25	9.7	17.4
Total	176	46	222		

Table 7.4 Summary of results from comparative and congeneric studies on traits promoting invasiveness in plants[a]

Group of traits	Trait	Comparative studies	No.	Congeneric studies	No.
Complex	Growth form and life form	Short-lived form promotes transport and in disturbed habitats, long-lived competitiveness in seminatural habitats	13	N.a.	
	Grime life strategy	Ambiguous	6	N.a.	
	Area of origin	Aliens of distant origin arrive later but become more abundant	4	N.a.	
	Use by humans	*Promotes early arrival and invasiveness*	4	N.a.	
Morphological	Biomass	N.a.		Ambiguous	15
	Plant height	*Promotes invasiveness*	10	*Promotes invasiveness*	6
	Vegetative spatial growth	*Promotes invasiveness, but the role is context-specific*	9	*Promotes invasiveness*	5
	Leaf number	N.a.		Ambiguous	10
	Leaf morphology, canopy structure	Ambiguous	6	N.a.	
Physiological	Photosynthetic rate/capacity	N.a.		*Promotes invasiveness*	15
	Water, N and P use efficiency	N.a.		*Promotes invasiveness*	9
	Chlorophyll contents	N.a.		Not enough studies	1
	Leaf N contents	N.a.		Not enough studies	1
	Leaf longevity	N.a.		Not enough studies	3
	Tissue construction costs	N.a.		Not enough studies	3
	Specific leaf area	Promotes invasiveness (only few studies)	2	*Promotes invasiveness*	11
	Leaf area ratio	N.a.		Tends to promote invasiveness	8
	Total leaf area	N.a.		Ambiguous	10
	Seedling relative growth rate	Not enough studies	3	Tends to promote invasiveness	8
	Growth rate, allocation to growth	N.a.		*Promotes invasiveness*	8

Reproductive	Self-compatibility	Not enough studies, but no support for it being a feature of aliens	3	Not enough studies	4
	Breeding system	Hermaphroditism and dioecy support invasiveness	5	N.a.	
	Pollen quality	N.a.		Not enough studies	2
	Pollen vector	No pattern	8	N.a.	
	Time of flowering	*Early and longer flowering promotes invasiveness*	11	*Early and longer flowering promotes invasiveness*	5
	Reproductive maturity (generation time)	N.a.		Not enough studies	4
	Fecundity	N.a.		*Promotes invasiveness*	13
	Propagule size	Not very clear pattern, the effect seems to be stage-related	10	Not enough studies and ambiguous	4
	Dispersal mode and efficiency	Ambiguous	11	*Efficient dispersal promotes invasiveness*	5
	Seed release, serotiny	N.a.		Not enough studies	3
	Germination ability	N.a.		*Promotes invasiveness*	18
	Seedling survival and establishment	N.a.		*Promotes invasiveness*	6
	Seed dormancy, seed bank longevity and size	Not enough studies	3	*Promotes invasiveness*	10
Response	Tolerance to herbivory/clipping/cutting	N.a.		*Promotes invasiveness*	11
	Tolerance to drought	N.a.		Not enough studies	4
	Tolerance to low nutrients	N.a.		Promotes invasiveness (but only few studies)	4
	Tolerance to fire	N.a.		*Promotes invasiveness*	5
	Habitat preferences	Ambiguous, trend for dry and warm habitats	7	N.a.	

[a] Robust patterns are shown in italics. Number of studies and species pairs used for comparison (see Tables 7.2 and 7.3) is indicated. *N.a.* Not addressed, in comparative studies mostly because data are not available for large sets of species, in congeneric studies mostly because complex traits are controlled by approach – species of the same life form and strategy, morphology, pollination syndrome, etc., are usually chosen for congeneric comparisons to reduce bias

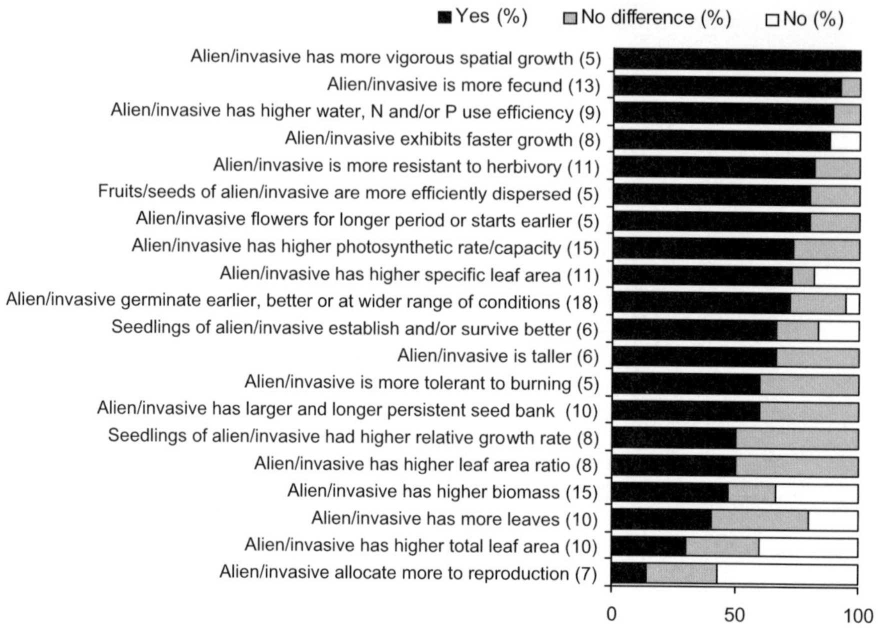

Fig. 7.2 Summary of the results of 59 studies comparing 64 aliens with their native congeners or related taxa, and alien congeners with different degree of invasiveness. Percentages of significant results supporting (*yes*) or rejecting (*no*) given statements, or yielding no difference are shown. Traits are listed according to decreasing unambiguousness of results. The number of species pairs on which a given trait was tested is given in parentheses following the statement; only traits tested on at least five species pairs are displayed. The following studies comparing traits of alien and native congeners or closely related taxa within a genus (the latter cases are where family name is given), and of alien congeners differing in the degree of invasiveness were used as dataset: *Acer* (Kloeppel and Abrams 1995), *Ageratum* (Baker 1965), *Agropyron* (Caldwell et al. 1981; Richards 1984; Black et al. 1994), *Agrostis* (Pammenter et al. 1986), *Amsinckia* (Pantone et al. 1995), Asteraceae (Smith and Knapp 2001), *Atriplex* (Mandák 2003), *Bidens* (Gruberová et al. 2001), *Bromus* (Kolb and Alpert 2003), *Carpobrotus* (Vilà and D'Antonio 1998a, b), *Celastrus*, *Parthenocissus*, *Polygonum* (Van Clef and Stiles 2001), *Centaurea* (Gerlach and Rice 2003), *Centaurea*, *Crepis* (Muth and Pigliucci 2006), *Cortaderia* (Lambrinos 2001, 2002), *Crataegus* (Sallabanks 1993), Cyatheaceae (Durand and Goldstein 2001), *Echium* (Forcella et al. 1986), *Eucalyptus* (Radho-Toly et al. 2001), *Eupatorium* (Baker 1965), Fabaceae (Smith and Knapp 2001), *Hakea* (Richardson et al. 1987), *Impatiens* (Perrins et al. 1993), *Lonicera* (Sasek and Strain 1991; Schierenbeck and Marshall 1993; Schierenbeck et al. 1994; Schweitzer and Larson 1999; Larson 2000), *Mikania* (Deng et al. 2004), *Oenothera* (Mihulka et al. 2003), Oleaceae (Morris et al. 2002), *Pinus* (Rejmánek and Richardson 1996; Grotkopp et al. 2002), *Plantago* (Matsuo 1999), Poaceae (Harris 1967; Baruch et al. 1985; Pyke 1986, 1987; Bilbao and Medina 1990; Williams and Black 1994; Baruch and Goméz 1996; Holmes and Rice 1996; Baruch and Bilbao 1999; Goergen and Daehler 2001a, b, 2002; Smith and Knapp 2001), Proteaceae (Honig et al. 1992), *Reynoutria* (Pyšek et al. 2003a; Bímová et al. 2003), *Rubus* (McDowell 2002; Lambrecht-McDowell and Radosevich 2005), *Senecio* (Radford and Cousens 2000, Sans et al. 2004), *Spartina* (Callaway and Josselyn 1992; Anttila et al. 1998), *Tradescantia*, *Commelina*, *Murdannia* (Burns 2004). Citation details can be found at www.ibot.cas.cz/personal/pysek

the seemingly contradicting conclusions of Daehler's review could be due to the fact that his review was not confined to congeners. In our analysis, phylogenetic bias is reduced.

In agreement with Daehler (2003), congeneric studies suggest that leaf area is important, although this was manifested as total leaf area in Daehler's dataset (which gave ambiguous results in ours), and as specific leaf area (SLA) in 11 studies in our review (Fig. 7.2). As SLA is positively associated with invasiveness in multispecies comparative studies as well (see above), it seems to be one of most robust indicators/predictors of invasiveness. Invasions are generally associated with disturbed habitats, and high SLA is typical of rapidly colonizing species. High SLA is correlated with short leaf retention and fast growth rate; this is associated with avoidance of investing biomass into long-lasting structures, which is, in turn, a critical precondition of success in disturbed habitats where fast growth is paramount (Grotkopp et al. 2002). Of other physiological traits, photosynthetic rate/capacity and water and/or resource use efficiency promote invasiveness, and this pattern is very robust and supported in 15 and nine studies, respectively.

The results of 80 % of studies that address flowering phenology accord with conclusions from multispecies comparative studies; early flowering and extended flowering period, compared to natives/non-invasives, provide invaders with an advantage. Reproductive traits in general appear important determinants of invasiveness, and these traits are identified by congeneric studies much more reliably than by multispecies comparisons, because sufficiently accurate data are mostly not available for large numbers of species. High fecundity and efficient dispersal of seeds promote invasiveness (Fig. 7.2, Table 7.4). Many studies compared features associated with seed germination, dormancy, and seed bank longevity (Fig. 7.2), and together they clearly indicate that easy germination, and long-term seed banks that allow species to extend germination over time and to wait for preferred conditions increase invasiveness.

7.4 Combining Approaches: Pooling the Evidence

From the above, it follows that each of the two main approaches discussed has its own strengths and weaknesses. It is symptomatic that the best progress to date toward a general theory of plant invasiveness has been achieved by pooling evidence from both approaches. Genera with enough invasive and non-invasive taxa to enable rigorous statistical analysis, and for which detailed autecological information is available are ideally suited for extracting robust generalizations. The genus *Pinus* provides the best example known to date.

Rejmánek and Richardson (1996) were able to explain invasiveness in *Pinus* species using only three traits (seed mass, length of juvenile period, and interval between good seed crops). They defined a discriminant function that successfully separated invasive and non-invasive species. This framework was expanded, by adding considerations relating to dispersal by vertebrates and characteristics of fruits, and successfully applied to predict invasiveness in other gymnosperms and woody angiosperms (Rejmánek and Richardson 1996; Richardson and Rejmánek 2004).

Rejmánek combined the results from congeneric studies on pines with robust patterns emerging from multispecies comparative studies to formulate his "Theory of seed plant invasiveness", the most ambitious attempt to create a unified scheme (Rejmánek 1996, 2000; Rejmánek et al. 2005). This theory posits that a low nuclear amount of DNA, as a result of selection for short generation time, membership to alien genera, and primary latitudinal range are major factors contributing to the invasiveness of seed plants. Large geographical range is often among the best predictors of invasion success (Rejmánek 1996; Goodwin et al. 1999). Widespread species are more likely to be adapted to a wider range of conditions, and have better chances to be dispersed because they occur in more locations (Booth et al. 2003). Although there are exceptions to this general rule reported for individual species (Richardson and Bond 1991), the same traits that allow a species to be widespread in the native range seem to be also favorable for a successful invasion (Booth et al. 2003). An additional study identified RGR as the most important predictor of invasiveness in disturbed habitats, and related invasiveness to physiological measures (Grotkopp et al. 2002).

7.5 Conclusions: Where Do we Stand?

The two main approaches to the role of plant traits in determining invasiveness (Table 7.4) provide complementary answers. The congeneric studies identified a higher number of important traits, because they are better focused and more detailed. Some of the traits simply cannot be addressed by multispecies studies, because this approach is too "coarse-grained". Methodologically, there is another difference between the two approaches. Congeneric/confamilial comparisons, by involving an invasive or at least naturalized alien, address later stages of invasion, while analyses of whole floras are in some cases biased by including casual species. Since different traits potentially influence different stages of invasion (Kolar and Lodge 2001; Pyšek et al. 2003; Perrings et al. 2005), this introduces a bias into multispecies comparisons that does not influence congeneric studies.

On the other hand, conclusions yielded by comparisons of whole floras are fairly robust, and often generally valid for all vascular plants. Detailed congeneric studies are sometimes difficult to compare directly because of the variety of methods used, these being specific and suited to a given species, region and the investigator's research priorities.

Cadotte et al. (2006b) recently reminded us of John Harper's contention that historically the field of plant ecology has been dominated by two major themes, i.e., description of vegetation, dealing with species assemblages and their classification, and autecological single-species descriptions (Harper 1977). Seeking traits associated with invasiveness has followed a similar two-pronged approach, with multispecies studies being somewhat analogous to vegetation description, and research on congeners comparable to autecological studies. As for plant ecology in general, both approaches yield unique and mutually enriching results.

When looking at the effect of traits on invasiveness, we must remember that different species were introduced at different times and are at different stages of naturalization/invasion. Studies that explicitly attempt to filter out such effects and other biases are extremely useful for revealing inherent trait-related determinants of invasibility. The role of plant traits in the invasion process is to a very large extent stage- and habitat-specific. Traits that confer an advantage at a given stage of the process, and in a particular habitat may be neutral or even detrimental at another phase and/or for a different habitat.

Most importantly, however, many traits have been tested repeatedly and often enough to allow us to draw fairly robust conclusions regarding their role. This review clearly indicates that successful invaders possess some traits that unsuccessful invaders do not have. Traits do matter! Unfortunately, crucial information is lacking for many species, and the challenge for the invasion-ecology community is to collate such information and to make it widely available.

Acknowledgments. We thank Mark Cadotte and Brad Murray for providing us with their unpublished papers, and Ivan Ostrý and Zuzana Sixtová for logistic support. P.P. was supported by the European Union FP 6 integrated project ALARM (GOCE-CT-2003-506675) and by the specific targeted research project DAISIE (SSPI-CT-2003-511202), grant no. 206/05/0323 from the Grant Agency of the Czech Republic, and by institutional long-term research plans no. AV0Z60050516 from the Academy of Sciences of the Czech Republic, and no. 0021620828 from the Ministry of Education of the Czech Republic. D.M.R. acknowledges support from the DST-NRF Centre of Excellence for Invasion Biology.

References

Alpert P, Bone E, Holzapfel C (2000) Invasiveness, invasibility and the role of environmental stress in the spread of non-native plants. Perspect Plant Ecol Evol Syst 3:52–66

Andersen UV (1995) Comparison of dispersal strategies of alien and native species in the Danish flora. In: Pyšek P, Prach K, Rejmánek M, Wade M (eds) Plant invasions: general aspects and special problems. SPB Academic, Amsterdam, pp 61–70

Baker HG (1965) Characteristics and modes of origin of weeds. In: Baker HG, Stebbins GL (eds) The genetics of colonizing species. Academic Press, New York, pp 147–172

Bellingham PJ, Duncan RP, Lee GW, Buxton RP (2004) Seedling growth rate and survival do not predict invasiveness in naturalized woody plants in New Zealand. Oikos 106:308–316

Blaney CS, Kotanen PM (2001) Effects of fungal pathogens on seeds of native and exotic plants: a test using congeneric pairs. J Appl Ecol 38:1104–1113

Booth BD, Murphy SD, Swanton CJ (2003) Weed ecology in natural and agricultural systems. CABI, Wallingford

Burns JH (2004) A comparison of invasive and non-invasive dayflowers (Commelinaceae) across experimental nutrient and water gradients. Diversity Distrib 10:387–397

Cadotte MW, Lovett-Doust J (2001) Ecological and taxonomic differences between native and introduced plants of southwestern Ontario. Ecoscience 8:230–238

Cadotte MW, Murray BR, Lovett-Doust J (2006a) Evolutionary and ecological influences of plant invader success in the flora of Ontario. Ecoscience 13:388–395

Cadotte MW, Murray BR, Lovett-Doust J (2006b) Ecological patterns and biological invasions: using regional species inventories in macroecology. Biol Invasions 8:809–821

Collingham YC, Wadsworth RA, Willis SG, Huntley B, Hulme PE (2000) Predicting the spatial distribution of alien riparian species: issues of spatial scale and extent. J Appl Ecol suppl 1 37:13–27

Crawley MJ (1987) What makes a community invasible? In: Gray AJ, Crawley MJ, Edwards PJ (eds) Colonization, succession and stability. Blackwell, Oxford, pp 429–453

Crawley MJ (1989) Chance and timing in biological invasions. In: Drake JA, Mooney HA, di Castri F, Groves RH, Kruger FJ, Rejmanek M, Williamson M (eds) Biological invasions: a global perspective. Wiley, Chichester, pp 407–424

Crawley MJ, Harvey PH, Purvis A (1996) Comparative ecology of the native and alien floras of the British Isles. Biol Trans R Soc B 351:1251–1259

Daehler CC (1998) The taxonomic distribution of invasive angiosperm plants: ecological insights and comparison to agricultural weeds. Biol Conserv 84:167–180

Daehler CC (2003) Performance comparisons of co-occurring native and alien invasive plants: implications for conservation and restoration. Annu Rev Ecol Syst 34:183–211

Daehler CC, Strong DR (1996) Status, prediction and prevention of introduced cordgrass *Spartina* spp. invasions in Pacific estuaries, USA. Biol Conserv 78:51–58

Daehler CC, Denslow JS, Ansari S, Kuo H-C (2004) A risk assessment system for screening out invasive pest plants from Hawai'i and other Pacific Islands. Conserv Biol 18:360–368

Dreyer GD, Baird LM, Fickler C (1987) *Celastrus scandens* and *Celastrus orbiculatus* comparisons of reproductive potential between a native and an introduced woody vine. Bull Torrey Bot Club 114:260–264

Gerlach JD, Rice KJ (2003) Testing life history correlates of invasiveness using congeneric plant species. Ecol Appl 13:167–179

Gilpin M (1990) Ecological prediction. Science 248:88–89

Goodwin BJ, McAllister AJ, Fahrig L (1999) Predicting invasiveness of plant species based on biological information. Conserv Biol 13:422–426

Grotkopp E, Rejmánek M, Rost TL (2002) Towards a casual explanation of plant invasiveness: seedling growth and life history strategies of 29 pine (*Pinus*) species. Am Nat 159:396–419

Hamilton MA, Murray BR, Cadotte MW, Hose GC, Baker AC, Harris CJ, Licari D (2005) Life-history correlates of plant invasiveness at regional and continental scales. Ecol Lett 8:1066–1074

Harper JL (1977) Population biology of plants. Academic Press, London

Harris GA (1967) Some competitive relationships between *Agropyron spicatum* and *Bromus tectorum*. Ecol Monogr 37:89–111

Harvey PH, Pagel MD (1991) The comparative method in evolutionary biology. Oxford University Press, Oxford

Harvey PH, Read AF, Nee S (1995) Why ecologists need to be phylogenetically challenged? J Ecol 83:535–536

Henery M, Westoby M (2001) Seed mass and seed nutrient contents as predictors of seed output variation between species. Oikos 92:479–490

Kolar CS, Lodge DM (2001) Progress in invasion biology: predicting invaders. Trends Ecol Evol 16:199–204

Lake JC, Leishman MR (2004) Invasion success of exotic plants in natural ecosystems: the role of disturbance, plant attributes and freedom from herbivores. Biol Conserv 117:215–226

Leishman MR, Thomson VP (2005) Experimental evidence for the effects of additional water, nutrients and physical disturbance on invasive plants in low fertility Hawkesbury Sandstone soils, Sydney, Australia. J Ecol 93:38–49

Lloret F, Médail F, Brundu G, Hulme P (2004) Local and regional abundance of exotic plant species on Mediterranean islands: are species traits important? Global Ecol Biogeogr 13:37–45

Lloret F, Médail F, Brundu G, Camarda I, Moragues E, Rita J, Lambdon P, Hulme PE (2005) Species attributes and invasion success by alien plants on Mediterranean islands. J Ecol 93:512–520

Mack RN (2003) Phylogenetic constraint, absent life forms, and preadapted alien plants: a prescription for biological invasions. Int J Plant Sci 164 suppl 3:185–196

Muth NZ, Pigliucci M (2006) Traits of invasives reconsidered: phenotypic comparisons of introduced invasive and introduced noninvasive plant species within two closely related clades. Am J Bot 93:188–196

Nadel H, Frank JH, Knight RJ (1992) Escapees and accomplices: the naturalization of exotic *Ficus* and their associated faunas in Florida. Florida Entomol 75:30–38

Noble I (1989) Attributes of invaders and the invading process: terrestrial and vascular plants. In: Drake JA, Mooney HA, di Castri F, Groves RH, Kruger FJ, Rejmanek M, Williamson M (eds) Biological invasions: a global perspective. Wiley, Chichester, pp 301–314

Perrings C, Dehnen-Schmutz K, Touza L, Williamson M (2005) How to manage biological invasions under globalization. Trends Ecol Evol 20:212–215

Perrins J, Fitter A, Williamson M (1993) Population biology and rates of invasion of three introduced Impatiens species in the British Isles. J Biogeogr 20:33–44

Prinzing A, Durka W, Klotz S, Brandl R (2002) Which species become aliens? Evol Ecol Res 4:385–405

Pyšek P (1997) Clonality and plant invasions: can a trait make a difference? In: de Kroon H, van Groenendael J (eds) The ecology and evolution of clonal plants. Backhuys, Leiden, pp 405–427

Pyšek P (1998) Is there a taxonomic pattern to plant invasions? Oikos 82:282–294
Pyšek P (2003) How reliable are data on alien species in Flora Europaea? Flora 198:499–507
Pyšek P, Hulme PE (2005) Spatio-temporal dynamics of plant invasions: linking pattern to process. Ecoscience 12:302–315
Pyšek P, Jarošík V (2005) Residence time determines the distribution of alien plants. In: Inderjit (ed) Invasive plants: ecological and agricultural aspects. Birkhäuser, Basel, pp 77–96
Pyšek P, Prach K, Šmilauer P (1995) Relating invasion success to plant traits: an analysis of the Czech alien flora. In: Pyšek P, Prach K, Rejmánek M, Wade M (eds) Plant invasions: general aspects and special problems. SPB Academic, Amsterdam, pp 39–60
Pyšek P, Sádlo J, Mandák B, Jarošík V (2003) Czech alien flora and a historical pattern of its formation: what came first to Central Europe? Oecologia 135:122–130
Pyšek P, Richardson DM, Rejmánek M, Webster G, Williamson M, Kirschner J (2004a) Alien plants in checklists and floras: towards better communication between taxonomists and ecologists. Taxon 53:131–143
Pyšek P, Richardson DM, Williamson M (2004b) Predicting and explaining plant invasions through analysis of source area floras: some critical considerations. Diversity Distrib 10:179–187
Rejmánek M (1996) A theory of seed plant invasiveness: the first sketch. Biol Conserv 78:171–181
Rejmánek M (2000) Invasive plants: approaches and predictions. Austral Ecol 25:497–506
Rejmánek M, Richardson DM (1996) What attributes make some plant species more invasive? Ecology 77:1655–1661
Rejmánek M, Richardson DM, Higgins SI, Pitcairn MJ, Grotkopp E (2005) Ecology of invasive plants: state of the art. In: Mooney HA, Mack RM, McNeely JA, Neville L, Schei P, Waage J (eds) Invasive alien species: searching for solutions. Island Press, Washington, DC, pp 104–161
Richardson DM, Bond WJ (1991) Determinants of plant-distribution – evidence from pine invasions. Am Nat 137:639–668
Richardson DM, Cowling RM (1992) Why is mountain fynbos invasible and which species invade? In: Van Wilgen BW, Richardson DM, Kruger FJ, van Hensbergen HJ (eds) Fire in South African mountain fynbos. Springer, Berlin Heidelberg New York, pp 161–181
Richardson DM, Pyšek P (2006) Plant invasions: merging the concepts of species invasiveness and community invasibility. Prog Phys Geogr 30:409–431
Richardson DM, Rejmánek M (2004) Conifers as invasive aliens: a global survey and predictive framework. Diversity Distrib 10:321–331
Richardson DM, Pyšek P, Rejmánek M, Barbour MG, Panetta FD, West CJ (2000a) Naturalization and invasion of alien plants: concepts and definitions. Diversity Distrib 6:93–107
Richardson DM, Allsopp N, D'Antonio CM, Milton SJ, Rejmánek M (2000b) Plant invasions: the role of mutualisms. Biol Rev 75:65–93
Roy J (1990) In search of characteristics of plant invaders. In: di Castri J, Hansen AJ, Debussche M (eds) Biological invasions in Europe and the Mediterranean Basin. Kluwer, Dordrecht, pp 335–352
Schierenbeck K, Mack RN, Sharitz RR (1994) Effects of herbivory on growth and biomass allocation in native and introduced species of *Lonicera*. Ecology 75:1661–1672
Smith MD, Knapp AK (2001) Physiological and morphological traits of exotic, invasive exotic, and native plant species in tallgrass prairie. Int J Plant Sci 162:785–792

Sutherland S (2004) What makes a weed a weed: life history traits of native and exotic plants in the USA. Oecologia 141:24–39

Thébaud C, Simberloff D (2001) Are plants really larger in their introduced ranges? Am Nat 157:231–236

Thompson K (1994) Predicting the fate of temperate species in response to human disturbance and global change. In: Boyle TJB, Boyle CEB (eds) Biodiversity, temperate ecosystems and global change. Springer, Berlin Heidelberg New York, pp 61–76

Thompson K, Band SR, Hodgson JG (1993) Seed size and shape predict persistence in soil. Funct Ecol 7:236–241

Thompson K, Hodgson JG, Rich TCG (1995) Native and alien invasive plants: more of the same? Ecography 18:390–402

Timmins SM, Williams PA (1987) Characteristics of problem weeds in New Zealand's protected natural areas. In: Saunders DA, Arnold G, Burbidge A, Hopkins A (eds) Nature conservation: the role of remnants of native vegetation. Surrey Beatty, Chipping Norton, pp 241–247

Williamson M (1993) Invaders, weeds and the risk from genetically manipulated organisms. Experientia 49:219–224

Williamson MH, Fitter A (1996) The characters of successful invaders. Biol Conserv 78:163–170

8 Do Successful Invaders Exist? Pre-Adaptations to Novel Environments in Terrestrial Vertebrates

Daniel Sol

8.1 Introduction

Central in invasion biology is to understand why alien species, whose initial populations are generally small and genetically depleted, can succeed to establish themselves in environments to which they have had no opportunity to adapt (Sax and Brown 2000). This paradox is usually resolved by invoking pre-adaptations of non-indigenous species to novel environments. The idea is that some species are successful invaders because they have attributes that pre-adapt them to survive and reproduce in novel environments (Mayr 1965). However, do we really have evidence that there exist properties of successful invaders?

The goal of this chapter is to evaluate to what extent establishment success of terrestrial vertebrates may be understood by the existence of pre-adaptations of species to novel environments. This implies answering two interrelated questions: (1) do species differ in their invasion potential? And if so, (2) what are the features of the species that identify some as successful invaders? Answering these questions is important not only to fully understand how animals respond to new environmental conditions, but also to help identify and prevent situations where the risk is high that a species becomes established and causes ecological impact when introduced in a novel region.

8.2 Framework

A population is considered to be established when it is able to develop a self-sustaining population, that is, a population that is maintained over time by reproduction without the need of additional introductions. To become estab-

Ecological Studies, Vol. 193
W. Nentwig (Ed.)
Biological Invasions

lished, the invader needs to find an appropriate niche to survive and reproduce in the novel environment, which should include environmental conditions (e.g., temperatures or precipitations) that the species may tolerate, resources (e.g., food or shelter) that are not monopolized by other species, and a pressure from enemies (e.g., parasites or predators) low enough to not compromise population growth. Consequently, any property of a species that increases the likelihood to find an appropriate niche in a variety of diverse environments will also make the species a potentially successful invader.

Yet, finding an appropriate niche in the novel region is not a guarantee of successful establishment. Introduced populations generally start in low numbers, which makes them vulnerable to extinction by demographic, environmental and genetic stochasticity (Legendre et al. 1999; Lockwood et al. 2005). Thus, the first stages in the invasion process are particularly critical in determining the chance that a species establishes a self-sustaining population. Not only is it necessary that the balance between births and deaths be positive, but also that the increase of population size is fast enough to reduce the period in which the population remains vulnerable to stochasticity.

The need to grow fast in numbers has four main implications for understanding the invasion process. First, the likelihood of establishment is likely to increase with the size of the founder population, an idea that is strongly supported by evidence in both birds and mammals (Lockwood et al. 2005). Second, a non-indigenous species that, due to the Allee effect, has a reduced per capita growth rate at low population densities will be particularly prone to extinction (Reed 1999). Third, a non-indigenous species with a life history that facilitates a higher intrinsic rate of population increase will be more likely to escape the critical founding stage, and hence have higher chances to develop a self-sustaining population. Finally, any property of the invader that reduces survival or reproduction in the novel environment will impair population growth, increasing the risk of population extinction.

Thus, the ideal invader should posses a combination of traits that allows it to pass three main barriers toward establishment (Fig. 8.1). Such a combination of traits is unlikely to be found in a same species, implying that for most species, establishing self-sustaining populations in novel environments should be inherently difficult. Indeed, Williamson and others have repeatedly noted that most past introductions of plants and animals have failed to establish self-sustaining populations (e.g., Williamson and Fitter 1996). Moreover, many species that establish successfully in new regions often do so only after having failed in multiple earlier introductions (Sax and Brown 2000).

Despite the inherent difficulty to invade a new region, some vertebrates appear to be extremely successful invaders, while others are not (Ehrlich 1989; Williamson and Fitter 1996). Such varying invasion success is often held to arise from differences in certain properties of species (Ehrlich 1989), yet a similar pattern could result from a variety of alternative processes. For example, differences in invasion success could be explained by simple neutral

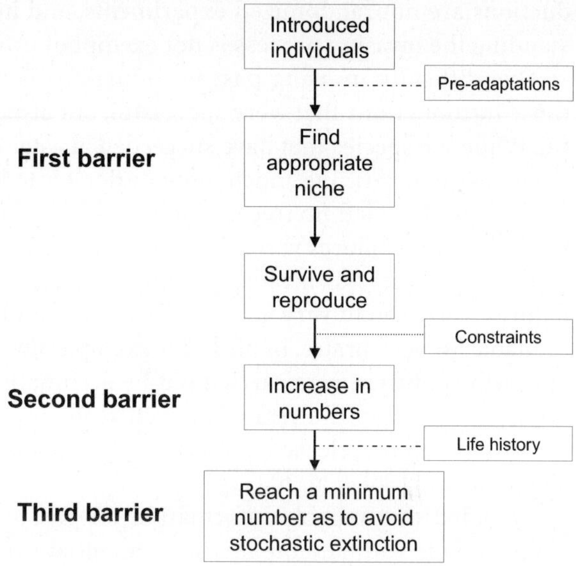

Fig. 8.1 Schematic representation of the three main barriers a species may pass to become established in a novel region. The species must first be able to find an appropriate niche to survive and reproduce, then have to reproduce at a rate high enough to counterbalance mortality, and finally must grow in number to reach a population size large enough to escape the effects of stochasticity. The success to pass each one of these barriers may mostly be affected by different attributes of the invaders

processes, such as differences in the number of individuals released, without the need to invoke differences in the properties of species. So, we must start asking whether differences between species really matter when it comes to determining invasion success.

8.3 Do Successful Invaders Exist?

The main difficulty in answering the above question is how to estimate the invasion potential for a given set of species. The invasion potential defines the ability of a species to establish itself in novel regions (Lonsdale 1999). The ideal approach to measure it would be to conduct experimental introductions under controlled conditions (e.g., releasing the same number of individuals of each species) in a variety of regions, and then use the rate of successes as an estimate of the invasion potential of each species. In terrestrial vertebrates, however, such large-scale experiments are not generally feasible for ethical and practical reasons.

As alternative, the invasion potential may be assessed using information of past historical introductions. The use of past introductions within a comparative framework has become the most widely used tool to study the success of vertebrate invasions, broadening our understanding of the invasion process, and providing general principles that are realistic enough to be used in risk assessments of future invaders (Duncan et al. 2003). Yet, past historical introductions are not randomized experiments, and hence their utility in understanding the invasion process is not exempt of problems.

One difficulty in using past introductions is the need to know not only those introductions that were successful, but also those that were unsuccessful. While the species that have succeeded at establishing themselves are relatively easy to determine, much more difficult is to know those that have failed, as they may have left no traces of their presence in the region. If the probability of detecting failures varies between species, then some species will appear to be very successful invaders simply because they have many unrecorded failures. Fortunately, very accurate records of both successes and failures are available for vertebrates. In birds, for example, many introductions in the 18th and 19th centuries were carried out by acclimatisation societies, which kept accurate records of the year of introduction, its outcome, and even the number of individuals released, providing high-quality data to estimate the invasion potential (Sol et al. 2005a).

Provided that reliable information on past introductions is available, a simple way to estimate the invasion potential would be to calculate the proportion of introductions that were successful. This method is nonetheless problematic when one uses historical introductions. It is well-known that introductions are non-randomly distributed across regions (Blackburn and Duncan 2001b), and that some species have been introduced in larger numbers than others (Cassey et al. 2004). In vertebrates, there is good evidence that success increases with the size of the founder population, so even successful invaders are expected to fail in some introductions if released in low numbers (Lockwood et al. 2005). Moreover, some regions may be easier to invade than others, as a result of their characteristics (e.g., species richness or degree of disturbance), as well as their ecological similitude with the native regions of the species introduced (Williamson 1996; Shea and Chesson 2002). In birds, for example, exotic species are more likely to fail on islands with species-rich mammalian predator assemblages (Cassey et al. 2005). Consequently, one species may seem to be a worse invader than another simply because it has been introduced in lower numbers, or in hard-to-invade regions. Clearly, differences in invasion potential between species must be evaluated under similar conditions of introduction.

One possibility suggested in the literature to obtain reliable measures of the invasion potential is the use of generalized linear mixed models (Blackburn and Duncan 2001a; Steele and Hogg 2003; Sol et al. 2005b). The idea is to estimate the magnitude of species (or higher taxonomic levels) differences in establishment success while accounting for the confounding effect of differences in the conditions of introduction. This is done by including species (or higher taxonomic levels) as random effect coefficients into a multivariate model that contains as co-variates the region of introduction, introduction effort, and other confounding variables. Thus, the random effect coefficient of each species provides a relative measure of the ability of the species to establish itself in a novel location, having controlled for region and introduction effort effects. The need for generalized linear mixed model is because the response variable is either success or failure of introductions, which has to be modeled with a binomial structure of errors.

The mixed model approach has provided evidence that, at least in birds, species differ in their invasion potential: once controlled for region and introduction effort, some species show a higher probability of establishment than others (Blackburn and Duncan 2001a; Sol et al. 2002). Interestingly, most variation in establishment success is evident at low, rather than at high taxonomic levels, indicating that even closely related species may differ substantially in their probability of establishment. Such a pattern may imply that the traits that affect establishment success primarily vary between closely related species. Nonetheless, the possibility that differences in invasion success also exist at higher taxonomic levels cannot be completely ruled out (Forsyth and Duncan 2001; Moulton et al. 2001; Sol et al. 2005b). For example, for a given number of individuals introduced, ungulates were more likely to succeed in New Zealand than were birds (Forsyth and Duncan 2001).

Mixed models have also revealed that species may not simply be separated as successful and unsuccessful invaders. The majority of species has intermediate levels of invasion potential, so they may either succeed or fail when introduced into novel regions (Fig. 8.2). This finding contradicts what Simberloff and Boecklen (1991) called an all-or-non pattern, where some species are particularly good invaders and so succeed everywhere they are introduced, whereas others are poor invaders and always fail, regardless of the characteristics of the recipient community (Moulton 1993; Duncan and Young 1999). Deviations from an all-or-non pattern have several possible interpretations (Duncan and Young 1999; Duncan et al. 2003). First, some species may

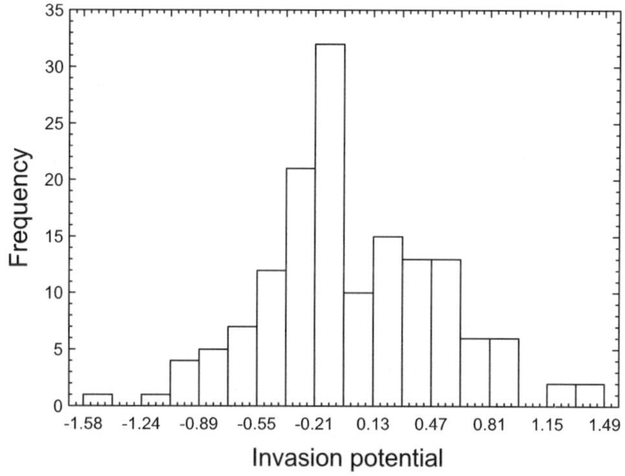

Fig. 8.2 Variation in invasion potential between avian species, when both introduction effort and region are accounted for by using generalized linear mixed models

have properties that make them good invaders when released in certain habitats, but not in others. Second, establishment success may be determined by a combination of properties that are unlikely to be all present in most introduced species. Finally, it is quite conceivable that even an exceptionally good invader might occasionally fail to establish itself due to a chance set of unfavorable circumstances, and that a poor invader may occasionally succeed under favorable circumstances. Thus, even though there exist differences in invasion potential at the species level, success still remains a very idiosyncratic event, limiting the utility of using past invasion success to assess the outcome of future introductions.

8.4 What Makes a Species a Successful Invader?

While in vertebrates there is evidence that species differ in their invasion potential, controversy still exists regarding the nature of the features best defining those species that are more successful. One important unresolved question is whether species are born as, or evolve to be successful invaders. Coming from distant regions, invaders are often confronted to sudden environmental changes to which they are unlikely to be fully adapted. Thus, evolutionary responses are likely to be important, at least in the long-term, to better fit the population into their new environment. Post-invasion evolutionary response has been suggested as one of the possible explanations for the time lag observed in many invasion events (Williamson 1996; Mooney and Cleland 2001), where the invader population remains at low numbers for a long time before starting an exponential population growth phase.

Post-invasion evolution has been shown in a number of studies (Mooney and Cleland 2001), but whether such evolutionary adjustments are important in determining differences between species in invasion potential remains unclear. This requires that successful invaders show a higher evolutionary potential over different ecological contexts (e.g., because of a higher genetic diversity) than less successful invaders, an aspect that still awaits empirical confirmation. Moreover, adaptive evolution is relatively slow in long-lived species such as vertebrates, and the evolutionary response of the population can be limited by insufficient genetic variation. This suggests that evolutionary adjustments should be more important in later transitions of the invasion process than during the process of establishment, at least in vertebrates.

The alternative to evolutionary responses is that successful invaders have some pre-adaptations that facilitate their establishment in novel environments. A large number of traits have been hypothesized to affect the invasion potential of terrestrial vertebrates (Table 8.1). Whether some of these features affect the invasion potential may be evaluated by measuring differences in

Table 8.1 Hypotheses proposed on the attributes that characterize successful invaders in terrestrial vertebrates[a]

Hypothesis	Description	Supporting evidence
Pre-adaptations to new environments		
Niche breadth	Generalist species should be better invaders than are specialists, because the former are more likely to find appropriate resources in a new environment[1,2]	Birds[3,4]
Behavioral flexibility	Species with larger brains and higher behavioral flexibility should be better invaders than less flexible ones, because they may behaviorally adapt to the new environment[1,5]	Birds[5–7]
Social behavior	Social species should be better invaders than solitary ones[1]. Social foraging may be advantageous for invaders because it can increase the probability of detecting a predator, locating food, and learning about new food sources. However, social species may also have difficulties to survive and/or reproduce when they are in low numbers due to the Allee effect, which may counterbalance the benefits	None
Pre-adaptations to specific environments		
Human commensalism	Human commensalists should be better invaders than non-commensalists[1,8], because introductions are generally carried out in human-modified habitats	Birds[7]
Traits that help avoid stochastic extinction		
Life histories and population growth	Species with life histories that increase intrinsic population growth rates are expected to have a better chance of surviving[9], because these species may attain large population size faster	Birds[10,11], but see[12]
Lifespan	Long-lived species should show a higher probability of establishing themselves in a new habitat[13,14], as they are less exposed to stochastic extinctions	None
Traits that constrain establishment		
Migratory behavior	Species that migrate within their native range are less likely to establish themselves than non-migratory species[15,16]. Long-distance migrants may be handicapped in invading novel regions by the incapacity to either develop novel migratory adaptations to reach suitable wintering habitats, or to adapt simultaneously to prevailing conditions in breeding and wintering areas	Birds[16]
Sexual selection	Compared to non-sexually selected species, sexually selected species should have lower introduction success[17]. Sexually selected species may be more vulnerable to extinction, because of production and maintenance costs of secondary sexual characters, and their reduced effective population size[3,17,18]	Birds[3,17,18], but see[4]

Table 8.1 (*Continued*)

Hypothesis	Description	Supporting evidence
Trophic level	Herbivores are predicted to invade new habitats more easily than are carnivores[19,20]. This is based on the idea that competition is the prime determinant of community structure, and that competition is less intense for herbivores than it is for carnivores	None
Nesting site	Ground nesters should have lower probabilities of establishing themselves in a new environment than would be the case for canopy, shrubs, or hole nesters[3,21]. This is because nest predation is generally higher in ground nesters, which may enhance the probability of extinction when the population is small	Birds[3], but see[7]

[a] Data sources: 1. Mayr (1965); 2. Ehrlich (1989); 3. Mclain et al. (1999); 4. Cassey et al. (2004); 5. Sol and Lefebvre (2000); 6. Sol et al. (2005b); 7. Sol et al. (2002); 8. Brown (1989); 9. Moulton and Pimm (1986); 10. Green (1997); 11. Cassey (2003); 12. Blackburn and Duncan (2001a); 13. Pimm (1989); 14. Legendre et al. (1999); 15. Thompson (1922); 16. Veltman et al. (1996); 17. Mclain et al. (1995); 18. Sorci et al. (1998); 19. Hairston et al. (1960); 20. Crawley (1986); 21. Reed (1999)

establishment success between species that differ in those traits (Newsome and Noble 1986). Although such a comparative approach has been extensively used in the last two decades, surprisingly, most of the traits found to be significantly associated with establishment success primarily explain why certain species repeatedly fail to invade a new environment, but say much less about why other species are successful invaders (Duncan et al. 2003; Sol et al. 2005a). For example, some studies have found that sexually dimorphic species are less likely to establish themselves in new regions than are sexually monomorphic species (Mclain et al. 1995, 1999; Sorci et al. 1998). This is consistent with sexual selection theory, which predicts a lower success in sexually selected species due to, among other reasons, the costs of producing and maintaining secondary sexual characters that promote male mating success at the expense of survival. However, while sexual selection theory may help understand why some species are bad invaders, it says nothing about why other species are so successful.

Life history has also been classically suggested to affect the ability of animals to establish themselves in new regions (Pimm 1991), although theoretical predictions are controversial. In general, "fast" life histories (i.e., small body size, fast body growth rate, early maturity, and short lifespan; Saether 1988) are thought to facilitate establishment success by promoting faster population growth, thereby reducing the period in which the population is small and highly vulnerable to extinction. However, small-bodied species also tend to be more vulnerable to environmental risks, and they tend to have more variable populations than do large-bodied ones (Pimm 1991). These contradictory theoretical predictions over how life history strategies affect the invasion potential have as yet not been resolved empirically (Duncan et al. 2003). While some studies have reported positive relationships between clutch size and establishment success (Green 1997; Cassey 2002), others have reported negative relationships, or no relationship at all (Veltman et al. 1996; Blackburn and Duncan 2001a). Despite the controversy, it is conceivable that life history affects establishment by shaping key demographic parameters. Nonetheless, life history does not explain why some species are better armed to find an appropriate niche in new regions, which is the first barrier in the invasion process.

Most ecologists would agree that the chances that a species find an appropriate niche in a new region should be higher if its environment is similar to that found in the native region of the invader, as then the invader will already be pre-adapted to it, an idea known as the "environmental matching" hypothesis. In vertebrates, the existence of pre-adaptations to specific habitats is suggested by several indirect lines of evidence. Establishment success in birds is significantly greater when the difference between a species' latitude of origin and its latitude of introduction is small (Blackburn and Duncan 2001a; Cassey 2002), when climatic conditions in the locations of origin and introduction are more similar (Duncan et al 2001), and when species are introduced to locations within their native biogeographical regions (Blackburn & Duncan 2001a). Direct evidence for the existence of pre-adaptations to specific habitats is less clear. In birds, species with a history of close association with humans tend to be successful invaders (Sol et al. 2002), which agrees with the fact that many avian introductions have taken places in human-modified habitats (Case 1996). By contrast, a comparison of avian introductions across convergent Mediterranean regions revealed that success was not higher for those originating from Mediterranean systems than for those from non-Mediterranean regions (Kark and Sol 2005).

Traits that pre-adapt species to specific habitats, however, cannot explain why some species are extremely successful at invading a variety of environments. Moreover, environments that may look very similar to us may actually show subtle differences in key aspects such as available food types or diversity of enemies, sometimes a matter of life and death. If successful invaders have in common some general attributes that help them to invade new regions, then these should be pre-adaptations to exploit novel niches under a wide diversity of environments. Yet, do we have evidence for such general attributes of successful invaders in vertebrates?

If features of successful invaders do exist, then they should not be multifaceted; the reason is that adaptations that are useful in some environments are often inappropriate for other environments. Thus, it would seem more likely to find pre-adaptations to invade specific habitats, rather than a wide variety of habitats. Yet, ecological theory suggests at least two classes of attributes of vertebrates that might predispose them to be successful invaders: niche breadth and behavioral flexibility.

The "niche breadth–invasion success" hypothesis represents the first attempt at generalization that species have attributes that make them successful invaders (Vasquez 2006). It suggests that species with broad niches ("generalists") are more likely to invade new regions than are species with narrower niches ("specialists"), because the former are more likely to find the necessary resources or conditions in the novel environment. Supporting evidence for the hypothesis is found in analyses showing that introduced birds that are either dietary or habitat generalists are more likely to establish successfully in new regions (Mclain et al. 1999; Cassey et al. 2004). Also consistent with the idea that niche breadth is important is the finding that, following introduction, bird species with larger geographic ranges are more likely to establish (Moulton and Pimm 1986; Blackburn and Duncan 2001a). Species may have large geographic ranges because they can exploit a broad range of conditions, although the alternative that they simply utilize conditions that are themselves widespread cannot be ruled out (Duncan et al. 2003).

Ecological generalism reflects the capacity of an animal to use a variety of resources, and is thus a static concept. However, coming from distant regions, invaders have to respond to dramatic changes in the environment, often facing what Schlaepfer et al. (2002) have termed "ecological traps". Ecological traps occur when invaders make wrong choices of resources, relative to conditions in their native environments. For example, the contrasting successes in North America of the common starling, *Sturnus vulgaris*, and the closely related Southeast Asian crested myna, *Acridotheres cristatellus*, have in part been attributed to the fact that mynas retained breeding habits appropriate to their homelands, but inappropriate in their new home (British Columbia; Ehrlich 1989).

Animals may in part compensate for such poor adaptive fit by means of behavioral changes (Klopfer 1962; Sol 2003). Behavioral flexibility may aid establishment through, for example, the ready adoption of new food resources (Lefebvre et al. 2004), the adjustment of breeding to the prevailing environmental conditions (Arcese et al. 1997), or rapid behavioral changes to avoid novel enemies (Berger et al. 2001). Because behaviorally flexible species are believed to be more exploratory (Greenberg and Mettke-Hofmann 2001) and ecologically generalist (Sol 2003), they may also have higher chances of discovering and adopting new habitats or new resources that may be important to survive and reproduce in the novel environment. The hypothesis that behavioral flexibility enhances establishment dates from Mayr (1965), but has only recently received empirical support for birds. This derives from observations that, compared to unsuccessful species, established birds tend to have a larger brain size, relative to their body mass, and to show more innovative behaviors in their region of origin (Sol and Lefebvre 2000; Sol et al. 2002, 2005b). The importance of behavioral adjustments to deal with novel ecological problems is also supported by experimental evidence. Martin II and Fitzgerald (2005) found that house sparrows from an invading population in North America tended to approach and consume novel foods more readily than those from a well-established population.

8.5 Conclusions and Future Directions

Progress in the last decades has provided firm evidence in vertebrates supporting that species differ in their invasion potential, and that such differences are associated with certain features that facilitate establishment in novel regions. A number of features appears to combine to affect the ability of species to cross the three barriers that lead to successful establishment, yet only two of these characteristics appear to provide general explanations in understanding why some vertebrates are so extremely successful invaders: a broad ecological niche, and a high degree of behavioral flexibility. These traits are presumed to facilitate that vertebrates find an appropriate niche in a variety of environments, even in those to which they have had no opportunity to adapt to.

The conclusion that species attributes influence invasion success is important for three main reasons. First, it informs us on the mechanisms that allow animals to invade novel environments, improving our understanding of the invasion process, and providing cues to identify and prevent situations where the risk is high that a species becomes established in a novel region. Second, it indicates that invasion success cannot simply be explained by neutral processes, such as the differences in introduction effort, but that properties of the species matter when it comes to understand invasion success. Finally, it suggests that animals differ in the way they respond to changes in their environment, which has obvious implications for the conservation of vertebrates in the face of environmental threats such as the destruction and fragmentation of habitats, and global climate change.

Despite our progress, we still have a long way to go to fully understand why some vertebrates are so successful invaders. Below, I highlight five issues that I envision as important avenues of future research.

First, the role of some traits in determining the invasion potential of vertebrates remains unclear. The reasons include that the theoretical basis is insufficiently developed, that the empirical evidence is contradictory, or simply that the effect of the trait has never been tested. Traits that remain insufficiently studied include life history strategies, human commensalisms, and social behaviour. Testing the importance of these and other factors requires large, representative samples of introduced species, adopting appropriate methodologies that take into account the non-random nature of past historical introductions (Duncan et al. 2003).

Second, previous work has largely ignored possible interactions between species attributes and the characteristics of the recipient community, even though this may contribute to better understanding the underlying mechanisms. For example, if behavioral flexibility affects establishment by enhancing the individual's response to novel environments, then we should expect this to play an even bigger role when the species is introduced into habitats that differ strongly from its original one – and hence demand greater behavioral adjustments – than would be the case for more similar habitats. Likewise, migratory behavior has been suggested to constrain establishment success in vertebrates on the grounds that, on isolated islands, species cannot develop migratory routes (Veltman et al. 1996). Such a mechanism should be tested – not simply assumed – for example, by assessing whether migratory species are particularly unsuccessful when introduced onto very isolated islands.

Third, although establishment success is undoubtedly central in the invasion process, the impact of the invader is determined mostly by the ability of the species to grow in numbers, and expand over large regions. Thus, it is critical that we understand what determines that a species is able to successfully spread and impact over ecosystems. A particularly intriguing question is whether those traits that have been found to affect establishment also influence spread, as then we would easily detect situations of high risk. Indeed, both niche breadth (Brooks 2001) and behavioral flexibility (Sol 2003) have been suggested to affect spread, although supporting evidence is still lacking.

Fourth, most past work on establishment success has been done in birds. We need to extend the results to other terrestrial vertebrates, particularly in mammals for which good data are available on many successful and failed introductions (Forsyth and Duncan 2001; Forsyth et al. 2004). Such studies will allow us to ascertain the generality of the hypothesis suggested to explain establishment success, and hence build a general framework that is common to all vertebrates.

Finally, if we wish to fully understand the mechanisms that allow some vertebrates to be so successful, then we should progress from the present focus on comparative approaches toward increasingly experimental approaches. While experimental introductions are not generally feasible in vertebrates, we can nevertheless study underlying mechanisms by using translocations of species, or by running experiments on both native and introduced populations. One good example of the type of work required is the study of Martin II and Fitzgerald (2005), who used common garden experiments to evaluate differences in behavioral flexibility between an invading and a well-established population of house sparrows (*Passer domesticus*). The use of experiments will serve to validate the evidence stemming from comparative analyses, and will help better understand the exact mechanisms that facilitate that a species can establish itself in an environment to which it is not well adapted.

Acknowledgements. I thank Louis Lefebvre, Richard Duncan, Tim Blackburn, Phill Cassey, Montse Vilà, Joan Pino, Salit Kark, Sven Bacher, Wojciech Solarz, Wolfgang Nentwig, Simon Reader, Andrea Griffin, Julie Morand-Ferron, Diego Vasquez, and Gray Stirling for fruitful discussions over the past years. This work was supported by a Ramón y Cajal fellowship and a Proyecto de Investigación (ref. CGL2005-07640/BOS) from the Ministerio de Educación y Ciencia (Spain), and European Union FP 6th integrated project ALARM grant GOCE-CT-2003-506675.

References

Arcese P, Keller LF, Cary JR (1997) Why hire a behaviorist into a conservation or management team? In: Clemmons JR, Buchholz R (eds) Behavioral approaches to conservation in the wild. Cambridge University Press, Cambridge, pp 48–71

Berger J, Swenson JE, Persson I-L (2001) Recolonizing carnivores and naive prey: conservation lessons from Pleistocene extinctions. Science 291:1036–1039

Blackburn TM, Duncan RP (2001a) Determinants of establishment success in introduced birds. Nature 414:195–197

Blackburn TM, Duncan RP (2001b) Establishment patterns of exotic birds are constrained by non-random patterns in introduction. J Biogeogr 28:927–939

Brooks T (2001) Are unsuccessful avian invaders rarer in their native range than successful invaders? In: Lockwood JL, McKinney ML (eds) Biotic homogenization. Kluwer/Plenum, New York, pp 125–155

Brown JH (1989) Patterns, modes and extents of invasions by vertebrates. In: Drake JA, Mooney HA, di Castri F, Groves RH, Kruger FJ, Rejmanek M, Williamson M (eds) Biological invasions: a global perspective. Wiley, Chichester, pp 85–109

Case TJ (1996) Global patterns in the establishment and distribution of exotic birds. Biol Conserv 78:69–96

Cassey P (2002) Comparative analysis of successful establishment among introduced landbirds. PhD Thesis, Griffith University, Australia

Cassey P (2003) A comparative analysis of the relative success of introduced land birds on island. Evol Ecol Res 5:1–11

Cassey P, Blackburn TM, Sol D, Duncan RP, Lockwood JL (2004) Global patterns of introduction effort and establishment success in birds. Proc R Soc Lond Ser B 271:405–408

Cassey P, Blackburn TM, Duncan RP, Gaston KJ (2005) Causes of exotic bird establishment across oceanic islands. Proc R Soc Lond Ser B 272:2059–2063

Crawley MJ (1986). The population biology of invaders. Philos Trans R Soc Lond Ser B 314:711–731

Duncan RP, Young JR (1999) The fate of passeriform introductions on oceanic islands. Conserv Biol 13:934–936

Duncan RP, Bomford M, Forsyth DM, Conibear L (2001) High predictability in introduction outcomes and the geographical range size of introduced Australian birds: a role for climate. J Anim Ecol 70:621–632

Duncan RP, Blackburn TM, Sol D (2003) The ecology of avian introductions. Annu Rev Ecol Evol Syst 34:71–98

Ehrlich PH (1989) Attributes of invaders and the invading processes: vertebrates. In: Drake JA, Mooney HA, di Castri F, Groves RH, Kruger FG, Rejmánek M, Williamson MH (eds) Biological invasions: a global perspective. Wiley, Chichester, pp 315–328
Forsyth DM, Duncan RP (2001) Propagule size and the relative success of exotic ungulate and bird introductions to New Zealand. Am Nat 157:583–595
Forsyth DM, Duncan RP, Bomford M, Moore G (2004) Climatic suitability, life-history traits, introduction effort, and the establishment and spread of introduced mammals in Australia. Conserv Biol 18:557–569
Green RE (1997) The influence of numbers released on the outcome of attempts to introduce exotic bird species to New Zealand. J Anim Ecol 66:25–35
Greenberg RS, Mettke-Hofmann C (2001) Ecological aspects of neophobia and neophilia in birds. In: Nolan V Jr (ed) Current ornithology. Springer, Berlin Heidelberg New York, pp 119–178
Hairston NG, Smith FE, Slobodkin LB (1960) Community structure, population control, and competition. Am Nat 94:421–425
Kark S, Sol D (2005) Establishment success across convergent Mediterranean ecosystems: an analysis of bird introductions. Conserv Biol 19:1519–1527
Klopfer PH (1962) Behavioral aspects of ecology. Prentice-Hall, London
Lefebvre L, Reader SM, Sol D (2004) Brains, innovations and evolution in birds and primates. Brain Behav Evol 63:233–246
Legendre S, Clobert J, Møller AP, Sorci G (1999) Demographic stochasticity and social mating system in the process of extinction of small populations: the case of passerines introduced to New Zealand. Am Nat 153:449–463
Lockwood JL, Cassey P, Blackburn TM (2005) The role of propagule pressure in explaining species invasions. Trends Ecol Evol 20:223–228
Lonsdale WM (1999) Global patterns of plant invasions and the concept of invasibility. Ecology 80:1522–1536
Martin LB II, Fitzgerald L (2005) A taste for novelty in invading house sparrows, *Passer domesticus*. Behav Ecol 16:702–707
Mayr E (1965) The nature of colonising birds. In: Baker HG, Stebbins GL (eds) The genetics of colonizing species. Academic Press, New York, pp 29–43
Mclain DK, Moulton MP, Redfearn TP (1995) Sexual selection and the risk of extinction of introduced birds on oceanic islands. Oikos 90:599–605
Mclain DK, Moulton MP, Sanderson JG (1999) Sexual selection and extinction: the fate of plumage-dimorphic and plumage-monomorphic birds introduced onto islands. Evol Ecol Res 1:549–565
Mooney HA, Cleland EE (2001) The evolutionary impact of invasive species. Proc Natl Acad Sci USA 98:5446–5451
Moulton MP (1993) The all-or-none pattern in introduced Hawaiian Passeriformes: the role of competition sustained. Am Nat 141:105–119
Moulton MP, Pimm SL (1986) Species introductions to Hawaii. In: Mooney HA, Drake JA (eds) Ecology of biological invasions in North America and Hawaii. Springer, Berlin Heidelberg New York, pp 231–249
Moulton MP, Miller KE, Tillman EA (2001) Patterns of success among introduced birds in the Hawaiian islands. Stud Avian Biol 22:31–46
Newsome AE, Noble IR (1986) Ecological and physiological characters of invading species. In: Groves RH, Burdon JJ (eds) Biological invasions. Cambridge University Press, Cambridge, pp 1–20
Pimm SL (1989) Theories of predicting success and impact of introduced species. In: Drake JA, Mooney HA, di Castri F, Groves RH, Kruger FG, Rejmanek M, Williamson M (eds) Biological invasions: a global perspective. Wiley, Chichester, pp 351–367
Pimm SL (1991) The balance of nature? University of Chicago Press, Chicago

Reed JM (1999) The role of behavior in recent avian extinctions and endangerments. Conserv Biol 13:232–241
Saether B-E (1988) Pattern of covariation between life-history traits of european birds. Nature 331:616–617
Sax DF, Brown JH (2000) The paradox of invasion. Global Ecol Biogeogr 9:363–371
Schlaepfer MA, Runge MC, Sherman PW (2002) Ecological and evolutionary traps. Trends Ecol Evol 17:474–480
Shea K, Chesson P (2002) Community ecology theory as a framework for biological invasions. Trends Ecol Evol 17:170–176
Simberloff D, Boecklen W (1991) Patterns of extinction in the introduced Hawaiian avifauna: a reexamination of the role of competition. Am Nat 138:327
Sol D (2003) Behavioural flexibility: a neglected issue in the ecological and evolutionary literature? In: Reader SM, Laland KN (eds) Animal innovation. Oxford University Press, Oxford, pp 63–82
Sol D, Lefebvre L (2000) Behavioural flexibility predicts invasion success in birds introduced to New Zealand. Oikos 90:599–605
Sol D, Timmermans S, Lefebvre L (2002) Behavioural flexibility and invasion success in birds. Anim Behav 63:495–502
Sol D, Blackburn TM, Cassey P, Duncan RP, Clavell J (2005a) The ecology and impact of non-indigenous birds. In: Del Hoyo J, Elliot A, Christie D (eds) Handbook of the Birds of the World, vol 10. Lynx, Barcelona, pp 13–35
Sol D, Duncan RP, Blackburn TM, Cassey P, Lefebvre L (2005b) Big brains, enhanced cognition, and response of birds to novel environments. Proc Natl Acad Sci USA 102:5460–5465
Sorci G, Møller AP, Clobert J (1998) Plumage dichromatism of birds predicts introduction success in New Zealand. J Anim Ecol 67:263–269
Steele BM, Hogg JT (2003) Measuring individual quality in conservation and behavior. In: Festa-Bianchet M, Apollonio M (eds) Animal behavior and wildlife conservation. Island Press, Washington, DC, pp 243–269
Thompson GM (1922) The naturalization of plants and animals in New Zealand. Cambridge University Press, Cambridge
Vazquez D (2006) Exploring the relationship between niche breadth and invasion success. In: Cadotte MW, McMahon SM, Fukami T (eds) Conceptual ecology and invasions biology: reciprocal approaches to nature. Springer, Berlin Heidelberg New York, pp 307–322
Veltman CJ, Nee S, Crawley MJ (1996) Correlates of introduction success in exotic New Zealand birds. Am Nat 147:542–557
Williamson MH (1996) Biological invasions. Chapman and Hall, London
Williamson MH, Fitter A (1996) The varying success of invaders. Ecology 77:1661–1666

Section III

Patterns of Invasion and Invasibility

Short Introduction

WOLFGANG NENTWIG

It is easy to state that the process of invasion by alien species has been occurring since time immemorial and has been slowly accelerating for the last few centuries. Now, in the age of globalization, this process is exploding and aliens are spreading all over. Again, in the following four chapters we search for patterns of invasion but no longer at the species level. Are there supra-species patterns which promote the success of alien species in general, or of the most successful invaders?

Globalization, i.e. a bundle of simultaneous changes which happen worldwide, contains alterations which definitely favour alien species. Agriculture, land management, and land use change are some of the obvious factors facilitating the occurrence of alien species (Chap. 9). Also, the invasibility of ecosystems has changed, with heavily disturbed ecosystems being much easier to invade. Consequently, modern anthropogenic modifications of natural ecosystems increase their invasibility, and the global restructuring of the ecosystems of the world for human purposes implicitly favours biological invasions (Chap. 11).

Pollution is certainly one of the factors promoting the spread of alien species. As a by-product of our increasing energy consumption and fertilizer use, nitrogen compounds flood our environment. Eutrophication of waters and soils has long received considerable attention but it has been largely overlooked that nitrogen-fixating plants and plants which profit from high nitrogen levels are among the most successful invaders. So, the question is justified as to whether nitrogen enrichment in general supports alien invaders (Chap. 10).

Globalization not only affects ecosystem structures and functions but is also one of the causes of global climate change. This strongly modifies temperature and humidity conditions, and completely alters the foundations of existing species assemblages. Since the establishment of alien species becomes easier, climate change is considered to promote alien plant invasions (Chap. 12).

9 Effects of Land Management Practices on Plant Invasions in Wildland Areas

MATTHEW L. BROOKS

9.1 Introduction

The alteration of natural ecosystems by humans and anthropogenic dispersal of plant propagules beyond their native ranges have facilitated the dramatic spread and increase in dominance of nonnative plants worldwide since the late 1800s (Hobbs 2000; Mack et al. 2000). The amount of ecosystem alteration is related to predominant land uses, which can be summarized into four categories of increasing impact: (1) conservation – nature reserves, wilderness; (2) utilization – pastoralism, non-plantation silviculture, recreation; (3) replacement – cropping agriculture, plantation silviculture; and (4) removal – urbanization, mining, industrial development (Hobbs and Hopkins 1990; Hobbs 2000). The rate at which propagules are dispersed into new regions is largely related to the frequency and intensity of human activities, which generally covary with the degree of ecosystem alteration among the four land use categories.

Compared to areas where replacement or removal land uses are the norm, the management of plant invasions tends to be more complicated where conservation or utilization land uses prevail. The latter two land uses emphasize the need to maintain the integrity of natural ecosystems, whereas the former two do not require that natural ecosystem properties be maintained, and in some cases involve replacing them with simpler ecosystems (e.g., cropping monocultures). Options for controlling invading plants are more limited when their potential negative effects on native ecosystems may preclude their usage. This chapter is focused on conservation and utilization land uses that occur where native ecosystems are largely present and functioning, otherwise known as wildland areas.

Management plans for wildland areas typically focus on defining a balance between conservation and utilization, while maintaining ecosystem integrity in the process. Each land use type is associated with a range of land manage-

Ecological Studies, Vol. 193
W. Nentwig (Ed.)
Biological Invasions

ment practices designed to achieve particular objectives. It is not so much the land use itself that affects plant invasions, but rather the collective effects of all associated land management practices. For example, pastoralism is often associated with the practices of road building to facilitate access to rangelands and infrastructure (grazing allotments, watering sites, corrals), specific grazing methods (rotational, deferred), and forage improvement techniques (seeding, burning), all of which may be tailored for specific types of livestock (sheep, cattle, goats). Collectively, these practices have a net effect on the invasion potential of grazed landscapes, but individually they may have somewhat differing effects, and differing solutions tailored to their specific effects. In the field of pastoralism, the decisions that land managers typically face are generally not related to choosing between practicing, or not practicing livestock grazing, but rather to determining how to conduct livestock grazing in a manner that maximizes its sustainability and minimizes its potential negative effects on ecosystem integrity. Plant invasions are one of the key variables that can hinder the attainment of both these objectives.

In this chapter, I present a conceptual framework that can be used to evaluate the mechanisms by which land management practices affect plant invasions in wildland areas. I also discuss some of the measures that can be implemented to reduce the potential for invasion. Although the discussions and examples are limited to terrestrial ecosystems, the approach and principles presented in this chapter may be applied to other types of ecosystems as well.

9.2 Factors that Affect Plant Invasions

Factors that promote plant invasions are only generally understood. Many factors have been associated with the ability of nonnative plants to invade new areas, including various types of disturbance, connectivity to already invaded sites via pathways and vectors, disruption of large-scale ecological processes or regimes, loss of pollinators or other keystone species, and fluctuating resource levels (Hobbs and Huenneke 1992; D'Antonio 1993; Maron and Connors 1996; Lonsdale 1999; Davis et al. 2000). However, much of this research has produced contradictory results as to the primary factors that promote plant invasions (Lonsdale 1999; Williamson 1999).

The difficulty in finding clear and predictable patterns may be due to the episodic nature of plant invasions. This is not captured in most theoretical constructs, which tend to focus more on the inherent susceptibility of landscapes to invasion based on more fixed characteristics (species diversity, vegetation type, land use type). The susceptibility of landscapes to invasion can alternatively be viewed as fluctuating. In this case, invasions are most likely to occur during "windows of opportunity" when barriers that would otherwise prevent them are lowered (Johnstone1986). These barriers can be lowered and raised over time, alternatively opening and closing windows of opportunity for invasion. Barriers can be specific intrinsic characteristics affecting the "invasibility" (sensu Lonsdale 1999) of the landscape in question, or related to the extrinsic types and amounts of potential invader progagules. A thorough assessment of invasion potential must take into account both of these factors.

Davis et al. (2000) incorporated the concepts of variable opportunities for invasion into a theoretical model that assumes the invasion of landscapes by new species is affected by variations in (1) resource availability; (2) propagule availability; and (3) the types of species invading. These factors can be further grouped into the intrinsic property of resource availability, which is related to the availability of each type of potentially limiting resource (light, moisture, nitrogen, etc.), and the extrinsic property of propagule pressure, which includes the amount of each potentially invading species. Collectively, resource availability and propagule pressure affect invasion potential. The basic premise of this theory is that when and where the availability of otherwise limiting resources is high, landscapes are more invasible than when and where resources are low, but only if propagule pressure is sufficiently high and comprised of species well suited to colonize and establish new populations under prevailing environmental conditions (Fig. 9.1). Thus, the invasion potential of a landscape is highly contextual, as are the relative levels of resource availability and propagule pressure, which both vary over space and time.

Plant resource availability is a function of the underlying supply of light, water, and mineral nutrients, and the proportions of these resources that are unused by the existing vegetation. Resource availability can increase due to

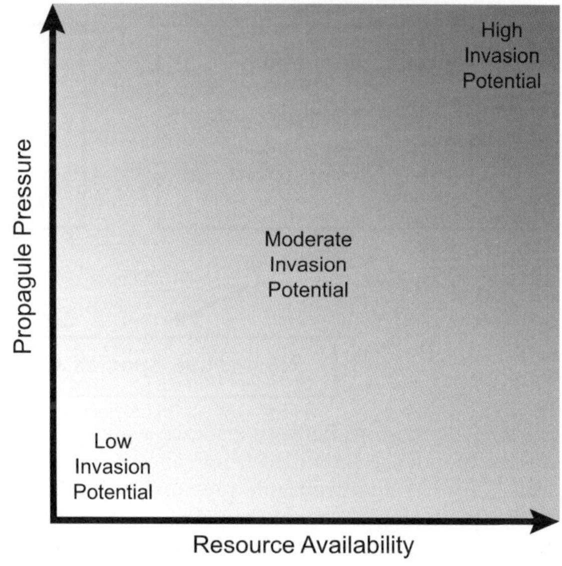

Fig. 9.1 Invasion potential of landscapes is related to both resource availability and propagule pressure

direct additions to the landscape (atmospheric nitrogen deposition, agricultural fertilization) or increased rates of production within the landscape (nutrient cycling rates), or by reduced rates of uptake following declines in competition from extant plants after they are thinned or removed (Fig. 9.2, link A). The rate of resource uptake is inversely related to disturbance levels, because disturbance typically reduces vegetation biomass, thus reducing the amount of resources used. Established populations of nonnative plants can also feed back to affect resource supply. This can occur by direct increases in nutrient supply (nitrifying plants), or indirect increases brought about by inhibiting the growth of other species through competition (e.g., Brooks 2000) or inhibition (e.g., Callaway and Aschehoug 2000; Fig. 9.2, link C). Conversely, processes that reduce plant resource availability (e.g., increased plant productivity) may dampen invasion potential.

Propagule pressure is related to the number of propagules available to establish and increase populations. Propagules can be introduced deliberately (seeding projects or ornamental plants) or accidentally (adhering to vehicles or as contaminant in hay; Fig. 9.2, link B). Once established, populations of nonnative plants can promote their dominance by adding to their own pool of available propagules (Fig. 9.2, link D). Propagule pressure can be negatively affected by predators (granivores) or diseases that reduce the rate of production of new propagules. The types of propagules that are present are also an important factor. Species or functional groups of plants with properties that confer an advantage in colonizing and establishing populations under prevailing environmental conditions will be more successful than those lacking such properties.

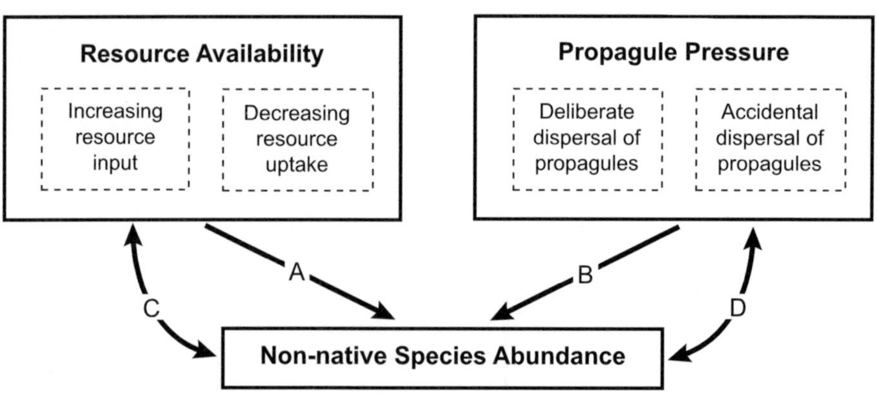

Fig. 9.2 Relationships between resource availability, propagule pressure, and nonnative species abundance (modified from concepts presented in Davis et al. 2000). Resource availability (*A*) and propagule pressure (*B*) both affect the abundance of nonnative species populations. Once these populations become established, they can themselves affect resource availability (*C*) and propagule pressure (*D*)

9.3 Linking Land Management Practices with Invasion Potential

Much has been written about various aspects of plant invasions, but to date a universal approach to evaluating the relationships between land management practices and invasion potential by plants has not been explicitly described. The theoretical framework presented in Figs. 9.1 and 9.2 can be used for such a purpose. Specifically, this framework can be applied to any land management practice to evaluate links between invasion potential, resource availability, and propagule pressure, as well as specific management actions that may reduce the invasion potential of landscapes. The fluctuating and episodic dynamics of this theoretical framework is similar in nature to land management practices, which tend to also occur as punctuated activities.

There are a number of management actions that can target specific parts of the conceptual model presented in Fig. 9.2. For example, resource availability may be minimized by reducing the rate of resource input and/or increasing its rate of uptake. Similarly, propagule pressure may be minimized by reducing the rates of intentional and/or unintentional dispersal. The specific actions that will generate the most benefit will vary on a case-by-case basis, depending on the relative importance of resource supply and propagule pressure. This is important to understand, because costly efforts may otherwise be wasted on reducing resource availability when propagule pressure is relatively low (i.e., there are few propagules available to respond to resource fluctuations), or reducing propagule pressure when resource availability is limited (i.e., there are few resources for propagules to respond to).

An example is provided below that illustrates some of the ways that a land management practice – in this case, the management of vehicular routes – may increase the invasion potential of landscapes, and how these effects may be mitigated by specifically managing resource availability and propagule pressure.

9.3.1 Vehicular Route Management

Vehicular routes are perhaps the single, most pervasive land use feature in wildland areas worldwide. Approximately 6.3 million km of roads have been reported in the United States alone (National Research Council 1997), which vastly underestimates the actual amount of vehicular routes, due to extensive networks of unimproved routes and trails that remain uninventoried (Forman et al. 2003). The ecological effects of vehicular routes can range from physical and chemical changes of ecosystems to alterations in the population and community structure of organisms (Forman and Alexander 1998; Spellerberg 1998, 2002; Forman et al. 2003), including the spread and dominance of nonnative plants (Brooks and Lair 2007).

Vehicular routes can be classified into three general categories: off-highway vehicle trails and unimproved local roads, improved local roads and collector roads, and arterial roads and limited-access highways (Brooks and Lair 2007). Of these, off-highway vehicle trails and unimproved local roads characterize the typical routes encountered in wildland areas.

Vehicular routes are part of larger transportation infrastructures that some classify as a form of removal land use (e.g., Hobbs 2000). From this perspective, the immediate footprints of routes are recognized for their direct amelioration of local ecosystem properties (vegetation cover, soil hydrology). This has also been referred to as the direct local effect of vehicular routes (Fig. 9.3a; Brooks and Lair 2007). However, vehicular routes also have significant indirect and diffuse effects that encompass much larger areas, and do not impact local ecosystem processes as severely (Fig. 9.3a). These latter effects are of primary concern to managers of wildlands, since they threaten ecosystem structure and processes beyond route corridors. In addition, vehicular routes are integral to most other types of land uses. Anywhere people need to travel to conduct activities associated with various land uses, they usually travel by vehicle upon some sort of vehicular route.

The ecological effects of vehicular routes stem from both the vehicles themselves, and the surfaces created to facilitate their travel (Brooks and Lair 2007). Both can affect resource supply and propagule pressure of invading plants, although the management of each requires differing approaches.

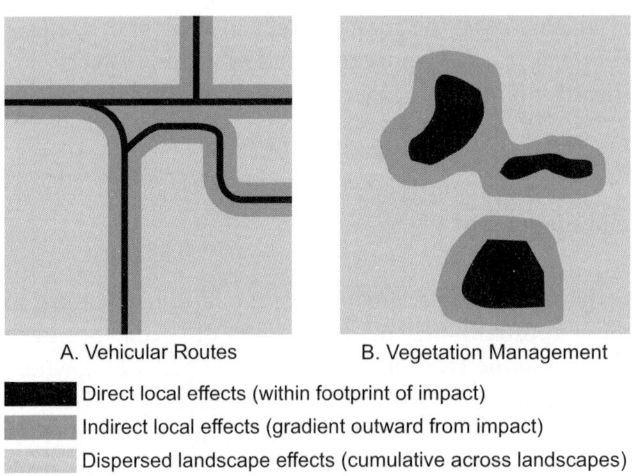

Fig. 9.3a, b Three primary scales of ecosystem impact of vehicular routes (**a**) and vegetation management treatments (**b**), modified from Brooks and Lair (2007)

9.3.1.1 Vehicles

Vehicles produce atmospheric pollution in the form of nitrogen oxides and other compounds. At a regional scale, these pollutants can produce nitrogen deposition gradients that increase soil nitrogen levels (e.g., Padgett et al. 1999), and can lead to increased dominance by nonnative plants (Brooks 2003; Allen et al. 2007). At a local scale, they can create deposition gradients radiating outward from individual roads (e.g., Angold 1997). Unfortunately, the reduction of nitrogen deposition rates is outside the scope of what local land managers can typically influence – this can occur primarily through the efforts of regional air quality management districts or based on national automobile emission standards. To some degree, limitations on local rates of vehicular travel may reduce local deposition rates associated with individual roads. However, local land managers still should understand how broad-scale atmospheric nitrogen deposition may affect the processes of plant invasions into the wildland areas they manage. Such information may help them monitor more efficiently for the arrival of new plant invaders, and predict where these new invaders may reach levels that negatively affect ecosystem properties (e.g., fire regimes in deserts; Brooks 2003, Allen et al. 2007).

Vehicles can also serve as vectors for the unintentional dispersal of nonnative plant propagules (Clifford 1959; Schmidt 1989; Lonsdale and Lane 1994). Propagules may adhere directly to vehicles, or be blown along by wind currents created by vehicular travel. Maintenance guidelines for vehicles used by land managers may help reduce propagule pressure if they stipulate that vehicles be periodically washed, or at least washed when they are moved from one region to another. Vehicle washing is especially important after they have been operated in the vicinity of populations of nonnative plants that are high priorities for containment and control. Management of private vehicles used by people visiting a management unit is more problematic. Beyond relatively simple efforts to reduce dispersal rates by land management vehicles, it may be more efficient to focus resources on early detection and eradication of colonizing nonnative plants, rather than on extensive efforts to reduce their dispersal by private vehicles into a management unit (Lonsdale and Lane 1994).

9.3.1.2 Vehicular Routes

Vehicular routes can have much greater effects on soil nutrient availability than do the vehicles themselves. For example, in arid and semi-arid environments, rainfall accumulation along roadsides can increase soil moisture levels, making conditions more conducive to plant growth (Brooks and Lair 2007). Even the tracks created by a single crossing of a motorcycle in desert soil can create microsites that facilitate the establishment and growth of non-native plants, as demonstrated in the deserts of Kuwait (Brown and Schoknecht 2001) and in the Mojave Desert of North America (Davidson and Fox 1974; Brooks 2007).

The maintenance and engineering of roads can also significantly affect resource availability. Where vegetation is removed along roadside verges, reduced competition may increase resource availability (Vasek et al. 1975), and thus invasion potential. The creation of roadside berms can improve soil conditions, making it more suitable for the establishment and growth of nonnative species (Gelbard and Belnap 2003), especially if the conditions they create are significantly different from those of the surrounding landscape (Brooks 2007). The abundance of nonnative plants may also increase where new soils are introduced to create roadbeds, such as clay and limestone soils in an otherwise sandy landscape (Greenberg et al. 1997). Contouring to reduce the prominence of berms, and the use of roadbed materials that do not increase the relative fertility of the soil may help reduce rates of establishment by invading plant species. If the underlying fertility of roadsides cannot be reasonably managed, then regular vegetation management to maintain bare soil, and repeatedly remove new invaders as they become established, may be another option.

The construction and maintenance of vehicular routes can also significantly affect propagule pressure of invading species. Roadbed materials often originate from quarries that contain significant stands of invasive plants (M. Brooks, personal observations). In some cases, roadsides are recontoured and the materials redistributed elsewhere along other roadsides, potentially spreading nonnative plant propagules. The results of these activities often lead to new populations of plant invaders establishing in the vicinity of major road construction (M. Brooks, personal observations). Careful monitoring and control of nonnative plants at sites from which road materials originate is required to reduce the rates of propagule dispersal onto roadsides.

The verges of vehicle routes are often revegetated if plant cover has been lost during the course of construction or maintenance activities, especially if soil conservation or aesthetic degradation is of concern. Seedings are the most common revegetation method, and in many cases nonnative species are used. Nonnatives are often chosen simply because they have been used in the past and are part of institutionalized practices, but also because they are relatively inexpensive, compared to native seeds, and are often bred for high establishment and rapid growth rates. This last factor is of particular concern because selection for these traits also improves the chances that seeded species will spread beyond their points of application into wildland areas, and potentially become problems for land managers. Economic incentives are required to promote the development of native seed stocks, and research is needed to identify those native species that are most appropriate for specific vegetation types and ecoregions.

9.4 Managing Established Populations of Invasive Plants

Once plants have invaded and naturalized, control efforts involve treatments to remove, or at least reduce, their populations. In some cases, other species may be introduced through the process of revegetation to hinder the reestablishment of invading species after they are removed from an area. These practices are typical of the broader field of vegetation management, which transcends most realms of land management. For example, silviculturalists manage forests to maximize lumber production, and range conservationists manage rangelands to maximize livestock production. Fire managers manage vegetation before and after fires to manipulate fuelbed characteristics that affect fire behavior and fire regimes. Law enforcement and cultural resource managers may manage vegetation to respectively facilitate the detection of illegal activities and to maintain historically significant vegetation stands. Natural resource managers manage vegetation to create and maintain habitat for wildlife (forage and cover), reduce rates of soil erosion (species that stabilize soils), and promote certain plant community characteristics (high diversity, healthy populations of rare species). Vegetation management may also be targeted directly at eradicating or controlling the dominance of nonnative plants. All of these land management activities can affect invasion potential by influencing resource availability and propagule pressure. Even efforts to manage specific nonnative plants may unintentionally promote the subsequent invasion and rise to dominance of other nonnative species.

Just as vehicular routes can have ecosystem impacts at various spatial scales, so too can vegetation management treatments (Fig. 9.3b). Areas directly within the footprint of the treatments can have direct local effects that are very obvious, such as a clearcut in a forest. Invasive plants can dominate these areas where competition for light and soil resources has been temporarily reduced (Hobbs and Huenneke 1992). There may also be gradients of resource availability and propagule pressure extending outward from areas where vegetation has been removed, and which can affect landscape invasibility (Zink et al. 1995). The cumulative effect of multiple direct and indirect local effects can have diffuse landscape impacts that influence much broader areas.

9.4.1 Effects of Vegetation Management on Resource Availability

The removal of vegetation has obvious implications for resource availability to species that may subsequently invade. When plants are removed, there can be an immediate release from competition for soil nutrients and light, and when plants are revegetated, the opposite may occur. However, the different methods used to remove or add vegetation may have additional effects on nutrient availability that warrant further examination.

Fire is perhaps the oldest and most widely used tool for vegetation management worldwide. Its application requires only an ignition source, and appropriate fuelbed and weather conditions. Its effects on vegetation vary depending on the life history strategies and phenologic stages of the plants, and the intensity and duration of the fire. Continuous fires that have high intensity, long duration, and high percent fuel consumption result in significant removal of plant tissue that can increase resource availability by reducing plant competition (Fig. 9.4). Fires can also alter the chemistry of soils, increasing rates of nutrient input, as long as fire intensities are not excessively high, in which case nutrients may be volatilized and lost from the local landscape (DeBano et al. 1998).

Mechanical treatments can be targeted to either selectively remove all, or selectively thin a proportion, of an individual species or group of species

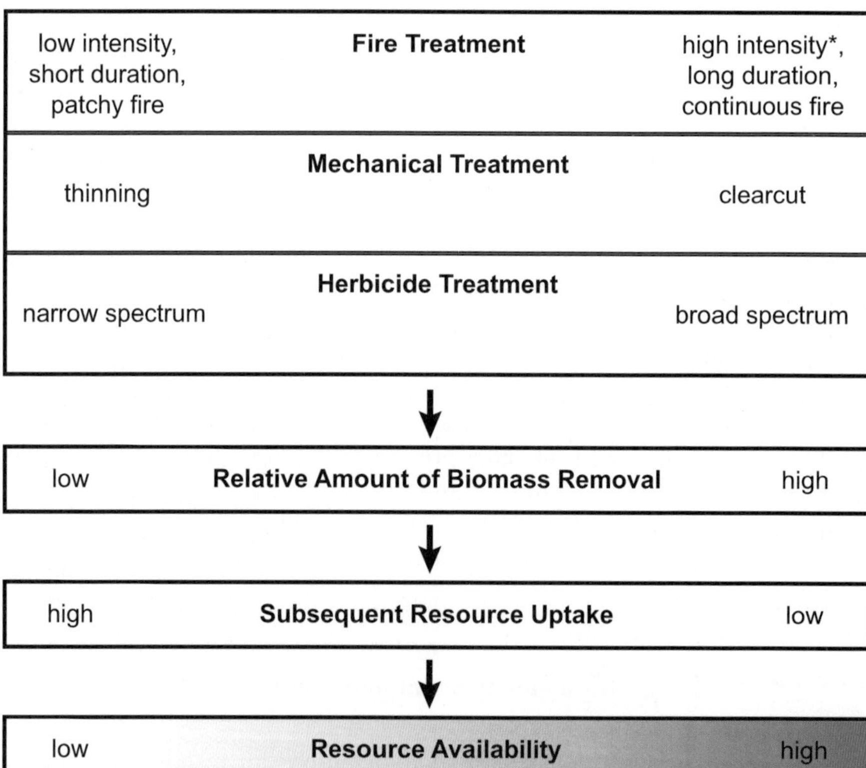

Fig. 9.4 Relative effects of different types of fire, mechanical, and herbicide vegetation removal treatments on resource availability (* unless of very high intensity, in which case soil nutrients may be volatilized)

present at a site. Another approach involves removing the aboveground biomass of all plant species through clearcutting or blading. As the level of biomass removal increases, the rate of resource uptake declines, and the relative availability of resources increases (Fig. 9.4). However, mechanical treatments also typically cause significant soil disturbance, which can reduce soil fertility if subsequent erosion removes topsoil (Edeso et al. 1999), or increase soil fertility due to the decomposition of freshly removed plants (McLellan et al. 1995).

Herbicides are designed to either target a specific subset of plants (narrow spectrum), or target all or most plants (broad spectrum). The more species and biomass that are removed, the lower the resource uptake and the greater the resource availability. Herbicide treatments do not typically cause significant soil disturbance, although the decomposing plant material, especially the roots, can over time increase rates of nutrient input and potentially increase nutrient availability (McLellan et al. 1995).

In general, biomass reductions can be relatively transient if individual plants are not removed, but merely defoliated or thinned in the process of vegetation removal. Incomplete vegetation removal is most common with narrow-spectrum herbicides that target only specific suites of species (e.g., grasses), mechanical treatments that target specific life forms (e.g., forest understory thinning), and fires that only partially consume vegetation, or leave unburned islands within their perimeters. Biomass reductions can persist longer if whole plants are removed, and their replacements are slow to reestablish. Broad-spectrum herbicide treatments, mechanical treatments that completely remove plant biomass (e.g., bladed fuelbreaks), and fires with high intensity and long duration can all result in high rates of plant mortality.

The different techniques used for revegetation can also affect resource availability. In some cases, revegetation efforts include fertilization or mulching treatments that can increase resource supply, and may improve conditions for plant invasions. The benefits of these treatments for the establishing plants should be weighed against their effects on invasion potential. Seedings are typically used over large areas, especially after major disturbances such as fires or floods. If these treatments are applied aerially, then they have virtually no direct impact on the soil. However, if they are followed by tillage, or are applied with ground-based equipment, then significant soil disturbance may occur that could increase rates of nutrient mineralization. Again, the comparison of tillage vs. non-tillage seeding methods should include consideration for the potential effects of soil disturbance.

9.4.2 Effects of Vegetation Management on Propagule Pressure of Invaders

Any time humans, their mechanized equipment, or domesticated animals pass through a landscape, there is a chance they will disperse propagules of invading plants. All land management activities have this potential, although the degree to which they affect propagule pressure can be variable.

On its own, fire does not directly promote the dispersal of plant propagules. However, the management activities associated with fire, such as fuels management, fire suppression, and post-fire emergency stabilization, rehabilitation, and restoration activities, can increase propagule pressure by either accidentally or deliberately introducing propagules. By contrast, mechanical and herbicide treatments tend to involve significant travel throughout the landscape, which can facilitate plant dispersal. In many cases, herbicides are applied aerially, but in most cases in wildlands they are applied using a vehicle or on foot. In the latter case, there is the chance of spread of propagules as the applicators traverse the landscape. These impacts themselves may affect resource availability and propagule pressure. Efforts to mitigate these impacts may include ensuring that propagules are not adhering to people and equipment (this may require periodic decontamination), avoiding passage through known stands of invasives, and traversing the landscape in the most efficient manner (i.e., covering minimal ground).

9.4.3 Predicting the Effects of Vegetation Management Treatments

Vegetation removal is targeted at undesirable, often nonnative, species. Revegetation is focused on promoting the dominance of desirable species, which can target either nonnative or native species, or sometimes both. The factors that determine which species are undesirable and targeted for removal, and which are desirable and targeted for revegetation, depend on the desired effects of treatments that have their roots in broad management objectives. For example, if a plant invasion has altered fuelbed characteristics to the point that fire behavior and fire regimes are affected, creating an invasive plant/fire regime cycle (Brooks et al. 2004), then an important management objective may be to restore pre-invasion fuel and fire regime characteristics. This would involve removal of the undesirable species causing the fuelbed change, and possible revegetation of the desirable species that are necessary to restore the pre-invasion conditions. Unfortunately, it can be very difficult to predict if specific vegetation management treatments will have the effects necessary to achieve the desired management objective. In this example, fuelbed and fire regime characteristics are higher-order effects, with a number of intermediate steps and interactions that can lead to variable results (Fig. 9.5). Similar

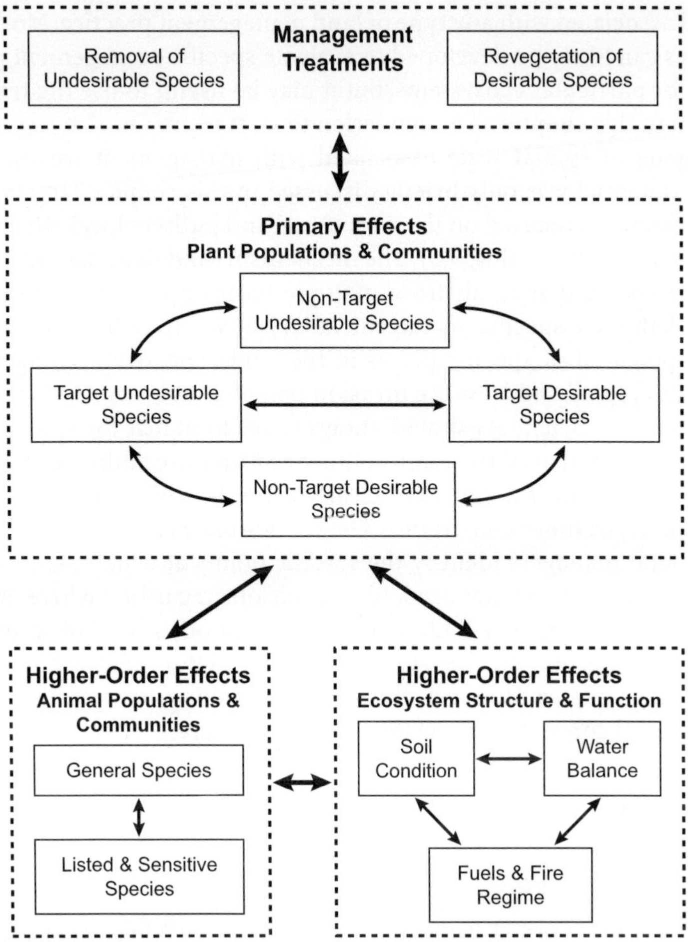

Fig. 9.5 Conceptual model linking vegetation management treatments with primary and higher-order effects

difficulties may be encountered when vegetation management is prescribed to benefit other ecosystem properties, or populations of listed and sensitive wildlife species.

9.5 Conclusions

The processes that affect plant invasions can vary widely, so much so that unifying principles have been difficult to identify. The guidelines presented in this chapter focus on two factors, resource availability and propagule pressure, which can be used to provide a coarse-scale assessment of the invasion potential associated with any type of land management practice. More precise guidelines can likely be developed to evaluate specific management practices tailored for particular ecosystems, but it may be useful to use the framework presented in this chapter as a foundation to start from.

The issue of spatial scale associated with management treatments and invasion potential was only briefly discussed in this chapter, largely because most research has focused on the direct local and indirect local effects of land management practices (Fig. 9.3). The mechanisms and dynamics of dispersed landscape effects that result from multiple local impacts need to be better studied. Although specific management objectives may be focused on the invasion potential of specific places in the landscape, overarching management goals typically address the invasion potential of broad landscapes (e.g., nature reserves). Scientists should always strive to match the spatial scale of their studies with that of the information need they are addressing.

Some management actions are clearly more feasible than others, due to financial costs or other constraints. The framework presented in this chapter can help land managers identify the specific points at which plant invasions may most effectively be managed. Final decisions regarding where and when to apply specific actions will ultimately require another level of scrutiny that involves social, economic, logistical, and other factors that are beyond the scope of this chapter. However, this decision-making process should begin with biological concepts such as those presented here.

References

Allen EB, Leela ER, Steers RJ, Bytnerowicz A, Fenn ME (2007) Impacts of atmospheric nitrogen deposition on Mojave Desert vegetation and soils. In: Webb RH, Fenstermaker LF, Heaton JS, Hughson DL, McDonald EV, Miller DM (eds) The Mojave Desert: ecosystem processes and sustainability. University of Arizona Press, Tucson (in press)

Angold PG (1997) The impact of roads on adjacent heathland vegetation: effects on plant species composition. J Appl Ecol 34:409–417

Brooks ML (2000) Competition between alien annual grasses and native annual plants in the Mojave Desert. Am Midl Nat 144:92–108

Brooks ML (2003) Effects of increased soil nitrogen on the dominance of alien annual plants in the Mojave Desert. J Appl Ecol 40:344–353

Brooks ML (2007) Spatial and temporal distribution of nonnative plants in the Mojave Desert. In: Webb RH, Fenstermaker LF, Heaton JS, Hughson DL, McDonald EV, Miller DM (eds) The Mojave Desert: ecosystem processes and sustainability. University of Arizona Press, Tucson (in press)

Brooks ML, Lair B (2007) Ecological effects of vehicular routes in a desert ecosystem. In: Webb RH, Fenstermaker LF, Heaton JS, Hughson DL, McDonald EV, Miller DM (eds) The Mojave Desert: ecosystem processes and sustainability. University of Arizona Press, Tucson (in press)

Brooks ML, D'Antonio CM, Richardson DM, Grace J, Keeley J, DiTomaso J, Hobbs R, Pellant M, Pyke D (2004) Effects of invasive alien plants on fire regimes. BioScience 54:677–688

Brown G, Schoknecht N (2001) Off-road vehicles and vegetation patterning in a degraded desert ecosystem in Kuwait. J Arid Environ 49:413–427

Callaway RM, Aschehoug ET (2000) Invasive plants versus their new and old neighbors: a mechanism for exotic invasion. Science 290:521–523

Clifford HT (1959) Seed dispersal by motor vehicles. J Ecol 47:311–315

D'Antonio CM (1993) Mechanisms controlling invasions of coastal plant communities by the alien succulent, *Carpobrotus edulis*. Ecology 74:83–95

Davidson E, Fox M (1974) Effects of off-road motorcycle activity on Mojave Desert vegetation and soil. Madroño 22:381–412

Davis MA, Grime PJ, Thompson K (2000) Fluctuating resources in plant communities: a general theory of invasibility. J Ecol 88:528–534

DeBano LF, Neary DG, Ffoliott PF (1998) Fire's effects on ecosystems. Wiley, New York

Edeso JM, Merino A, Gonzalez MJ, Marauri P (1999) Soil erosion under different harvesting managements in steep forestlands from northern Spain. Land Degrad Develop 10:79–88

Forman RTT, Alexander LE (1998) Roads and their major ecological effects. Annu Rev Ecol Syst 29:207–231

Forman RTT, Sperling D, Bissonette JA, Clevenger AP, Cutshall CD, Dale VH, Fahrig L, France R, Goldman CR, Heanue K, Jones JA, Swanson FJ, Turrentine T, Winter TC (2003) Road ecology: science and solutions. Island Press, Washington, DC

Gelbard JL, Belnap J (2003) Roads as conduits for exotic plant invasions in a semiarid landscape. Conserv Biol 17:420–432

Greenberg CH, Crownover SH, Gordon DR (1997) Roadside soil: a corridor for invasion in xeric scrub by nonindigenous plants. Nat Areas J 17:99–109

Hobbs RJ (2000) Land-use changes and invasions. In: Mooney HA, Hobbs RJ (eds) Invasive species in a changing world. Island Press, Washington, DC, pp 55–64

Hobbs RJ, Hopkins AJM (1990) From frontier to fragments: European impact on Australia's vegetation. Proc Ecol Soc Am 16:93–114

Hobbs RJ, Huenneke LF (1992) Disturbance, diversity and invasion: implications for conservation. Conserv Biol 6:324–337

Johnstone IM (1986) Plant invasion windows: a time-based classification of invasion potential. Biol Rev 61:369–394

Lonsdale WM (1999) Global patterns of plant invasions and the concept of invasibility. Ecology 80:1522–1536

Lonsdale WM, Lane AM (1994) Tourist vehicles as vectors of weed seeds in Kakadu National Park, Northern Australia. Biol Conserv 69:277–283

Mack RN, Simberloff D, Lonsdale WM, Evans H, Clout M, Bazzaz FA (2000) Biotic invasions: causes, epidemiology, global consequences, and control. Ecol Appl 10:689–710

Maron JL, Connors PG (1996) A native nitrogen-fixing shrub facilitates weed invasion. Oecologia 105:302–312

McLellan AJ, Fitter AH, Law R (1995) On decaying roots, mycorrhizal colonization, and the design of removal experiments. J Ecol 83:225–230

National Research Council (1997) Toward a sustainable future: addressing the long-term effects of motor vehicle transportation on climate and ecology. National Academy Press, Washington, DC

Padgett PE, Allen EB, Bytnerowicz A, Minnich RA (1999) Changes in soil inorganic nitrogen as related to atmospheric nitrogenous pollutants in southern California. Atmospheric Environ 33:769–781

Schmidt W (1989) Plant dispersal by motor cars. Vegetatio 80:147–152
Silveri A, Dunwiddie PW, Michaels HJ (2001) Logging and edaphic factors in the invasion of an Asian woody vine in a mesic North American forest. Biol Invasions 3:379–389
Spellerberg IF (1998) Ecological effects of roads and traffic: a literature review. Global Ecol Biogeogr Lett 7:317–333
Spellerberg IF (2002) Ecological effects of roads. Science Publishers, Plymouth
Vasek FC, Johnson HB, Brum GD (1975) Effects of power transmission lines on vegetation of the Mojave Desert. Madroño 23:114–130
Williamson M (1999) Invasions. Ecography 22:5–12
Zink TA, Allen MF, Heindl-Tenhunen B, Allen EB (1995) The effects of a disturbance corridor on and ecological reserve. Restor Ecol 3:304–310

10 Nitrogen Enrichment and Plant Invasions: the Importance of Nitrogen-Fixing Plants and Anthropogenic Eutrophication

Michael Scherer-Lorenzen, Harry Olde Venterink, and Holger Buschmann

10.1 Introduction

The invasion of natural ecosystems by exotic species is an important component of global environmental change, and poses a major threat to biodiversity. Other drivers of global change – such as alteration of the atmospheric composition and associated climate change, changing patterns of land use that fragment habitats and alter disturbance regimes, and increasing levels of airborne nitrogen deposition – also influence resource dynamics and species composition of ecosystems (Sala et al. 2000). Consequently, they all have the potential to interact with biological invasions and to accelerate this process, for which evidence is accumulating (Dukes and Mooney 1999; Mooney and Hobbs 2000). In addition, biological invasions themselves can alter the biogeochemistry of ecosystems through particular traits of the invading species (Ehrenfeld and Scott 2001). If we wish to understand and eventually predict the ecological impacts of invasive species, it is thus of particular importance to reveal the many complex interactions between all elements of global change, and their effects on ecosystem processes. In this chapter, we focus on alterations of the nitrogen cycle of terrestrial ecosystems by exotic invasions, and how nitrogen deposition may influence the success of invaders.

It is not yet possible to predict which exotic species will become invasive, though successful invaders are often associated with a particular suite of traits: high seed output, high relative growth rate, high specific leaf area, low leaf construction costs, high phenotypic plasticity, and high nutrient concentrations (Rejmanek and Richardson 1996; Daehler 2003; see also Chap. 7). Species sharing these traits may be specifically capable to capitalize on the various elements of environmental change (Dukes and Mooney 1999). What

Ecological Studies, Vol. 193
W. Nentwig (Ed.)
Biological Invasions

makes a community invasible appears related to, among others, disturbance regimes, and resource availabilities (Lonsdale 1999; Davis et al. 2000; see Chaps. 9, 11, 12), stressing the relevance of global change drivers for invasibility. Some consequences of exotic invasions are clear: for example, they may have a devastating effect on communities of native plants and animals (see Chaps. 13–17), resulting in a strong decline in biodiversity (Mack et al. 2000; Sala et al. 2000). Other consequences – for example, effects on ecosystem biogeochemistry – may be initially less conspicuous, but are of great importance in the long term.

10.2 Alterations of the N-Cycle by Exotic Invaders

Some exotic species alter soil communities and nutrient cycling processes, often increasing nutrient availabilities for plants (Ehrenfeld 2003). Since exotic invaders usually benefit more from higher nutrient availability than do indigenous species, this may ultimately cause a positive feedback of invasion (D'Antonio et al. 1999), although negative feedbacks of invasions have also been reported (Ehrenfeld 2003). For a few exotic invaders, such as pyrophytic grasses (e.g., Mack and D'Antonio 2003) and the N_2-fixing shrub *Myrica faya* (Vitousek et al. 1987), the ecosystem effect and feedback mechanism appear to be well understood. By contrast, less information is available about the consequences for the invaded ecosystem, partly because the time course of the invasion is often unknown, but also because ecosystem effects may not become evident for many decades (Ehrenfeld 2003).

Mack et al. (2001) distinguished two types of alterations to nutrient cycles by exotic invasions, which they characterized as (1) dramatic, and (2) gradual. Dramatic alterations occur when exotic invasive species introduce a new functional trait into the ecosystem, making resources available from new sources (e.g., the introduction of N_2-fixing plants into ecosystems with no such symbioses). Gradual alterations occur when plant functional traits – such as size, growth rate, tissue quality – of invasive species differ quantitatively from those of indigenous species, causing different rates of nutrient uptake or nutrient turnover (e.g., the cycle involving plant-available nutrients in soil–nutrients in plants–nutrients in litter–plant-available nutrients in soil). Here, we focus on the first type: dramatic alterations by the introduction of nitrogen-fixing species.

10.2.1 Nitrogen-Fixing Species Among Invasives and Natives

Some of the world's most problematic invaders of natural ecosystems are N_2-fixing legumes (e.g., *Acacia*, *Albizia*, and *Leucaena* species), and actinorhizal

N_2-fixing species (e.g., *Casuarina equitesifolia* and *Myrica faya*; e.g., Richardson et al. 2000). There are examples of invasions by some of these species in Hawaii (*Myrica faya*), South African fynbos (*Acacia* species), the American continent (*Casuarina* species, *Lupinus arboreus*, *Robinia pseudoacacia*), Europe (*Lupinus polyphyllus*, *Robinia pseudoacacia*), New Zealand and Chile (*Ulex europaeus*), and several tropical islands (*Albizia falcataria*).

Approximately 10 % of the introduced and invasive flora of the United States are possible nitrogen-fixers (USDA 2006), and for highly invasive species the proportion could even be higher (e.g., six of 31 species in Florida; Gordon 1998). Hence, the ability to fix N_2 is not an uncommon feature of invasive species in North America, but we do not have the data to evaluate whether this proportion differs from that of native species. For Germany, such data are available from FloraWeb (BfN 2006). The proportion of N_2-fixers among the native and invasive flora of Germany is rather similar (natives 4 %, invasives 5 %). Moreover, only two (*Lupinus polyphyllus*, *Robinia pseudoacacia*) of 30 species listed by Kowarik (2003) as strong invaders with serious economical and ecological impacts have symbiotic rhizobia. Thus, it seems that nitrogen fixation is generally more pronounced in the introduced invasive flora of the new world than in Europe. There are, however, no clear indications that nitrogen fixation is more frequent in invasive than in native species.

10.2.2 Nitrogen Input by N_2-Fixing Invasive Species

Despite the apparent importance of invading exotic N_2-fixers, the roles of nitrogen-fixing symbioses have been surprisingly ignored in the invasion literature (Richardson et al. 2000). The outstanding exception is the study by Vitousek and coworkers in Hawaii, of N_2 fixation by the exotic *Myrica faya* in a N-limited forest where N_2-fixers were previously absent (Vitousek et al. 1987). In this study, the mean nitrogen input into the forest ecosystem due to *Myrica faya* was 18 kg N ha^{-1} $year^{-1}$, whereas other N inputs were only 5.5 kg N ha^{-1} $year^{-1}$. This study has become *the* example of a dramatic alteration of a nutrient cycle by an invasive plant species; by March 2005, it had been cited 390 times (ISI Web of Science; i.e., the papers from Vitousek et al. 1987, and Vitousek and Walker 1989). We screened the abstracts of these 390 papers, and found that N_2 fixation by invading exotic species had been quantified in only one other region (cf. two exotic *Acacia* species in two South African forests, Stock et al. 1995). Additionally, seven of the 390 studies reported an increased soil N status (higher N contents, higher mineralization rate) after invasions with exotic N_2-fixing species, while one study reported a decreased soil N status (Wolf et al. 2004). It should be noted, however, that the cases of higher soil N availability are not necessarily due to an enhanced N input, since at least six of the 390 studies reported similar increases following invasion by non-N_2-fixing species. Thus, this could also result from increased decomposition rates, since legumes often have relatively high N concentrations, which might stimulate litter decomposition rates (Ehrenfeld 2003).

10.2.3 Major Invasive Nitrogen-Fixing Species

Also other researchers came to the conclusion that there are hardly any data available about the N input into the ecosystem by N_2 fixation of exotic invading species. One reason for this may be the lack of precise methods for quantifying N_2-fixation rates. Nevertheless, there are some data available for some of these species when growing in their native range or in plantations. The N input values from these studies might give an indication of the potential importance of N_2 fixation when these species are invading. We note that rates in plantations could be higher than those in natural areas because of more optimal growth conditions in the former. Below, we will shortly present examples of some important invading N_2-fixing species, and their possible influence on N cycling and related ecosystem processes.

Myrica faya. As mentioned above, *Myrica faya* increased annual N input in young volcanic soils about fourfold. This additional nitrogen cycles rapidly within the system, increases nitrogen availability, enhances plant productivity, and alters community structure by differentially affecting survival and growth (Vitousek and Walker 1989; Walker and Vitousek 1991). It is interesting to note that these volcanic soils are poor in nitrogen, rich in phosphorus, and have scarce vegetation. Hence, these young soils are nitrogen-limited. *M. faya* does not invade on older, P-limited soils (Vitousek 2004).

Albizia falcataria (=*Falcataria moluccana* or *Paraserianthes falcataria*). There are several studies investigating the influence of N_2-fixing *Albizia falcataria* and non-N_2-fixing *Eucalyptus saligna* on soil properties and productivity of tree plantations (e.g., Binkley 1997; Garcia-Montiel and Binkley 1998; Binkley et al. 2003). Both species were introduced on Hawaii. Compared with *E. saligna* monocultures, mixed stands of *E. saligna* and *A. falcataria* may produce more biomass, contain larger aboveground nutrient pools, and cycle more nutrients through litterfall. *A. falcataria* can increase N pools in tropical soils with N limitation almost threefold. There are also indications that *A. falcataria* is able to acquire more P from the soil than is the case for *E. saligna*, leading to reduced soil P pools. As the effects of *A. falcataria* have been measured only in plantations, it is unknown whether it also causes N enrichment in natural ecosystems. This seems a very relevant question, since *A. falcataria* is an important invasive tree in some tropical areas.

Acacia spp. Invasive *Acacia* species have been intensively investigated in South African ecosystems. The results of these studies are unequivocal and ecosystem-dependent. In sand plains in the South African fynbos, alien nitrogen-fixing *Acacia* species increase soil nitrogen and phosphorus contents, and enhance nitrogen return in litterfall (e.g., Yelenik et al. 2004). Remarkably, they enhance soil nitrogen mineralization rates only at nutrient-rich sites, while phosphorus mineralization was increased in *Acacia* stands of both nutrient-rich and nutrient-poor ecosystems (Witkowski and Mitchell 1987; Stock et al. 1995). It has been suggested that these effects are only in part the result of nitrogen fixation, and rather due to associations with mycorrhiza (Versfeld and van Wilgen 1986). There are indications that enhanced soil nitrogen caused by *Acacia* species promotes the invasion of other plant species. For example, a bioassay experiment showed that the alteration of N availability by *Acacia* increases growth rates of the weedy grass *Ehrharta calycina* (Yelenik et al. 2004).

Lupinus arboreus. Pickart et al. (1998) studied the invasion of *Lupinus arboreus* in parts of North America where it does not occur naturally. Ammonium and nitrate levels in the soil of nutrient-poor coastal dunes of northern California were found to increase at higher abundance of *L. arboreus*. In addition, by removing *L. arboreus* from restoration plots, other invasive species were reduced. Pickart et al. (1998) suggested that *L. arboreus* invasion results in both direct and indirect soil enrichment, as a result of the associated encroachment of other nonnative species, particularly grasses.

Robinia pseudoacacia. According to Boring and Swank (1984), *Robinia pseudoacacia* is the second most abundant deciduous tree in the world, and one of the most aggressive invaders worldwide. They reported that N_2 fixation of this species can amount to 75 kg N ha^{-1} $year^{-1}$. Rice et al. (2004) investigated the influence of this nitrogen-fixing species on nitrogen cycling, when it was invading a nutrient-poor sand plain ecosystem in temperate North America. They carried out a comparison between pine-oak stands and 20–35 year-old *R. pseudoacacia* stands in terms of soil nutrient contents, soil nitrogen transformation rates, and annual litterfall biomass and nitrogen concentrations. Compared with soils of pine-oak stands, soils of *R. pseudoacacia* stands had 1.3–3.2 times greater nitrogen concentration. They also showed elevated net nitrification rates (25–120 times higher than in pine-oak stands), higher total net N mineralization rates, and elevated soil P pools.

Ulex europaeus. In New Zealand, Egunjobi (1969) studied nine ecosystems between 1 and 57 years of age and invaded by nitrogen-fixing *Ulex europaeus*. In vegetation stands up to 10 years old, *U. europaeus* was dominant, and dry matter accumulation and nitrogen contents of dead litter and soil were highest. In later stages of succession, non-fixing species were dominant, and dry matter accumulation and nitrogen contents declined with vegetation age, indicating that *U. europaeus* can increase soil nitrogen levels.

Despite such examples of N enrichment by exotic N_2-fixers, there are various reasons why invasion by such species may not result in N enrichment of the ecosystem. Firstly, not all leguminous species are able to form nodules and fix atmospheric N_2 (Sprent 2001). Secondly, the extent of nodulation and fixation is controlled largely by environmental factors, of which soil N and P availabilities are the most prominent. Indeed, nodulation and N_2 fixation are negatively affected by high soil N availability and by low soil P availability (Vitousek et al. 2002; Binkley et al. 2003). Hence, there may be a negative feedback between exotic invasions of N_2-fixing species and N enrichment of the soil, as well as a positive feedback between N_2-fixing species and factors enhancing soil P availability. The latter factors may include interactions with other invasive exotic species that are able to enhance P availability (Simberloff and Von Holle 1999). On the one hand, enhanced P availability may promote N_2 fixation, and on the other, increased N availability may stimulate root phosphatase activity and/or P mineralization, and hence may increase P availability for plants (e.g., Johnson et al. 1999). Although P is the main growth-limiting nutrient, N enrichment may stimulate production and cause shifts in species composition. As mentioned above, in some cases the introduction of an N_2-fixer is accompanied by a decrease of soil P pools (e.g., Binkley 1997), in others by an increase (e.g., Rice et al. 2004). Versfeld and Van Wilgen (1986) suggested that the increased P pools could be due to interactions with symbiotic mycorrhiza. Hence, the impact of invading and potentially N_2-fixing species is site-specific, and depends on local soil N and P conditions.

10.2.4 Facilitated Secondary Invasion

The consequences, and in particular, the long-term effects of N_2 fixation on ecosystem properties and vegetation composition are fairly unknown. Not only does N_2 fixation influence primary productivity, but this may have cascading effects on successional patterns, community composition, and disturbance regimes (Rice et al. 2004). For example, the increased availability of nitrogen following the invasion of nitrogen-fixing species might be an important pathway by which invaders alter community structure, possibly favoring the invasion of more exotic species. In coastal grasslands of California invaded by *Lupinus arboreus*, the impact of N enrichment on secondary invasion was most severe after senescence (Pickart et al. 1998; Maron and Jefferies 1999). Similar effects were documented after the death of *Myrica faya* (Adler et al. 1998). Hence, nitrogen enrichment may impede efforts to restore native plant communities and ecosystem functions at sites previously occupied by exotic nitrogen-fixers.

10.2.5 Nitrogen Fixation Suppressed by Invasion

Nitrogen fixation could also be negatively affected by plant invasions. For the invasion of African grasses in Hawaii, Ley and D'Antonio (1998) demonstrated that these species eliminate the fire-sensitive native tree *Metrosideros polymorpha* by promoting fire events. Non-symbiotic N_2-fixing bacteria are active on the litter of *M. polymorpha*, but not on that of the invasive grasses. Hence, elimination of these native species led to reduced nitrogen contents in invaded soils. In addition, invasive species that do not themselves fix N_2 may also indirectly affect the rate of fixation in co-occurring N_2-fixing species, possibly through an allelopathic mechanism (Ehrenfeld 2003). For example, in glasshouse and pasture studies in New Zealand, Wardle et al. (1994) found evidence that decomposing leaves of the invasive thistle *Carduus nutans* inhibit nitrogen fixation by the introduced *Trifolium repens*. *C. nutans* has invaded many areas of western North America. Therefore, it might be possible that this thistle also adversely affects growth and nitrogen fixation of native legumes.

10.3 Nitrogen Deposition and Exotic Invasions

10.3.1 N Deposition and Eutrophication in Natural Ecosystems

The input of nitrogen compounds into terrestrial ecosystems is a major component of global environmental change, and can have substantial effects on ecological processes and the biogeochemistry of these systems (Vitousek et al. 1997). Current estimates of airborne, mineral nitrogen deposition show that large portions of North America, Europe, and Asia have rates >75 kg N ha^{-1} $year^{-1}$, and some regions receive >100 kg N ha^{-1} $year^{-1}$ (Galloway et al. 2004), exceeding the natural background rate by two orders of magnitude. Anthropogenic emissions of nitrogen compounds (NH_x and NO_x) from fossil fuel combustion, fertilizer use, and animal husbandry are the underlying cause for this elevated deposition of biologically active N. Beside the relatively well-documented effects on biogeochemical cycles, N deposition can also have severe economic impacts, such as increased costs for the management or purification of drinking water (Wamelink et al. 2005).

The most direct ecological effect is a subtle, but continuous increase in nitrogen pools and fluxes, and hence in nitrogen availability for plant growth and productivity in N-limited ecosystems (e.g., Olde Venterink et al. 2002). The eutrophying influence of N deposition may be associated with an acidification of soils, mineral imbalances in plant nutrition resulting in reduced plant vitality and increased sensitivity against stress, and enhanced nitrate leaching into groundwater (Schulze 2000; Nadelhoffer 2001). Ecosystems that are characterized by low natural levels of nitrogen availability are most vulnerable – for example, ombrothrophic bogs, heathlands, temperate and boreal forests, or nutrient-poor grasslands. Critical loads for acidifying and eutrophying nitrogen for these ecosystems lie in the range of 5–20 kg N ha^{-1} $year^{-1}$, currently exceeded in many parts of the world (EEA 2005). These alterations of growing conditions also have direct consequences for the competitive success of species, associated with often dramatic changes in community composition and decrease of species richness, especially at N-limited oligo- and mesotrophic sites such as in northern temperate forests (Bobbink et al. 1998; Sala et al. 2000): slow-growing plants adapted to nutrient-poor soils are outcompeted by faster-growing species originating from nutrient-rich sites. Responsiveness of species to N deposition is thus highly related to plant traits typical for high-nutrient environments.

Since invasive species are often associated with a particular suite of traits that are characteristic for plants of nutrient-rich sites (see Sect. 10.1), one might assume that N deposition could also influence the abundance patterns of native vs. invasive plant species. In addition, the invasibility of ecosystems is most often also related to the degree of resource availability (Davis et al. 2000), and invasive species generally occur more frequently at nutrient-rich sites, as shown for the floras of Germany (Scherer-Lorenzen et al. 2000), and the Czech Republic (Pyšek et al. 1995). One could therefore hypothesize that invasive species are more successful, and therefore more abundant in areas of high atmospheric N deposition than in areas of low deposition.

10.3.2 A Short Note on Mechanisms

As Scherer-Lorenzen et al. (2000) have discussed, the mechanism of competitive exclusion of species adapted to low nutrient levels by faster-growing, more nitrophilic species is independent of whether the invading species is either alien or native. There are numerous examples of successful native species invading species-rich communities under high N deposition (Bobbink et al. 1998). Particularly well documented is the case of ombrotrophic bogs, an ecosystem type among the most sensitive to N enrichment because these bogs receive most of their nutrients from the atmosphere (e.g., Tomassen et al. 2004). Other examples include community changes in wet heathlands (e.g., Aerts and Berendse 1988), or after acidic deposition-induced forest dieback. In the latter case, native grasses responsive to nitrogen invade the understorey of the forests, replacing species adapted to less fertile soils. Examples from mountain spruce forests in Europe include the replacement of the dwarf shrubs *Vaccinium myrtillus* and *Calluna vulgaris* or the grass *Deschampsia flexuosa* by the grass *Calamagrostis villosa* (Scherer-Lorenzen et al. 2000). In all cases, combinations of specific traits, such as high relative growth rate and higher rates of nutrient uptake associated with a higher demand for soil nutrients, result in competitive advantages for the more nitrophilic species. As discussed above, the same suites of traits are also often found in alien invasive species, which may at least partly explain their success in formerly nutrient-poor habitats subsequently exposed to high levels of eutrophication related to N deposition.

10.3.3 Evidence for Effects of N Deposition on Plant Invasions?

10.3.3.1 Spatial Correlations

A first, correlative approach is to compare the distribution patterns of N deposition with those of invasive species. Obviously, this comparison has to be done on different spatial scales, ranging from global to local. The major problem with such an approach, however, is the difference in data availability and quality between, on the one hand, spatially explicit information on N deposition, and on the other, scanty information on plant invasions. Indeed, data on N deposition are now available both at global and regional scales – for example, global: Galloway et al. (2004), Europe: EMEP – European Monitoring and Evaluation Programme (EEA 2005), and USA: National Atmospheric Deposition Program/National Trends Network (NADP 2006). By contrast, a mapping of plant invasions on similar scales currently does not exist. Even readily accessible global datasets such as the ISSG Global Invasive Species Database as yet do not allow us to compile maps with a robustness and resolution comparable to those for N deposition. Despite this limitation, interesting insights can still be derived from such a correlative approach.

At the global scale, highest N deposition with values greater than 20 kg ha^{-1} $year^{-1}$ (or 2,000 mg $m^{-2} yr^{-1}$) occurs in densely populated regions in the vicinity of large urban agglomerations, or in intensively managed agricultural areas and their downwind neighborhoods (Fig. 10.1, Galloway et al. 2004), in central and eastern Europe, eastern North America, the Indian subcontinent, and eastern Asia. Smaller spots of relatively high deposition occur in South America and central Africa. Still unaffected are large areas of the circum-boreal zone, the western coasts of both Americas, the Amazon basin, Patagonia, Saharan Africa, central and southeast Asia, Australia and New Zealand, and the oceans. From the knowledge we have about global patterns of invasion (Chap. 11), it is clear that present-day N deposition levels cannot be assigned as a crucial factor determining the invasibility of ecosystems by alien, invasive species at the global scale: the hotspots of invasion, such as all oceanic islands, Australia, New Zealand, Chile, and South Africa, clearly lie outside of the hotspots of deposition – either because of low human population densities and agricultural N losses, or because of low levels of industrialization.

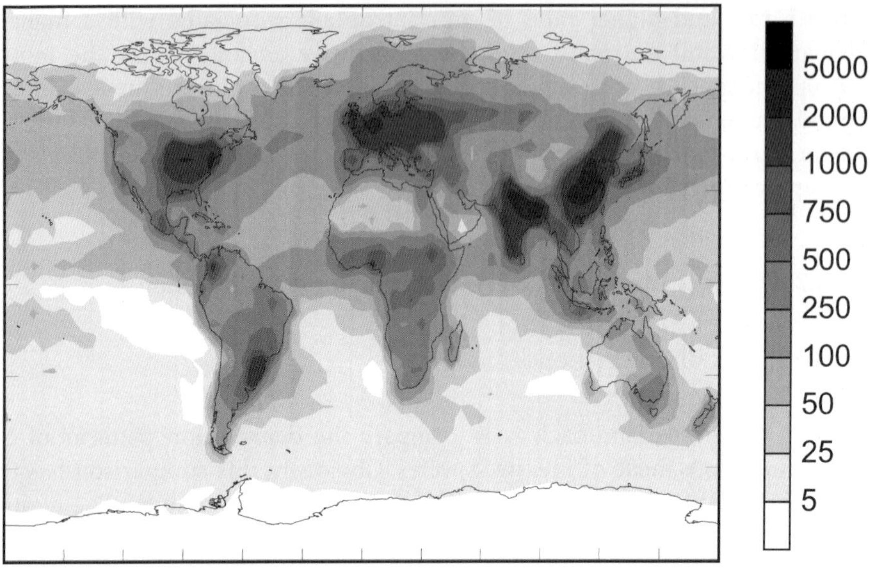

Fig. 10.1 Spatial pattern of total inorganic nitrogen deposition in the early 1990s (mg N m^{-2} $year^{-1}$; reprinted from Galloway et al. 2004, with permission from the author)

At a continental or regional scale, however, we found some evidence for a positive correlation between N deposition and invasive species abundance. We combined data on inorganic nitrogen wet deposition from nitrate and ammonium, extracted from the US National Atmospheric Deposition Program/National Trends Network (NADP 2006), with estimates of nonnative plant species abundance in North American ecoregions, provided by the PAGE project of the World Resources Institute (WRI 2000, based on data of the US World Wildlife Fund WWF). This shows that the proportion of exotic species in forest ecoregions of the USA clearly increases at higher levels of wet N deposition (Fig. 10.2). By contrast, no such pattern emerges for corresponding data from grassland ecoregions (data not shown). Of course, the relationship depicted in Fig. 10.2 is no absolute proof of a causal link between N enrichment and invasibility, because other factors determining invader success do also change in the vicinity of centers of N emission, such as disturbance, traffic, or seed input. Nevertheless, it convincingly demonstrates the potentially accelerating effects on invasibility resulting from complex interactions between various drivers of environmental change.

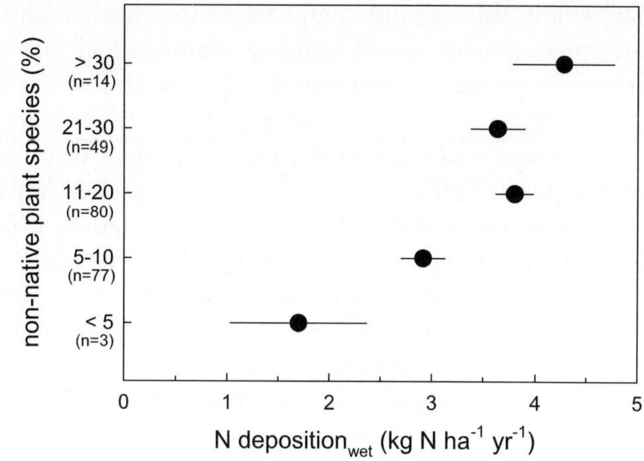

Fig. 10.2 Correlation between total N wet deposition as nitrate and ammonium and the proportion of nonnative plant species in forest ecoregions of the USA. Data taken from the US National Atmospheric Deposition Program (NADP 2006) and from the World Resources Institute (WRI 2000). Wet deposition represents means±s.e. of annual averages of 223 monitoring sites located in forest systems

10.3.3.2 Observational Studies

More reliable data on the relationship between N deposition and invader success are available only on a local or plot scale, i.e., from a few square kilometers to several square meters. Theory predicts that a plant community should become more invasible if there is an increase in the amount of unused resources (Davis et al. 2000), i.e., if the use of resources by resident species declines, or if resource supply increases at a rate faster than the resident species can make use of. The latter may occur in the case of eutrophication, e.g., by N deposition. There are indeed several reports of promoted invasion on more fertile soils. For instance, in New Zealand's *Nothofagus* forests, areas invaded by *Hieracium* species had higher N availabilities as well as higher P, Ca, and Mg concentrations than those recorded in uninvaded areas (Wiser et al. 1998). Likewise, in southern Utah, exotic plant invasions were highly correlated to soils high in C, N, and P (Bashkin et al. 2003). Although variations in soil fertility were of natural origin in these studies, the authors concluded that because habitats containing fertile soils appeared more vulnerable to exotic invasions than those with less-fertile soils, shifts in soil conditions induced by N deposition could shift the balance for native and exotic species locally. Similarly, the clonal-growing species *Arundo donax*, which is known to be very responsive to N enrichment, invades riparian areas in the USA preferentially at sites of high nitrogen availability (Decruyenaere and Holt 2005). Its higher success under such conditions is attributable to year-round activity, greater overall ramet flux, and higher foraging. Although eutrophication of riparian ecosystems is related more to nutrient enrichment of surface waters than to atmospheric N deposition, this example emphasizes the mechanisms of responsiveness to eutrophication of clonal-growing plants, which are known to show high growth responses to nutrient enrichment and which can be highly invasive (e.g., *Solidago canadensis* and *S. gigantea* or *Helianthus tuberosus* in Europe).

Much work on ecological effects of N deposition on plant invasions has been done in California, USA, especially in the chaparral and coastal sage scrub ecosystems invaded by alien grasses (Allen 2004; also see Fenn et al. 2003, and references therein). Gradients of N deposition are well reflected in soil nitrogen concentrations, and increased soil N concentrations may contribute to the growth of invasive grasses. A variety of underlying mechanisms have been tested, including (1) direct growth responses due to high responsiveness of these species to nutrient enrichment, (2) higher N uptake rates of invasive grasses, (3) changes in mycorrhizal fungal community composition and functioning, and (4) indirect alterations of the fire cycle due to increased fire fuel loads. It seems that direct growth responses are less important than expected, because in addition to invasive species, also native shrub species responded strongly to fertilization, while changes in mycorrhizal community composition, higher N uptake rates, and especially positive feedbacks via fire may be responsible for the observed shifts in community composition caused by N deposition. Phenology also plays a role here, as the exotic grasses germinated more rapidly than native species in response to rain during the first winter growing season, and they therefore may have had a week of growth advantage in terms of N uptake. In species-rich Californian serpentinitic grasslands, there are indications that dry N deposition from smog near urban areas is responsible for invasions by annual grasses (mainly *Lolium*, *Bromus*, and *Avena*) that ultimately lead to crashes in rare butterfly populations (Weiss 1999). A moderate, well-managed cattle grazing is now needed to prevent dominance of exotic grasses, and to maintain native plant and insect diversity.

10.3.3.3 Nutrient Addition Experiments

There are numerous nitrogen addition experiments confirming that N enrichment stimulates the dominance of alien plants, and decreases the overall performance, abundance, and diversity of native species. For example, increasing nutrient levels of Californian serpentine grasslands linked with the use of N, NP or NPK fertilizers facilitated the rapid invasion and dominance of nonnative annual grasses in patches originally dominated by native forbs (Hobbs et al. 1988; Huenneke et al. 1990). Compared to N deposition levels (100 kg N ha^{-1} $year^{-1}$, Huenneke et al. 1990), levels of fertilization in these studies were rather high (313 kg N ha^{-1} $year^{-1}$, Hobbs et al. 1988). Still, atmospheric deposition of even less than 10–15 kg ha^{-1} $year^{-1}$ at these sites (Weiss 1999) may have accumulated nitrogen in plants and microbes over several years in these strongly N-limited serpentinitic grasslands that are highly retentive of N. Thus, under strong N limitation, even low (albeit chronic) levels of N deposition may be sufficient to induce dramatic changes in community composition associated with shifts from native annual forbs to invasive grasses. Similarly, Wedin and Tilman (1996) showed, for N-limited prairie ecosystems in Minnesota, USA, that simulated gradients of N deposition resulted in decreasing levels of plant species diversity, and shifts in dominance from native C_4 to exotic C_3 grasses, even at N input rates comparable to current deposition levels. For deserts, still among the least invaded of all ecosystems and possibly because of very low levels of soil nitrogen, Brooks (2003) were able to demonstrate that enhancing N load at rates comparable to published deposition rates around urban areas in the Mojova Desert, USA, increased the density and biomass of exotic annual species. Native species density, biomass, and richness responded negatively in years of high annual productivity, presumably due to increased competition for soil water and other nutrients caused by the increased biomass production of the invaders. The higher standing biomass, in turn, represents higher fuel load, so that fire frequency could be increased in those systems affected by N deposition (see also Sect. 10.3.4).

Compared to control plots in the shortgrass steppe of Colorado, USA, Milchunas and Lauenroth (1995) found large increases of exotic species after addition of nitrogen or nitrogen plus water. Interestingly, these changes were much more pronounced after a period of 7 years without any further experimental addition of resources, emphasizing the importance of time lags in response to enrichment stressors.

Evidence for superiority of alien invasive plants over native ones under high soil N availability also comes from restoration ecology. Daehler (2003) summarizes examples from nutrient reduction experiments via the addition of carbon supplements to the soil, which increases microbial N immobilization. In many cases, this treatment increased the competitive performance of native species vis-à-vis invasives, which could even lead to the transformation of alien-dominated communities into those closely resembling the natural vegetation.

10.3.4 Interaction of N Deposition with Other Drivers of Environmental Change

Clearly, drivers of environmental change other than N deposition have large effects on plant invasion. Many of these will interact in complex ways, presumably accelerating biological invasions and their ecological impact (Dukes and Mooney 1999; Mooney and Hobbs 2000).

A particularly important interaction might exist between N deposition and increasing levels of disturbance by fire. Indeed, there is evidence from several studies that N deposition increases biomass production, contributing to greater fuel loads and thus altering the fire cycle in a variety of ecosystems (cf.

Fenn et al. 2003). Particularly prone to such interactions between N enrichment and fire seem to be ecosystems invaded by grasses, because of their generally strong positive response in biomass production under N loading, and compared with other plant functional types, their higher inflammability. In addition, grasses are able to recover relatively quickly after fire, producing a positive feedback loop of grass and fire until an annual grassland is stabilized under a high-frequency fire cycle (D'Antonio and Vitousek 1992). There are many examples from other ecosystems where plant invasion has caused alterations of the fire regime (Brooks et al. 2004); it would be interesting to see whether nitrogen enrichment could also interact with the process of invasion and the fire regime in these systems.

10.4 Future Challenges

The few nitrogen-fixing species that have to date been intensively investigated show partly high impacts on community structure and ecosystem processes, but it remains unclear whether these results are applicable for all invasive N_2-fixers: many species have not yet been investigated perhaps simply due to an absence of obvious impact on the ecosystem. Levine et al. (2003) criticized that most studies have been carried out in environments where effects would be most likely (e.g., nutrient-poor soils with sparse vegetation), and that nitrogen-rich and densely vegetated ecosystems would not be expected to show similar impacts. Thus, general statements about the influence of invasive nitrogen-fixers may have been biased by the choice of species, and of study sites. Evidently, there is high need for further detailed studies about effects of other invasive N_2-fixing species on N cycling and ecosystem properties under various site conditions.

It is very difficult to draw any general conclusion about the impact of nitrogen-fixing invasives, largely because of the paucity of comparisons with native nitrogen-fixers, and with non-invasive exotic nitrogen-fixers. Furthermore, essentially nothing is known about whether N_2-fixers change their fixing capacity if they become invasive, or about the compatibility of these plants with local symbionts in the invaded range. To investigate these questions, comparative studies between the native and invaded range are needed. A comprehensive approach would entail comparing the provenances of species in the invaded and native ranges by means of experimental studies.

Although we have shown that there is now some basic knowledge about the effects of invasive species on N dynamics, and about the impact of N deposition on invasibility, there are still many open questions to be resolved. An interesting research project would be a compilation of the occurrence and proportions of native and alien species responsive to N addition in local floras, and to what extent this is reflected in the invader success along gradients of N deposition. Particularly important in this context would be a high-resolution mapping of invasive species at local scales. Such a study should be complemented by experimental N addition experiments under various levels of soil fertility and/or soil disturbance, because performance comparisons between native and invasive species are often context-dependent (Daehler 2003). Ideally, other aspects of environmental change could be added in a combinatorial approach.

The significance of interactions between N enrichment and other environmental factors, such as global warming, hydrological changes, P enrichment, or fire, in plant invasions could be investigated by means of full-factorial experiments. These may provide insight in the relative importance of these interactions, and how they may be related to regional or ecosystem characteristics. For example, further study is needed of whether N enrichment as such leads to a real loss of species, since many endangered species appear to persist under P-limited conditions, and hence seem particularly sensitive to P, rather than to N enrichment (Wassen et al. 2005).

In conclusion, our literature review on the impact of N deposition on plant invasibility of ecosystems supports the hypothesis formulated by Dukes and Mooney (1999): depending on the occurrence of species responsive to nitrogen in the native flora, the impact of invaders may be minimal, e.g., in European temperate forests. However, if most responsive species are aliens, e.g., in many parts of North America, then the impact of N deposition (or other forms of eutrophication) will be much larger. Thus, N deposition effects on invasibility will certainly vary from region to region.

References

Adler P, D'Antonio CM, Tunison JT (1998) Understorey succession following a dieback of *Myrica faya* in Hawaii Volcanoes National Park. Pacific Sci 52:69–78

Aerts R, Berendse F (1988) The effect of increased nutrient availability on vegetation dynamics in wet heathlands. Vegetatio 76:63–69

Allen EB (2004) Restoration of *Artemisia* shrublands invaded by exotic annual *Bromus*: a comparison between southern California and the intermountain region. USDA For Ser Proc 31:9–17

Bashkin M, Stohlgren TJ, Otsuki Y, Lee M, Evangelista P, Belnap J (2003) Soil characteristics and plant exotic species invasions in the Grand Staircase – Escalante National Monument, Utah, USA. Appl Soil Ecol 22:67–77

BfN (2006) Floraweb. Bundesamt für Naturschutz, Bonn, http://floraweb.de/neoflora

Binkley D (1997) Bioassays of the influence of *Eucalyptus saligna* and *Albizia falcataria* on soil nutrient supply and limitation. For Ecol Manage 91:229–234

Binkley D, Senock R, Cromack K Jr (2003) Phosphorus limitation on nitrogen fixation by *Facaltaria* seedlings. For Ecol Manage 186:171–176

Bobbink R, Hornung M, Roelofs JGM (1998) The effects of air-borne nitrogen pollutants on species diversity in natural and semi-natural European vegetation. J Ecol 86:717–738

Boring LR, Swank WT (1984) Symbiotic nitrogen fixation in regenerating black locust (*Robinia pseudoacacia* L.) stands. Forest Sci 30:528–537

Brooks ML (2003) Effects of increased soil nitrogen on the dominance of alien annual plants in the Mojave Desert. J Appl Ecol 40:344–353

Brooks ML, D'Antonio CM, Richardson DM, Grace JB, Keeley JE, DiTomaso JM, Hobbs RJ, Pellant M, Pyke D (2004) Effects of invasive alien plants on fire regimes. BioScience 54:677–688

Daehler CC (2003) Performance comparisons of co-occurring native and alien invasive plants: implications for conservation and restoration. Annu Rev Ecol Evol Syst 34:183–211

D'Antonio CM, Vitousek PM (1992) Biological invasions by exotic grasses, the grass/fire cycle, and global change. Annu Rev Ecol Evol Syst 23:63–87

D'Antonio CM, Dudley TL, Mack M (1999) Disturbance and biological invasions: direct effects and feedbacks. In: Walker LR (ed) Ecosystems of disturbed ground. Elsevier, Amsterdam, pp 413–452

Davis MA, Grime JP, Thompson K (2000) Fluctuating resources in plant communities: a general theory of invasibility. J Ecol 88:528–534

Decruyenaere JG, Holt JS (2005) Ramet demography of a clonal invader, *Arundo donax* (Poaceae), in Southern California. Plant Soil 277:41–52

Dukes JS, Mooney HA (1999) Does global change increase the success of biological invaders? Trends Ecol Evol 14:135–139

EEA (2005) The European environment. State and outlook 2005. European Environment Agency, Copenhagen

Egunjobi JK (1969) Dry matter and nitrogen accumulation in secondary successions involving gorse (*Ulex europaeus* L.) and associated shrubs and trees. N Z J Sci 12:175–193

Ehrenfeld JG (2003) Effects of exotic plant invasions on soil nutrient cycling processes. Ecosystems 6:503–523

Ehrenfeld JG, Scott N (2001) Invasive species and the soil: effects on organisms and ecosystem processes. Ecol Appl 11:1259–1260

Fenn ME, Baron JS, Allen EB, Rueth HM, Nydick KR, Geiser L, Bowman WD, Sickman JO, Meixner T, Johnson DW, Neitlich P (2003) Ecological effects of nitrogen deposition in the western United States. BioScience 53:404–420

Galloway JN, Dentener FJ, Capone DG, Boyer EW, Howarth RW, Seitzinger SP, Asner GP, Cleveland CC, Green PA, Holland EA, Karl DM, Michaels AF, Porter JH, Townsend AR, Vörösmarty CJ (2004) Nitrogen cycles: past, present, and future. Biogeochemistry 70:153

Garcia-Montiel D, Binkley D (1998) Effect of *Eucalyptus saligna* and *Albizia falcataria* on soil processes and nitrogen supply in Hawaii. Oecologia 113:547–556

Gordon DR (1998) Effects of invasive, non-indigenous plant species on ecosystem processes: lessons from Florida. Ecol Appl 8:975–989

Hobbs RJ, Gulmon SL, Hobbs VJ, Mooney HA (1988) Effects of fertiliser addition and subsequent gopher disturbance on a serpentine annual grassland community. Oecologia 75:291–295

Huenneke LF, Hamburg SP, Koide R, Mooney HA, Vitousek PM (1990) Effects of soil resources on plant invasion and community structure in Californian serpentine grassland. Ecology 71:478–491

Johnson D, Leake JR, Lee JA (1999) The effects of quantity and duration of simulated pollutant nitrogen deposition on root-surface phosphatase activities in calcareous and acid grasslands: a bioassay approach. New Phytol 141:433–442

Kowarik I (2003) Human agency in biological invasions: secondary releases foster naturalisation and population expansion of alien plant species. Biol Invasions 5:293–312

Levine JM, Vila M, D'Antonio CM, Dukes JS, Grigulis K, Lavorel S (2003) Mechanisms underlying the impacts of exotic plant invasions. Proc R Soc Lond B 270:775–781

Ley RE, D'Antonio CM (1998) Exotic grass invasion alters potential rates of N fixation in Hawaiian woodlands. Oecologia 113:179–87

Lonsdale WM (1999) Global patterns of plant invasions and the concept of invasibility. Ecology 80:1522–1536

Mack MC, D'Antonio CM (2003) Exotic grasses alter controls over soil nitrogen dynamics in a Hawaiian woodland. Ecol Appl 13:154–166

Mack RN, Simberloff D, Lonsdale WM, Evans H, Clout M, Bazzaz FA (2000) Biotic invasions: causes, epidemiology, global consequences, and control. Ecol Appl 10:689–710

Mack MC, D'Antonio CM, Ley RE (2001) Alteration of ecosystem nitrogen dynamics by exotic plants: a case of C_4 grasses in Hawaii. Ecol Appl 11:1323–1335

Maron JL, Jefferies RL (1999) Bush lupine mortality, altered resource availability, and alternative vegetation states. Ecology 80:443–454

Milchunas DG, Lauenroth WK (1995) Inertia in plant community structure: state of changes after cessation of nutrient-enrichment stress. Ecol Appl 5:452–458

Mooney HA, Hobbs RJ (eds) (2000) Invasive species in a changing world. Island Press, Washington, DC

Nadelhoffer KJ (2001) The impacts of nitrogen deposition on forest ecosystems. In: Follett RF, Hatfield JL (eds) Nitrogen in the environment: sources, problems and management. Elsevier, Amsterdam, pp 311–331

NADP (2006) National Atmospheric Deposition Program (NRSP-3). NADP Program Office, Champaign, http://nadp.sws.uiuc.edu

Olde Venterink H, Pieterse NM, Belgers JDM, Wassen MJ, de Ruiter PC (2002) N, P and K budgets along nutrient availability and productivity gradients in wetlands. Ecol Appl 12:1010–1026

Pickart AJ, Miller LM, Duebendorfer TE (1998) Yellowbush lupine invasion in northern California coastal dunes. I. Ecological impacts and manual restoration techniques. Restor Ecol 6:59–68

Pyšek P, Prach K, Smilauer P (1995) Relating invasion success to plant traits: an analysis of the Czech alien flora. In: Pyšek P, Prach K, Rejmánek M, Wade M (eds) Plant invasions – general aspects and special problems. Academic Publishing, Amsterdam, pp 39–60

Rejmanek M, Richardson DM (1996) What attributes make some plant species more invasive? Ecology 77:1655–1661

Rice SK, Westerman B, Federici R (2004) Impacts of the exotic, nitrogen-fixing black locust (*Robinia pseudoacacia*) on nitrogen-cycling in a pine–oak ecosystem. Plant Ecol 174:97–107

Richardson DM, Allsop N, D'Antonio CM, Milton SJ, Rejmanek M (2000) Plant invasions – the role of mutualisms. Biol Rev 75:65–93

Sala OE, Chapin FS, Armesto JJ, Berlow E, Bloomfield J, Dirzo R, Huber-Sanwald E, Huenneke LF, Jackson RB, Kinzig A, Leemans R, Lodge DM, Mooney HA, Oesterheld M, Poff NL, Sykes MT, Walker BH, Walker M, Wall DH (2000) Biodiversity – global biodiversity scenarios for the year 2100. Science 287:1770–1774

Scherer-Lorenzen M, Elend A, Nöllert S, Schulze E-D (2000) Plant invasions in Germany: general aspects and impact of nitrogen deposition. In: Mooney HA, Hobbs RJ (eds) Invasive species in a changing world. Island Press, Washington, DC, pp 351–368

Schulze E-D (ed) (2000) Carbon and nitrogen cycling in European forest ecosystems. Springer, Berlin Heidelberg New York

Simberloff D, Von Holle B (1999) Positive interactions of nonindigenous species: invasional meltdown? Biol Invasions 1:21–32

Sprent JI (2001) Nodulation in Legumes. Royal Botanic Gardens, Kew

Stock WD, Wienand KT, Baker AC (1995) Impacts of invading N_2-fixing *Acacia* species on patterns of nutrient cycling in two Cape ecosystems: evidence from soil incubation studies and ^{15}N natural abundance values. Oecologia 101:375–382

Tomassen HBM, Smolders AJP, Limpens J, Lamers LPM, Roelofs JGM (2004) Expansion of invasive species on ombrotrophic bogs: desiccation or high N deposition. J Appl Ecol 41:139–150

USDA (2006) The PLANTS database, version 3.5. National Plant Data Center, Baton Rouge, http://plants.usda.gov

Versfeld DB, van Wilgen BS (1986) Impact of woody aliens on ecosystem properties. In: Macdonald IAW, Kruger FJ, Ferrar AA (eds) The ecology and management of biological invasions in southern Africa. Oxford University Press, Cape Town, pp 239–246

Vitousek PM (2004) Nutrient cycling and limitation; Hawai'i as a model system. Princeton University Press, Princeton

Vitousek PM, Howarth RW (1991) Nitrogen limitation on land and in the sea: how can it occur? Biogeochemistry 13:87–115

Vitousek PM, Walker LR (1989) Biological invasion by *Myrica faya* in Hawaii: plant demography, nitrogen fixation, ecosystem effects. Ecol Monogr 59:247–265

Vitousek PM, Walker LR, Whiteaker LD, Mueller-Dombois D, Matson PA (1987) Biological invasion by *Myrica faya* alters ecosystem development in Hawaii. Science 238:802–804

Vitousek PM, Aber JD, Howarth RW, Likens GE, Matson PA, Schindler DW, Schlesinger WH, Tilman DG (1997) Human alteration of the global nitrogen cycle: sources and consequences. Ecol Appl 7:737–750

Vitousek PM, Cassman K, Cleveland C, Crews T, Field CB, Grimm NB, Howarth RW, Marino R, Martinelli L, Rastetter EB, Sprent JI (2002) Towards an ecological understanding of biological nitrogen fixation. Biogeochemistry 57:1–45

Walker LR, Vitousek PM (1991) An invader alters germination and growth of a native dominant tree in Hawaii. Ecology 72:1449–1455

Wamelink GWW, de Jong JJ, Van Dobben HF, Van Wijk MN (2005) Additional costs of nature management caused by deposition. Ecol Econ 52:437–451

Wardle DA, Nicholson KS, Ahmed M, Rahman A (1994) Interference effects of the invasive plant *Carduus nutans* L. against the nitrogen fixation ability of *Trifolium repens* L. Plant Soil 163:287–297

Wassen MJ, Olde Venterink H, Lapshina ED, Tanneberger F (2005) Endangered plants persist under phosphorus limitation. Nature 437:547–550

Wedin D, Tilman D (1996) Influence of nitrogen loading and species composition on the carbon balance of grasslands. Science 274:1720–1723

Weiss SB (1999) Cars, cows, and checkerspot butterflies: nitrogen deposition and management of nutrient-poor grasslands for a threatened species. Conserv Biol 13:1476–1486

Wiser SK, Allen RB, Clinton PW, Platt KH (1998) Community structure and forest invasion by an exotic herb over 23 years. Ecology 79:2071–2081

Witkowski ETF, Mitchell DT (1987) Variations in soil phosphorus in the fynbos biome, South Africa. J Ecol 75:1159–1171

Wolf JJ, Beatty SW, Seastedt TR (2004) Soil characteristics of Rocky Mountain National Park grasslands invaded by *Melilotus officinalis* and *M. alba*. J Biogeogr 31:415–424

WRI (2000) World resources 2000–2001: people and ecosystems. World Resources Institute, Washington, DC

Yelenik SG, Stock WD, Richardson DM (2004) Ecosystem level impacts of invasive *Acacia saligna* in the South African fynbos. Restor Ecol 12:44–51

11 From Ecosystem Invasibility to Local, Regional and Global Patterns of Invasive Species

Ingolf Kühn and Stefan Klotz

11.1 Introduction

Distribution patterns of species are a consequence of long evolutionary histories. Biogeographical barriers have resulted in separate developments of biota with specific adaptations to their native ecosystems and associated environmental conditions. Especially during the past centuries, human activity has helped species to surmount these natural barriers, so that present-day patterns of alien species result from natural drivers as well as man's history of land exploitation and construction of traffic routes. Humans created new pathways of species introductions (Chaps. 2 and 3), and also new habitats. Introduced species were thus able to invade both (semi-)natural and human-made habitats, which differ considerably in their proportion of alien species (e.g. Chytrý et al. 2005).

With the arrival of aliens in a novel environment, interactions between resident species are disrupted, and interactions among resident and invading species have to be newly established. Though unplanned and mostly unwanted, biological invasions are considered to be an important ecological experiment, well suited for ecological studies. Because many aspects are better known in alien species than in native ones (e.g. time of isolation from the original gene pool, and we have replications by introductions into multiple localities), species invasions provide a unique opportunity to test general ecological theories as an alternative approach to focused experimental manipulations which might be more constrained by time, space, research budgets, etc. (Rice and Sax 2005).

Here, we employ this approach by using habitat availability, and the fact that habitats differ in their proportion of alien species across a multitude of ecosystems and spatial scales to investigate the question of spatial patterns of alien species distribution, and the consequences of invasions for communities and ecosystems. Many of these ideas were outlined initially by Elton (1958),

Ecological Studies, Vol. 193
W. Nentwig (Ed.)
Biological Invasions

who provided a set of key observations and hypotheses in this field of research. In this chapter, we will discuss the invasibility of ecosystems as well as local, regional and global patterns of alien species occurrence as a tool to understand two different niche theories and general biogeographical patterns. In this context, alien species are here defined as exotic (non-native) species which have been intentionally or unintentionally introduced into an area after the discovery of the Americas by Columbus.

11.2 Background

In his famous book 'The ecology of invasions by animals and plants', Elton (1958) laid the foundation for modern invasion biology. He describes why relatively simple communities are "more easily upset than richer ones; that is, more subject to destructive oscillations in populations, especially of animals, and more vulnerable to invasions" (Elton 1958: 145). The idea that species-rich communities are more resistant to invasions than are species-poor communities has challenged generations of ecologists.

Particularly important for invasion biology is the question whether ecological communities are saturated or unsaturated, and whether the more invaded ones are less saturated. This aspect is directly coupled with that of the ecological niche. The concept of the 'ecological niche' is considered to be among the most important in ecology (Cherrett 1989). However, there are essentially at least two different niche concepts, based on the ideas of Grinnell (1928), on the one hand, and Hutchinson (1957) on the other. Grinnell (1928) used the term 'niche' to characterise species-specific requirements. Besides habitat, these can be nutrients, mating places or other resources and requisites associated with a species' occurrence (Brandl et al. 2001). Grinnell (1928) defines a niche as an ..."ultimate distributional unit, within which each species is held by its structural and instinctive limitations, these being subject only to exceedingly slow modification down through time". According to this concept, the niche has an autecological character – any community to which a species belongs is of less importance. However, even Grinnell had pondered on whether all niches are necessarily occupied within a community (Grinnell and Swarth 1913). Grinnell's theory was mainly used to describe and to understand changes in species distributions in relation to environmental variability. Jäger (1988) directly applied Grinnell's idea to invasion problems. He characterised the introduced range of a species by the properties of its native range. This idea also forms a basis for species distribution modelling using climate envelopes.

Hutchinson (1957) introduced a new concept to niche theory. He stated that basic autecological (environmental) factors are not the only ones determining niche dimensions. The role of a species in its community is an important additional factor influencing the presence or absence of a species at a given site. The difference between the Grinnell and the Hutchinson approach can be viewed as that between an address (Grinnell) and a profession (Hutchinson). Hutchinson (1957) described the niche as an n-dimensional hyper-volume characterised by several niche dimensions such as nutrients and other resources. The fitness of a species may vary along these niche axes, which may lead to a reduction in available niche space. He thus distinguished between the fundamental niche and the realised niche. The former covers all potential resources and requisites, the latter only the resources and requisites available to a species within a given community.

The main differences between these two concepts are

1. Grinnell highlights the distributional range (geography) of a species to characterise the niche (regional concept). Hutchinson stresses the use of resources within a given community (local concept).
2. Following Hutchinson, the niche of a species depends on other species in a community. Grinnell's concept emphasises the fundamental niche, Hutchinson's the realised niche.
3. Hutchinson defined the niche in terms of species characteristics and community structure. By definition, there exists no vacant niche within a community.

One conceptual problem with Hutchinson's niche concept is the idea that 'vacant niches' do not occur. Indeed, it is obvious that, in some systems, there may be possibilities for species to exist which, due to evolutionary constraints, are simply not made use of. A prominent example are large herbivores represented in African savannahs by ungulates, in Australian grasslands by marsupials but which were absent on the pre-Columbian South American pampas.

In addition, both these classical concepts consider species as static entities, and assume that communities are saturated and therefore in equilibrium (i.e. species gains are compensated by losses). However, these assumptions are both usually not met. Species evolve continuously, and microevolution can occur over short time periods. Furthermore, an ecosystem is rarely in equilibrium, and this for several reasons: for example, systems in temperate regions may not have reached their full set of species after the last glaciation (i.e. they are unsaturated), the climate changes continuously, systems are disturbed more or less frequently by natural or human processes, and propagule pressure ensures a steady influx of new species.

Nevertheless, these two niche concepts facilitate a wider understanding of scale-dependent processes in biological invasions, as their underlying processes work at different scales. Grinnell's concept is more regional and describes a species' potential impact whereas Hutchinson's concept is more local and describes a species' existing requirements within a community.

11.3 Case Studies on Ecosystem Invasibility

There are several case studies shedding light on the patterns and processes of biological invasions. Ecosystems can be invaded if there is a 'vacant niche', i.e. resources which are not utilised. A good example for this is the Central European aquatic mammal community (Brandl et al. 2001): the water vole (*Arvicola terrestris*) and European beaver (*Castor fiber*) are native herbivores of these inland waters, where humans successfully introduced the North American muskrat (*Ondatra zibethica*) and South American coypu (*Myocastor coypus*). Why was this possible – was there a 'vacant niche'? It has been suggested that, in vertebrates, resource use correlates with body mass (Brown 1975). The difference in resource use between the beaver (over 20 kg body mass) and water vole (ca. 0.1 kg) is substantial: a beaver feeds on trees, a water vole on grass. Between these two extremes are some resources which are evidently unutilised and thus available for the muskrat (ca. 1 kg body mass) and coypu (ca. 7–8 kg). This shows that increasing species numbers can lead to more complete resource use. Therefore, more species would mean less 'vacancies'.

This example shows why we may expect a negative relationship between species richness and invasion resistance, but leaves room for other explanations as well. Indeed, it is useful to review some more patterns which (1) derive from a larger sample size, (2) are robust to statistical testing and (3) encompass different groups of organisms and different spatial scales. Gido (in Brown and Lomolino 1998) found a significant negative correlation between native and introduced fish species in North American rivers. A similar observation was made by Case and Bolger (1991) for reptiles on islands. They reported that, on islands with only few native reptile species, there were more invasive reptile species than on islands with many native reptile species. However, in both these fish and reptile datasets, native–invasive relationships were not linear, species-poor communities showing a higher variability in the number of invading species. This means that alien species which can potentially invade do not necessarily do so. An analysis of macrozoobenthos of German waterways again showed a significant negative correlation between native and alien species numbers (reported by Brandl et al. 2001) but the proportion of alien species decreased with increasing number of native species (see also Chap. 15). By contrast, Welter-Schultes and Williams (1999) found no significant relationship between species richness of native and alien species for molluscs of the Aegean islands. For plant species, most of the published studies report that species-rich habitats were also more strongly invaded, i.e. the rich became richer (Stohlgren et al. 2003). As an example, Stadler et al. (2000) analysed native and alien tree species richness in Kenya, reporting a positive correlation between native and alien tree species numbers. In a more complex approach, Chytrý et al. (2005) investigated over 20,000 vegetation plots, ranging in size from 1–100 m^2 in 32 habitats of the Czech Republic. They found no significant relationship between native and alien plant species numbers when analysing across all habitats but, within habitat types, mostly positive relationships were recorded.

Using both an observational and an experimental approach, Levine (2000) found two different patterns in native–alien relationships in his analysis of riverine Californian plant communities. He investigated tussocks of the sedge *Carex nudata*, which can host more than 60 native and three alien plant species (*Agrostis stolonifera, Plantago major, Cirsium arvense*). As a first approach, he counted the proportion of native tussocks in which the alien species occurred (i.e. invader incidence), finding a significant positive relationship with plant species richness (excluding the invaders). In a second approach, he manipulated the tussocks; specifically, he removed all species from randomly selected tussocks and assigned these to one of five species richness treatments. Then, he added 200 seeds of each alien plant species to the surface of the experimental tussocks. Here, with increasing native species richness, the number of alien seeds which germinated and survived two growing seasons decreased. Thus, in a controlled experiment but in a natural setting, Levine (2000) was able to separate effects of species richness from covarying effects of natural heterogeneity.

The examples above do not show a clear pattern. It therefore is necessary to have a closer look at the different factors driving the invasibility of habitats and ecosystems. This can help to understand which processes are relevant for invasibility, and to decide which niche concepts are best able to explain small- to large-scale invasion patterns.

11.4 Scale Dependence of Invasibility and the Importance of Environmental Factors

On a local scale, the main factors identified to date in explaining habitat invasibility are evolutionary history, disturbance, propagule pressure, abiotic stress, and community structure (Alpert et al. 2000).

Local patterns of invasibility differ strongly around the globe. For example, Europe is less affected by biological invasions whereas regions of North America, Australia and especially oceanic islands can be heavily affected. This can be explained by different evolutionary histories. It is argued that species and habitats which have shared a long co-evolutionary history with human land uses (such as agriculture) are better (pre-)adapted to biological invasion. Thus, these species had already been selected for their tolerance to disturbance. Therefore, regions such as the Mediterranean, with a long history in agriculture, may be less prone to biological invasions than others (di Castri 1990). By contrast, natural grasslands such as the North American prairies evolved under a regime of only little disturbance by native grazers and, therefore, under today's strong grazing pressure, are particularly susceptible to invasions (Mack and D'Antonio 1998). Likewise, islands which have been most strongly affected by invasions were often least disturbed before human colonisation. Island biota are evolutionary distinct from mainland biota, and have evolved very specific community structures and species traits. For one, oceanic islands are considered to host habitats showing relaxed selection for competitive ability (Loope and Mueller-Dombois 1989). In the presence of invading species, specific interactions are therefore more heavily affected in island biota than in mainland biota, which makes islands much more susceptible to invasions than mainland areas.

In numerous studies, disturbance is considered to be a key factor in biological invasions, but one which can act in several ways: disturbance can remove native competitors, facilitate a flush of surplus resources (Davis et al. 2000), such as light or nutrients, or can create completely new habitats. In many cases, therefore, increasing disturbance promotes invasibility. However, invasions can occur also without disturbance, and there are cases when suppressing disturbance can even increase invasions. Thus, it can be hypothesised that it is not necessarily disturbance per se which increases invasibility but rather the deviation from a typical disturbance regime (Alpert et al. 2000).

Lonsdale (1999) argues that invasion patterns are a function not only of habitat invasibility – as an idiosyncratic characteristic – but also of different propagule pressures (see also Williamson 1996). Therefore, the number of exotic species would be a function of the frequency and magnitude of intentional and unintentional introduction events, and of the ability of these species to successfully reproduce. The importance of propagule pressure for invasion patterns has been invoked in several analyses (e.g. Pyšek et al. 2003; Thuiller et al. 2005). Locally, propagule pressure can explain exotic species cover better than can environmental factors (Rouget and Richardson 2003). For birds, it is known that invasion success is higher when more species have been introduced into a target region (Duncan et al. 2003).

As another factor, environmental stress is hypothesised to be important for ecosystem invasibility. Stress can be caused by specific factors which are limiting for plant growth, such as the availability of nutrients, water and light, by the presence of toxins (incl. saline soils) or by other extreme conditions. The majority of studies found that ecosystem invasibility decreases with increasing stress or that invasion increases when limiting resources, such as nutrients, are provided (Alpert et al. 2000). Therefore, adding nutrients such as nitrogen or phosphorus can raise invasibility and promote a smaller number of faster-growing species. Depending on interactions with other factors, a complete shift of community structure was observed in several directions (Alpert et al. 2000 and references therein). For one, there is evidence of some interaction between stress and disturbance. When stress is low (i.e. resource availability is high), only little change in a typical disturbance regime is needed to facilitate invasions whereas when stress is high, a high deviation from the typical disturbance regime is needed (Alpert et al. 2000).

In terms of specific community structure, ecosystem invasibility involves several processes. Still, the basic concept behind this idea is that of the niche. The realised niche of a species may be altered by specific members of a community within the potential given by the fundamental niche. One classical example is Ellenberg's (1953) experiment, where he showed that several grass species (amongst others, *Bromus erectus*, *Arrhenatherum elatius* and *Alopecurus pratensis*) had the same optimal growth along a water gradient in single-species plots but displayed a considerable shift in multi-species plots (e.g. *Bromus erectus* towards dryer sites and *Alpoecurus pratensis* towards moister sites). Also, different members of a community can have very strong interactions which may either inhibit invasions (e.g. due to the depletion of resources) or facilitate these (e.g. nitrogen-fixing acacias, Holmes and Cowling 1997; see also Chap. 10).

Besides effects on resources, community structure can also determine the availability of natural enemies, thus creating a 'natural enemy escape opportunity' (Shea and Chesson 2002). This is explicitly explained by two important hypotheses – the enemy release hypothesis (ERH; Keane and Crawley 2002), and the evolution of increased competitive ability (EICA) hypothesis (Blossey and Nötzold 1995). The former states that plant species in their introduced range should experience a decrease in regulation by herbivores and other natural enemies when their specific enemies are absent. This would result in higher abundances and, thus, wider distributions of alien species in their introduced range. The latter hypothesis states that introduced species do not need to invest resources in the defence against enemies. They can therefore invest these resources in the evolution of increased competitive ability (Blossey and Nötzold 1995). There are examples both corroborating and rejecting these hypotheses (Chap. 6).

Important – though long overlooked – interactions exist between soil micro-organisms and macro-organisms. These seem to play an important role in the invasibility of ecosystems. Callaway and Aschehoug (2000) found that *Centaurea diffusa*, a noxious alien weed in North America, had much stronger negative effects on grass species from North America than on closely related grass species from communities to which *Centaurea* is native. On sterile soils, these differences disappeared. They argue that *Centaurea*'s advantage against North American species appears to be due to differences in the effects of its root exudates, indicating that micro-organisms are responsible for the invasion success of the species in its introduced range. More recently, Klironomos (2002) was able to show several interactions between soil micro-organisms and plant species. Rare native plant species cultivated in their own soils were smaller than those cultivated in soils of other species. Invasive species, on the other hand, grew better (cf. relative increase in growth) in their home soils than in soils of other species. Klironomos also found that rare native plant species accumulated species-specific pathogens quickly in their own soils and, therefore, maintained low densities. By contrast, invasive species benefited from interactions with mycorrhizal fungi.

The examples above show that biodiversity plays a major role in community structure and community susceptibility to biological invasions. Indeed, following the ideas of Elton (1958), biological diversity is considered to be a key element of invasion resistance. In a recent review, Levine at al. (2002) showed that in most experimentally assembled systems, species diversity enhances invasion resistance whereas those studies examining natural invasion patterns more often reported positive correlations between natural species diversity and invasion, rather than negative ones. This apparent contradiction has been widely discussed in the literature, and has spawned some idiosyncratic views of invasion processes and invaded systems. Nevertheless, this contradiction can be resolved within a general conceptual framework by distinguishing between local factors affecting biodiversity and those factors associated with diversity patterns across communities, i.e. on a larger scale (Shea and Chesson 2002; Levine et al. 2002). In the model of Shea and Chesson (2002), negative relationships between alien and native species numbers can be observed in each case for groups of locations where a given group shows little variation in environmental factors. When these data are combined across several groups together spanning highly variable environmental factors, the result is a large-scale positive relationship (Fig. 11.1). Levine at al. (2002) consider that small-scale diversity as such causes resistance against

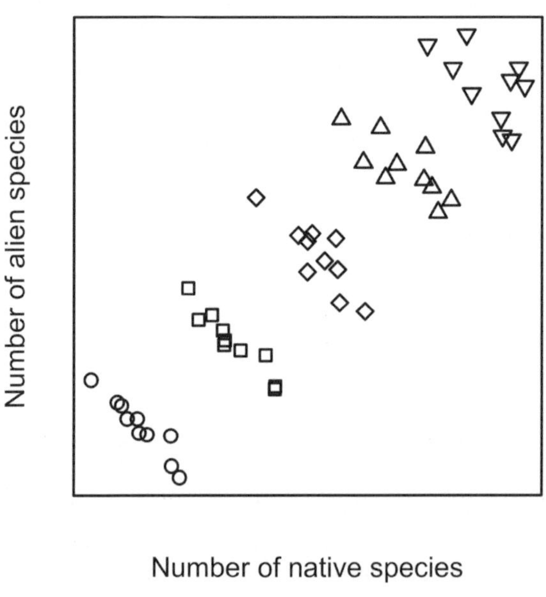

Fig. 11.1 Hypothesised relationship between native and alien species richness at different scales. At a local scale with little environmental variation within communities, a negative relationship between alien and native species richness can be observed due to small-scale neighbourhood processes such as competition. Across these communities, environmental heterogeneity increases and affects alien and native species richness in similar ways, through covarying factors (after Shea and Chesson 2002, using randomly generated data)

biological invasions. However, this could also be a consequence of small-scale ecological processes such as competition (even if it were simply competition for space). The positive correlation between diversity and invasion success across communities would result from the combined effects of these local factors and additional covarying factors (Levine et al. 2002). The latter may act at larger scales, such as gradients in disturbance regime, climate, soil properties, and dispersal (propagule pressure). Therefore, such larger-scale processes drive not only native species richness but, to a large extent, also alien species richness (Kühn et al. 2003), and can dominate over small-scale species interactions or neighbourhood effects. In Fig. 11.2, we present a causal framework to summarise these different scale-dependent processes, acting in the same direction on both alien and native species richness at larger scales but in opposite directions through well-documented local-scale neighbourhood processes such as competition.

Within the framework, we combine ideas of Brandl et al. (2001), Levine at al. (2002) and Shea and Chesson (2002) which can reconcile the niche concepts of Grinnell (1928) and Hutchinson (1957) discussed above, and the seemingly contrasting patterns of alien and native species richness on local and larger scales. Large-scale geographic gradients act mainly on more regional processes, especially as constraints for specific climates, soils, habitats, etc. Due to biogeographic constraints, however, there are direct influences of large-scale gradients on species distributions, e.g. through individual

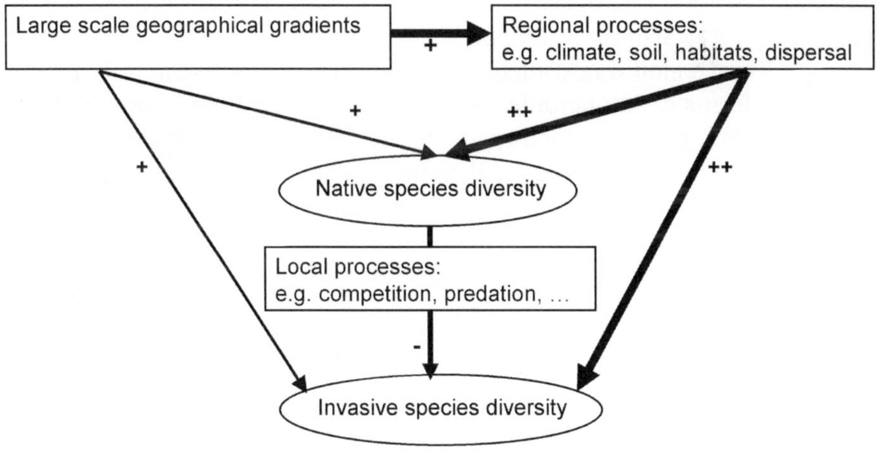

Fig. 11.2 A conceptual framework to reconcile small-scale neighbourhood processes (as explained by Hutchinson's (1928) niche concept) and large-scale environmental processes (as explained by Grinnell's (1957) niche concept). *Thick arrows* represent strong effects, *thin arrows* weak effects. *Plus symbols* indicate effects in the same direction (either both positive or both negative), and the *minus symbol* indicates effects in opposite directions

evolutionary histories, dispersal, movements or recolonisation after the last glaciation. These processes, of course, work at more local scales but nevertheless are constrained by large-scale processes which may hinder natural new species occurrences far outside a species' range. On a regional scale, those processes determining native species richness, such as resource availability (e.g. temperature, water, nutrients, habitats) act also on alien species diversity, these being the ones relevant in Grinnell's (1928) niche concept. It can therefore be expected that species richness patterns of native and alien species are positively correlated at larger scales. It is only at a very local scale that neighbourhood effects and other local-scale processes, inferred by Hutchinson (1957), come into play, so that native species richness can increase resistance to biotic invasions. These local patterns, however, are often much more weakly expressed than are the larger-scale patterns.

11.5 Local, Regional and Global Patterns

As described above, the relationship between species richness and ecosystem invasibility is scale-dependent. It should therefore be possible to recognise these patterns, and some possible turning point, in a nested analysis. We analyzed 30 plots, each 1 × 20 m in size, in tall herb communities along the river Elbe in Saxony, Germany, in 2002. Within each of these randomly selected plots, we used five point estimates by counting the number of native and alien species which touched a stake regularly put to the ground. We also noted all species present within each of the 30 plots. A species inventory at a landscape scale for the Elbe River region was available from the atlas of Hardtke and Ihl (2000), with a resolution of 5′ longitude and 3′ latitude (i.e. ca. 30 km^2). A major axis regression on log-transformed species numbers clearly exhibited a negative relationship at the point scale, no relationship at the 20 m^2 scale, and a positive relationship at the ca. 30 km^2 scale (Fig. 11.3). Thus, we were able to demonstrate the scale dependence of the relationship between native and alien species number for a single observatory frame within a restricted region.

Our study did not show any clear relationship at a resolution of 20 m^2. However, other studies have reported significant positive relationships at even smaller scales. Plots with sizes of 1 m^2 showed weak positive relationships (Stohlgren et al. 2003) or significant positive and negative relationships between alien and native plant species richness in grasslands of the USA (Stohlgren et al. 1999) whereas only positive correlations were observed by, for example, Sax (2002) at all scales between 1 and 400 m^2 in scrub communities of Chile and California.

Within Germany, we were able to show that the positive relationship between alien and native plant species was caused by a similar set of environmental factors, thus corroborating the notion of common large-scale

Fig. 11.3a–c Relationships between native species richness and alien species richness at three different, nested scales along the river Elbe in Saxony, Germany. **a, b** Stumpf, Klotz and Kühn (unpublished data; overlaying data points have been slightly shifted for better visualisation). **c** Hardtke and Ihl (2000)

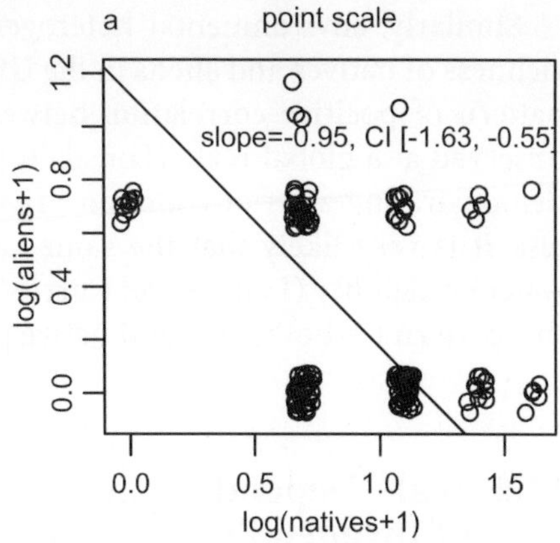

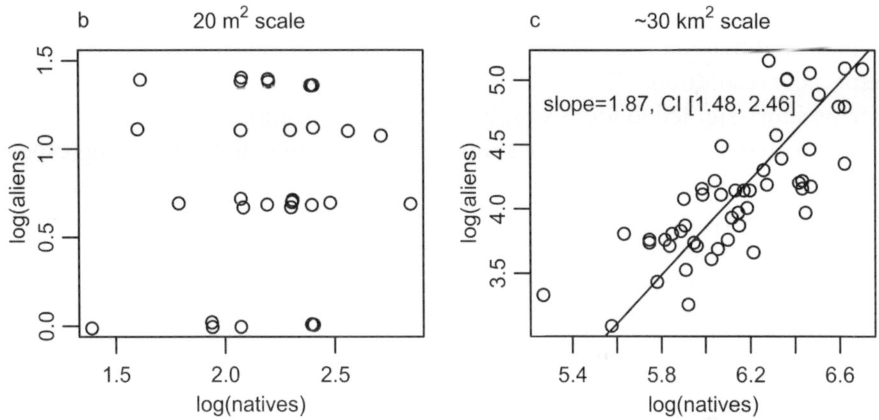

environmental factors driving both native as well as alien plant species richness at all but neighbourhood scales. Analysing 40 randomly selected plots of size 250 × 250 m in an urban and an agricultural landscape near Halle, Wania et al. (2006) confirmed the expected positive correlation, and showed that especially habitat diversity was able to explain both native and alien plant species richness. At a slightly larger scale, 5′ longitude and 3′ latitude in the district of Dessau (central Germany), Deutschewitz et al. (2003) explained increases in native and alien plant richness in terms of moderate levels of natural and/or anthropogenic disturbances, coupled with high levels of habitat and structural heterogeneity in these urban, riverine, and small-scale rural ecosystems. For Germany (at a scale of 10′ longitude and 6′ latitude, i.e. ca. 130 km²), the diversity of geological substrates proved to be the best predictor for both alien and native plant species richness (Kühn et al. 2003). Nevertheless, native plant species richness was further explained by other natural parameters whereas alien plant species richness was additionally explained by urban land cover.

Similarly, environmental heterogeneity was able to account for species richness of natives and aliens in the USA (Stohlgren et al. 2006). Although this pattern of positive correlation between native and exotic species was also observed at a global scale (Lonsdale 1999), we are not aware of any analysis demonstrating a set of common drivers behind this relationship. Nevertheless, it is very likely that the same variables, i.e. energy (temperature) and water availability (Francis and Currie 2003), are able to at least largely explain this pattern for both alien and native plant species.

11.6 Scale-Dependent Consequences for Biodiversity of Invaded Ecosystems

We showed that patterns of ecosystem invasibility changed with spatial scale, especially resolution. What will the consequences of this be for biodiversity? It seems short-sighted to focus simply on biodiversity and disregard other well-documented impacts of biological invasions involving nutrient cycling (especially by nitrogen fixers such as *Acacia* or *Myrica faya*, the fayatree, Chap. 10), water table depletion (*Acacia* or *Tamarix ramosissima*, the salt cedar), alteration of soil structure through salt accumulation (*Mesembryanthemum crystallinum*, the ice plant) or soil perturbation by digging (the feral pig *Sus scrofa domestica*), which additionally disperse seeds of alien plant species and fertilise the soil (Williamson 1996). However, it is biodiversity or rather, its elements (i.e. species) which largely drive ecosystem processes. Still, most conservation actions are concerned with species as such, not with the goods and services they provide as integral parts of an ecosystem.

We discussed several studies showing an increase of alien species at higher native species levels at larger spatial scales. At the global scale, however, alien species are considered to be among the major causes of species extinctions (e.g. Diamond 1989; Sala et al. 2000; Chaps. 13, 15, 16). This impact seems inevitable, given that global extinction rates cannot be compensated by speciation rates. At a global scale, the introduction of species into a new habitat or biogeographical region does not add to biodiversity but the loss of a single species due to this introduction decreases biodiversity. Within regions (i.e. areas which are intermediate in size between those of the globe and small study plots), Sax and Gaines (2003) show for a variety of groups of organisms and across many different parts of the world that the net gain of species due to biological invasions is higher than the loss of species. As an example from Europe, the German Red List of endangered vascular plant species (Korneck et al. 1996) lists 47 taxa as extinct and 118 as threatened. On the other hand, 470 vascular plant species are considered to be naturalised aliens (neophytes; Klotz et al. 2002), and alien species are not among the major causes for species extinctions in Germany (Korneck et al. 1998). Indeed, theoretical approaches (Rosenzweig 2001) and palaeontological records (Vermeij 1991) suggest that diversity increases after faunal mixing of formerly separated biota.

At the local scale, extirpations of rare native species have been observed. More common, however, are shifts in abundance. Sax and Gaines (2003) reported that the diversity of intact systems has often increased locally but can decrease or remain unchanged as well.

What are the consequences for formerly distinct biota? The introduction of alien species across biogeographical barriers into previously isolated regions was termed a 'new Pangaea' (Rosenzweig 2001). The idea is that formerly distinct biota become more similar, a process termed biotic homogenisation (McKinney and Lockwood 1999). Again, at a global scale, biotic homogenisation is the predictable result in the short term. At local or regional scales, however, patterns of homogenisation but also of differentiation can be observed. Which of these patterns predominates is again scale-dependent: at a local scale, differentiation seems predominant whereas, at a more regional scale, homogenisation can become important. Also, it seems that alien species from less-distant areas tend to promote homogenisation whereas species from more-distant areas tend to promote differentiation (Kühn et al. 2003; McKinney 2004; 2005; Kühn and Klotz 2006).

To better understand the consequences of biological invasions, and to be able to provide plausible scenarios for the future, it is not only necessary to study the problem at an appropriate scale. It is also necessary to use appropriate assumptions of future biodiversity in modelling ecosystem responses. However, most concepts postulate a decrease in biodiversity at all scales – actually, it would be meaningful to also examine the effects of biodiversity increase on ecosystems.

11.7 Conclusions

Patterns of ecosystem invasibility are scale-dependent. Though it seems obvious, we showed that it is indeed necessary to use the appropriate scale to analyse invasibilty. This choice of scale, however, is crucial not only in investigating relationships between biotic and abiotic factors but also for the selection of an appropriate theoretical framework and, hence, to understand a system correctly. We discussed that, at smallest scales, high native species richness enhances the invasion resistance of ecosystems through various neighbourhood interactions and processes, consistent with Hutchinson's (1957) niche concept. At larger scales, environmental heterogeneity increases and native as well as alien species richness is determined by largely the same environmental factors, and therefore covary. These larger-scale relationships can be explained by Grinnell's (1928) niche concept.

To increase the quality of future scenarios for invasive species, it is essential to fully comprehend the exact causal relationship between native and invasive species richness at relevant scales. For this, it is also crucial to use correct assumptions about the direction of future (native and invasive) species richness in a system which is also scale-dependent. To date, many analyses of invasibilty have been too descriptive or correlative, and lack a true mechanistic understanding of processes at different scales. This gap in our knowledge can probably be minimised by joint research programmes combining observational, experimental and mechanistic approaches across spatial scales.

Acknowledgements. Support by funding of the European Union within the FP 6 integrated project ALARM (GOCE-CT-2003-506675) and by the specific targeted research project DAISIE (SSPI-CT-2003-511202) is gratefully acknowledged.

References

Alpert P, Bone E, Holzapfel C (2000) Invasiveness, invasibility and the role of environmental stress in the spread of non-native plants. Perspect Plant Ecol Evol Syst 3:52–66
Blossey B, Nötzold R (1995) Evolution of increased competitive ability in invasive non-indigenous plants: a hypothesis. J Ecol 83:887–886
Brandl R, Klotz S, Stadler J, Auge H (2001) Nische, Lebensgemeinschaft und biologische Invasionen. In: Reichholf JH (ed) Gebietsfremde Arten, die Ökologie und der Naturschutz. Friedrich Pfeil, München, Rundgespräche der Kommission für Ökologie 22:81–98
Brown JH (1975) Geographical ecology of desert rodents. In: Cody ML, Diamond JM (eds) Community ecology. Harvard University Press, Cambridge, pp 314–341
Brown JH, Lomolino MV (1998) Biogeography. Sinauer, Sunderland
Callaway RM, Aschehoug ET (2000) Invasive plants versus their new and old neighbors: a mechanism for exotic invasion. Science 290:521–523
Case TJ, Bolger DT (1991) The role of introduced species in shaping the distribution and abundance of island reptiles. Evol Ecol 5:272–290
Cherrett JM (1989) Key concepts: the results of a survey of our members' opinion. In: Cherret JM (ed) Ecological concepts. Blackwell, Oxford, pp 1–161
Chytrý M, Pyšek P, Tichy L, Knollová I, Danihelka J (2005) Invasions by alien plants in the Czech Republic: a quantitative assessment across habitats. Preslia 77:339–354
Davis MA, Grime JP, Thompson K (2000) Fluctuating resources in plant communities: a general theory of invasibility. J Ecol 88:528–534
Deutschewitz K, Lausch A, Kühn I, Klotz S (2003) Native and alien plant species richness in relation to spatial heterogeneity on a regional scale in Germany. Global Ecol Biogeogr 12:299–311
di Castri F (1990) On invading species and invaded ecosystems: the interplay of historical chance and biological necessity. In: di Castri F, Hansen AJ, Debussche G (eds) Biological invasions in Europe and the Mediterranean Basin. Kluwer, Dordrecht, pp 3–16
Diamond JM (1989) The present, past and future of human-caused extinctions. Philos Trans R Soc Lond B Biol Sci 325:469–477

Duncan RP, Blackburn TM, Sol D (2003) The ecology of bird introductions. Annu Rev Ecol Evol Syst 34:71–98
Ellenberg H (1953) Physiologisches und ökologisches Verhalten derselben Pflanzenarten. Ber Dt Bot Ges 65:351–362
Elton C (1958) The ecology of invasions by animals and plants. Methuen, London
Francis AP, Currie DJ (2003) A globally consistent richness-climate relationship for angiosperms. Am Nat 161:523–536
Grinnell J (1928) Presence and absence of animals. Univ California Chronicle 30:429–450
Grinnell J, Swarth H (1913) An account of the birds and mammals of the San Jacinto area of Southern California. Univ California Publ Zool 10:197–406
Hardtke H-J, Ihl A (2000) Atlas der Farn- und Samenpflanzen Sachsens. Materialien zu Naturschutz und Landschaftspflege. Sächsisches Landesamt für Umwelt und Geologie, Dresden
Holmes PM, Cowling RM (1997) The effects of invasion by *Acacia saligna* on the guild structure and regeneration capabilities of South African fynbos shrublands. J Appl Ecol 34:317–332
Hutchinson GE (1957) Concluding remarks. Cold Spring Harbor Symp Quant Biol 22:427
Jäger EJ (1988) Möglichkeiten der Prognose synanthroper Pflanzenausbreitungen. Flora 180:101–131
Keane RM, Crawley MJ (2002) Exotic plant invasions and the enemy release hypothesis. Trends Ecol Evol 17:164–170
Klironomos JN (2002) Feedback with soil biota contributes to plant rarity and invasiveness in communities. Nature 417:67–70
Klotz S, Kühn I, Durka W (2002) BIOLFLOR – eine Datenbank zu biologisch-ökologischen Merkmalen der Gefäßpflanzen in Deutschland. Bundesamt für Naturschutz, Bonn, Schriftenreihe für Vegetationskunde 38
Korneck D, Schnittler M, Vollmer I (1996) Rote Liste der Farn- und Blütenpflanzen (Pteridophyta et Spermatophyta) Deutschlands. In: Ludwig G, Schnittler M (eds) Rote Liste gefährdeter Pflanzen Deutschlands. Bundesamt für Naturschutz, Bonn, Schriftenreihe für Vegetationskunde 28:21–187
Korneck D, Schnittler M, Klingenstein F, Ludwig G, Takla M, Bohn U, May R (1998) Warum verarmt unsere Flora? Auswertung der Roten Liste der Farn- und Blütenpflanzen Deutschlands. In: Klingenstein F, Ludwig G (eds) Ursachen des Artenrückgangs von Wildpflanzen und Möglichkeiten zur Erhaltung der Artenvielfalt. Bundesamt für Naturschutz, Bonn, Schriftenreihe für Vegetationskunde 29:299–444
Kühn I, Klotz S (2006) Urbanization and homogenization – comparing the floras of urban and rural areas in Germany. Biol Conserv 127:292–300
Kühn I, May R, Brandl R, Klotz S (2003) Plant distribution patterns in Germany – will aliens match natives? Feddes Repert 114:559–573
Levine JM (2000) Species diversity and biological invasions: relating local process to community pattern. Science 288:852–854
Levine JM, Kenedy T, Naeem S (2002) Neighborhood scale effects of species diversity on biological invasions and their relationship to community patterns. In: Loreau M, Naeem S, Inchausti P (eds) Biodiversity and ecosystem functioning: synthesis and perspectives. Oxford University Press, Oxford, pp 114–124
Lonsdale WM (1999) Global patterns of plant invasions and the concept of invasibility. Ecology 80:1522–1536
Loope LL, Mueller-Dombois D (1989) Characteristics of invaded islands. In: Drake J, diCastri F, Groves R, Kruger F, Mooney H, Rejmanek M, Williamson M (eds) Biological invasions: a global perspective. Wiley, Chichester, pp 257–280

Mack MC, D'Antonio CM (1998) Impacts of biological invasions on disturbance regimes. Trends Ecol Evol 13:195–198

McKinney ML (2004) Do exotics homogenize or differentiate communities? Roles of sampling and exotic species richness. Biol Invasions 6:495–504

McKinney ML (2005) Species introduced from nearby sources have a more homogenizing effect than species from distant sources: evidence from plants and fishes in the USA. Diversity Distrib 11:367–374

McKinney ML, Lockwood JL (1999) Biotic homogenization: a few winners replacing many losers in the next mass extinction. Trends Ecol Evol 14:450–453

Pyšek A, Pyšek P, Jarosik V, Hajek M, Wild J (2003) Diversity of native and alien plant species on rubbish dumps: effects of dump age, environmental factors and toxicity. Diversity Distrib 9:177–189

Rice WR, Sax DF (2005) Testing fundamental evolutionary questions at large spatial and demographic scales – species invasions as an underappreciated tool. In: Sax DF, Stachowicz JJ, Gaines SD (eds) Species invasions: insights into ecology, evolution, and biogeography. Sinauer, Sunderland, pp 291–308

Rosenzweig ML (2001) The four questions: what does the introduction of exotic species do to diversity? Evol Ecol Res 3:361–367

Rouget M, Richardson DM (2003) Inferring process from pattern in plant invasions: a semimechanistic model incorporating propagule pressure and environmental factors. Am Nat 162:713–724

Sala OE, Chapin FS, Armesto JJ, Berlow E, Bloomfield J, Dirzo R, Huber-Sanwald E, Huenneke LF, Jackson RB, Kinzig A, Leemans R, Lodge DM, Mooney HA, Oesterheld M, Poff NL, Sykes MT, Walker BH, Walker M, Wall DH (2000) Global biodiversity scenarios for the year 2100. Science 287:1770–1774

Sax DF (2002) Native and naturalized plant diversity are positively correlated in scrub communities of California and Chile. Diversity Distrib 8:193–210

Sax DF, Gaines SD (2003) Species diversity: from global decreases to local increases. Trends Ecol Evol 18:561–566

Shea K, Chesson P (2002) Community ecology theory as a framework for biological invasions. Trends Ecol Evol 17:170–176

Stadler J, Trefflich A, Klotz S, Brandl R (2000) Exotic plant species invade diversity hot spots: the alien flora of northwestern Kenya. Ecography 23:169–176

Stohlgren TJ, Binkley D, Chong GW, Kalkhan MA, Schell LD, Bull KA, Otsuki Y, Newman G, Bashkin M, Son Y (1999) Exotic plant species invade hot spots of native plant diversity. Ecol Monogr 69:25–46

Stohlgren TJ, Barnett DT, Kartesz JT (2003) The rich get richer: patterns of plant invasions in the United States. Frontiers Ecol Environ 1:11–14

Stohlgren TJ, Barnett D, Flather C, Fuller P, Peterjohn B, Kartesz J, Master LL (2006) Species richness and patterns of invasion in plants, birds, and fishes in the United States. Biol Invasions 8:427–447

Thuiller W, Richardson DM, Pysek P, Midgley GF, Hughes GO, Rouget M (2005) Niche-based modelling as a tool for predicting the risk of alien plant invasions at a global scale. Global Change Biol 11:2234–2250

Vermeij GJ (1991) When biotas meet – understanding biotic interchange. Science 253:1099–1104

Wania A, Kühn I, Klotz S (2006) Biodiversity patterns of plants in agricultural and urban landscapes in Central Germany – spatial gradients of species richness. Landscape Urban Planning 75:97–110

Welter-Schultes FW, Williams WR (1999) History, island area and habitat availability determine snail species richness of Aegean islands. J Biogeogr 26:239–249

Williamson M (1996) Biological invasions. Chapman & Hall, London

12 Will Climate Change Promote Alien Plant Invasions?

WILFRIED THUILLER, DAVID M. RICHARDSON, and GUY F. MIDGLEY

12.1 Introduction

Invasive alien plant species pose significant challenges to managing and maintaining indigenous biodiversity in natural ecosystems. Invasive plants can transform ecosystems by establishing viable populations with growth rates high enough to displace elements of the native biota (Rejmánek 1999) or to modify disturbance regimes (Brooks et al. 2004), thereby potentially transforming ecosystem structure and functioning (Dukes and Mooney 2004). Because the numbers of invasive plant species and the extent of invasions are increasing rapidly in many regions, concern has grown about the stability of these novel, emerging ecosystems (Hobbs et al. 2006). The question of how climate change will interact in this global process of ecosystem modification is becoming highly relevant for natural resource management.

Although many studies have addressed the potential threats to ecosystems from invasive alien plants and climate change separately, few studies have considered the interactive and potentially synergistic impacts of these two factors on ecosystems (but see Ziska 2003). Climatic and landscape features set the ultimate limits to the geographic distribution of species and determine the seasonal conditions for establishment, recruitment, growth and survival (Rejmánek and Richardson 1996; Thuiller et al. 2006b). Human-induced climate change is therefore a pervasive element of the multiple forcing functions which maintain, generate and threaten natural biodiversity.

A widely stated view is that climate change is likely to enhance the capacity of alien species to invade new areas, while simultaneously decreasing the resistance to invasion of natural communities by disturbing the dynamic equilibrium maintaining them. Links between invasion dynamics and climate change are, nevertheless, particularly difficult to conceptualize (Fig. 12.1). The determinants of plant invasiveness per se are extremely complex (Rejmánek et al. 2005). Consequently, efforts to combat plant invasions

Ecological Studies, Vol. 193
W. Nentwig (Ed.)
Biological Invasions

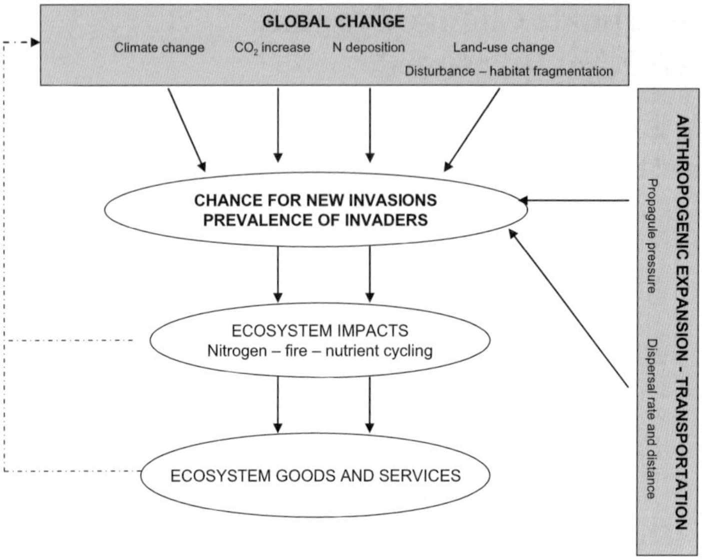

Fig. 12.1 Impacts of global change on invasions and the associated processes

have been largely reactive in nature: another species becomes invasive, and a plan must then be developed to combat it. Similarly, the question of whether invasibility is positively or negatively related to community diversity (defined in various ways) is still under debate (Lonsdale 1999). Given the level of uncertainty around determinants of invasiveness and invasibility, even without the additional complexity introduced by changed climatic conditions, it is clear that precise forecasts of the dynamics of invasions with climate change is a very tall order. Therefore, while climate change is a relatively slow ongoing process, human-induced fragmentation and disruption of disturbance regimes have probably a much greater impact on the dynamic of both native and alien species than does climate change in the short to medium term (Bond and Richardson 1990). Still, does this pragmatic view conceal underlying pressures which could eventually thwart current management approaches to alien species?

Changing climate affects natural communities, ecosystems and habitats in many ways (Parmesan and Yohe 2003) but most immediately through shifts in geographic range. Not all ecosystems seem equally vulnerable to global environmental change (Walther et al. 2002). For instance, arctic-alpine communities might see their distributions substantially reduced to the benefit of more temperate ones (Grabherr et al. 1994), and early work on climate impacts on ecosystems focused on potential geographic range shifts of species assemblages or even of biomes. However, given the idiosyncratic responses of species to past climate change (e.g. Prentice 1986), we can be sure that ongoing and future climate change will not harm or benefit all component species in any assemblage to the same extent. Shifts in the range of individual indigenous species under climate change involve processes closely akin to those driving the spread of alien species; the two can thus be addressed using a similar theoretical approach. This theory relates both to the demographic impacts of climate change as a consequence of individual physiological constraints, and also to changes in the outcomes of interactions between species – and not simply between species at the same trophic level (e.g. plants, but also plant–animal interactions; Davis et al. 1998). Finally, regional changes in community and ecosystem structure have the potential to influence both micro- and regional climate, providing a complex feedback effect which might exacerbate or retard the rate of ecosystem change.

Is it possible to simplify this plethora of impacts and interactions? As a starting point, it seems likely that the response of a few species (positive or negative) or functional groups will determine ecosystem resistance to biological invaders and changes in ecosystem functioning (Zavaleta and Hulvey 2004). Also, recent empirical evidence has provided support for the biomass ratio hypothesis (Grime 1998), showing that biogeochemical pools and fluxes are controlled by the expression of individual traits of the dominant plant species. This is especially the case in stressful environments such as alpine and artic ecosystems, where fewer than 10 species of higher plants make up more than 90 % of the vascular-plant biomass (Chapin III and Körner 1996). If such keystone species suffer population declines under climate change, then this is highly likely to increase the susceptibility of these communities to invasion by aliens, and to the disruption of key processes (Zavaleta and Hulvey 2004). Understanding how changes in species distributions resulting from global change may cascade to changes in ecosystem processes therefore requires advancing simultaneously our understanding of the processes determining community assembly and of the mechanisms through which species influence ecosystem functioning.

Climate change may, however, favour a subset of species with certain sets of traits (particularly those related to dispersal abilities and tolerance of disturbance) or species well adapted to, or tolerant of, warmer and/or drier environments and responsive to elevated atmospheric CO_2 levels. There are a number of pathways by which invasive species and climate change can interact (Fig. 12.1, Dukes and Mooney 1999). The system of complex interactions can be considered transient because the main players – anthropogenic environment, invasive species, and the components of the "host" ecosystem – are all dynamic (Sutherst et al. 2000). Climate change will likely increase this dynamism and transience, leading to a substantial impact on an already complex and almost certainly non-equilibrium relationship between invasive species and the host ecosystem. For example, climate change has been predicted to lead to greatly increased rates of species turnover (exceeding 40 %) in local communities in Europe (Thuiller et al. 2005a). Such species turnover will undoubtedly lead to severe ecological perturbation (disruption/disturbance) of these communities. Disturbance is a crucial mechanism in mediating the establishment, persistence and dramatic expansion (explosion in numbers) of invasive alien species (hence, ecosystem invasion; Brooks et al. 2004). Additionally, these new communities and ecosystems will have unknown properties, including the likely presence of species which exhibit all the traits of invasive alien species. This is the concern of an emerging concept, that of "novel ecosystems" (Hobbs et al. 2006), which we discuss below within a climate change context.

We first review the current knowledge of the potential synergies between climate change and invasive alien species, and then provide some perspectives and areas of research needed to manage biological invasions in the face of climate change.

12.2 Current and Emerging Knowledge

It is widely considered that climate change will enhance the success of invasive alien species (Dukes and Mooney 1999; Mooney and Hobbs 2000). For instance, the Union of Concerned Scientists (2001), synthesizing the views of many scientists in the USA, issued a press release stating that "Climate change could potentially favor invasive non-native species by either creating more favourable environmental conditions for them, e.g., increasing fire frequency, or by stressing native species to the point of being unable to compete against new invasives." Climate change could alter almost very facet of invasion dynamics and every interaction between different factors. In this section, we discuss some important aspects. Invasions generally have two distinct phases, separated by a time lag ranging from decades to a century: a quiescent phase, during which ranges shift only slightly, followed by a phase of active population growth and expansion. Numerous factors potentially act as trigger to start the rapid growth phase, notably natural or human-induced disturbances (Mooney and Hobbs 2000). Changing climate could provide new triggers, or fine-tune existing triggers, for instance, by creating disturbance events which open opportunities for previously quiescent alien species, e.g. by facilitating reproduction, survival, or enhancing competitive power.

Sutherst et al. (2000) suggest that climate change may impact on the overall invasion process by affecting three significant constraints: sources of invasive species, pathways of dispersal, and the invasion process in host ecosystems. Of these, we suggest that the last is by far the most important and relevant constraint, since climate change is not likely to generate additional sources of invasive species, and pathways of dispersal are overwhelmingly defined by human economic and trade activity. How, then, might climate change affect the process of invasion? This topic has been explored most succinctly by Dukes and Mooney (1999), and we build on elements of their argument. Essentially, we argue that climate change is likely to affect patterns of alien plant invasions through its effect on three overarching aspects: the invasibility of the host ecosystem, the invasive potential of the alien species, and climate impacts on indigenous species. Synergistic combinations between two or, in a worst-case scenario, all three elements is likely to lead to significantly increased vulnerability to climate change. What are the probabilities that positive synergies (i.e. increased invasiveness) will result? In the next section, we review these possibilities by assessing the potential impacts of important elements of climate change.

12.2.1 Elevated Carbon Dioxide

12.2.1.1 Observations and Experimental Findings

The significance of the direct CO_2 effect on vegetation and on invasive alien species is important – ambient CO_2 levels are currently 30 % higher than the pre-industrial level, and are higher than at any time in at least the last 750,000 years. There is general consensus on the direct physiological impact of increasing CO_2 on plant photosynthesis and metabolism (Ainsworth and Long 2005). Increasing CO_2 stimulates growth and development significantly in hundreds of plant species (Drake et al. 1997). Increasing atmospheric CO_2 generally increases the resource-use efficiency of plants (Drake et al. 1997), due to direct stimulation of photosynthetic CO_2 uptake rate or a reduction in stomatal conductance. Thus, more carbon is fixed per unit of water or nitrogen used in the process of fixation. This effect is evident in plants with both C_3 and C_4 photosynthetic pathways; species with these pathways dominate the world's flora. Preliminary work suggests that CAM plants also respond to higher CO_2 levels by increasing carbon uptake and metabolism (Dukes 2000).

The benefits of CO_2 stimulation predicted purely by photosynthetic theory and single-species experiments are difficult to extrapolate to multi-species communities. Nonetheless, the few experiments which have been done on invasive alien species suggest a strong positive response to elevated CO_2 (Dukes and Mooney 1999). However, it is difficult to tease species-specific effects from effects on native species vs. invasive species. For example, Hattenschwiler and Körner (2003) show that two indigenous European temperate forest species had a muted response to elevated CO_2 whereas an indigenous ivy, an indigenous deciduous species, and the invasive alien *Prunus laurocerasus* showed significant responses. By contrast, Nagel et al. (2004) show a clear CO_2 stimulation of an invasive grass species, and lack of response in a co-occurring native species. This is corroborated by Smith et al. (2000) who found that aboveground production and seed rain of an invasive annual grass increase more at elevated CO_2 than in several species of native annuals in the deserts of western North America. They also suggest that this increase in production would have the potential to accelerate the fire cycle, reduce biodiversity, and alter ecosystem function in these arid ecosystems. In some cases, species-specific effects are maintained in multi-species communities (Polley et al. 2002) but, to date, the number of case studies is too small to be able to make general conclusions.

Should we expect invasive alien species to have different responses than native species? It is possible that faster-growing species may benefit more than slower growers in more productive environments. Since rapid growth rate is a typical characteristic of invasive plant species, this may underpin the stronger response of the aliens. Combined with a higher reproductive output (Nagel et al. 2004) and possibly greater seedling survivorship (Polley et al. 2002), elevated CO_2 may well provide significant advantages to fast-growing alien species within the context of the host ecosystem, especially for invasive woody plants. Recently, Ziska (2003) suggested that increases in atmospheric CO_2 during the 20th century may have been a factor in the selection of six plant species widely recognised as among the most invasive weeds in the continental United States.

Ecosystem feedbacks through changes in the fire, water and nutrient cycles also complicate the issue. Recently, Kriticos et al. (2003) concluded that the invasive potential of the woody species *Acacia nilotica* in Australia may be enhanced by elevated CO_2, through improvements in plant water-use efficiency. Woody species invasions, which have the potential to subdue fire regimes in currently grass-dominated ecosystems, could drive rapid switches in ecosystem structure and function, with significant implications for the biodiversity of both flora and fauna. For example, invasion of grasslands by *Quercus macrocarpa* seedlings in the American Midwest may necessitate focused fire management strategies (Danner and Knapp 2003). Alternatively, it is conceivable that CO_2-driven increases in flammable woody or herbaceous plants will accelerate fire regimes in fire-prone systems (Grigulis et al. 2005).

12.2.1.2 Future Expectations

Looking at effects on host ecosystems, some simulations suggest that ecosystem structure might be significantly altered by elevated CO_2 (Bond et al. 2003), leading to switches in plant-functional type dominance and the opportunity for increased success of woody invaders, for example. In his experiments, Ziska (2003) argued that the average stimulation of plant biomass among invasive species from current (380 $\mu mol\ mol^{-1}$) to future (719 $\mu mol\ mol^{-1}$) CO_2 levels averaged 46 %, with the largest response (+72 %)

observed for Canada thistle *Cirsium arvense*. This study suggests that the CO_2 increase during the 20th century has selected for invasive alien species based on their positive response, and that further CO_2 increase into the future, as predicted by IPCC, will enhance the invasive potential of these recognised weeds.

In a more theoretical perspective, Gritti et al. (2006) using LPJ-GUESS, a generalized ecosystem model based on dynamic processes describing establishment, competition, mortality and ecosystem biogeochemistry, simulated the vulnerability of Mediterranean Basin ecosystems to climate change and invasion by exotic plant species. They simulated the vegetation dynamics using a set of native plant-functional types based on bioclimatic and physiological attributes (tree and shrub) and two invasive plant-functional types, an invasive tree type and invasive herb type, according to two climate change and CO_2 increase scenarios projected for 2050. The major point of relevance here is that these simulations suggested that the effect of climate change alone is likely to be negligible in several of the simulated ecosystems. The authors pointed out that the simulated progression of an invasion was highly dependent on the initial ecosystem composition and local environmental conditions, with a particular contrast between drier and wetter parts of the Mediterranean, and between mountain and coastal areas. They finally concluded that, in the longer term, almost all Mediterranean ecosystems will be dominated by exotic plants, irrespective of disturbance rates. Although there is no way of validating such projections, they do shed light on the extreme complexity of attempts to predict invasion success, especially when invoking synergies between climate change, CO_2 increase, disturbance regimes, and initial conditions.

12.2.2 Changing Climate with Respect to Temperature and Rainfall

There is overwhelming evidence of individual species responses to changing temperature regimes over the past century, the vast majority of range shift responses having been recorded in insect, bird and marine species (Parmesan and Yohe 2003). By contrast, plant responses to temperature increases over the past century have been mainly phenological (i.e. a change in timing of growing season). Changes in moisture regime are far more difficult to attribute to anthropogenic climate change, and therefore studies of these have been mainly experimentally based, rather than focused on historical trends and their impacts. With a few exceptions (e.g. tundra invasion by indigenous woody species, alpine range shifts, tree invasion in boreal regions, "laurophyllization" of European forests (Walther et al. 2002), plant range shifts appear unsurprisingly much slower that those of animals. The implications of this lag between animals and plants are most obvious in North American forests, where an indigenous insect species (pine bark beetle) appears to have extended its poleward range limit, with devastating consequences for indigenous forest tree species. Therefore, in addition to alien species invading new habitats/countries as a result of climate change, concern has also been raised over the potential of those species currently causing problems in managed ecosystems to becoming more widespread and damaging (Cannon 1998). These impacts are best observed on sub-Antarctic islands, where invasive plant species currently benefit from increasing temperature and decreasing rainfall trends, in synergy with enhanced success of invasive small mammals (Frenot et al. 2005).

12.2.3 Future Expectations

The impacts of temperature and rainfall change on plant species and ecosystems have been extensively investigated using modelling approaches, which fall into two main groups: mechanistically based models which simulate simplified, abstract versions of ecosystems, and statistically based (i.e. niche-based) models which match individual species to their ecological niches and simulate potential changes in range. Such models seem poorly capable of projecting the complex interactions observed in the natural world (cf. above). Nevertheless, they may provide guidelines on which ecosystems are vulnerable to the development of such interactive effects.

In general, dynamic global vegetation models predict quite significant changes in vegetation structure and function at a global scale (Cramer et al. 2001). This type of approach yields some insights into the potential structural and functional changes accompanying climate change. For instance, Gritti et al. (2006) projected that, although climate change alone could enhance exotic invasion in Mediterranean landscapes, the interaction with the direct effect of CO_2 was the most important driver controlling the invasion by shrubs. This interaction between climate change and CO_2 to drive vegetation distribution and structure was corroborated by Harrison and Prentice (2003), who showed that both climate change and CO_2 controlled the global vegetation distribution during the last glacial maximum.

Alternatively to dynamic global vegetation models, niche-based models (Guisan and Thuiller 2005) have been the tools of choice when addressing the biodiversity implications of climate change (Thuiller et al. 2005a) and invasion potentials (Welk et al. 2002; Thuiller et al. 2005b). Niche-based models simulate quite strong negative effects of climate change on species range sizes in specific ecosystems such as Alpine environments (Guisan and Theurillat 2000), dry and hot areas (Thuiller et al. 2006a) or Mediterranean ecoregions (Midgley et al. 2003).

Such negative impacts on native ecosystems are likely to trigger and promote invasion. However, very few studies have investigated potential climate change impacts on exotic species ranges. Interestingly, potential range contractions are projected for a number of invasive alien species in South Africa (Richardson et al. 2000), in a pattern matching projections for indigenous species.

Given the strong interactions between plants and insect "pests", it is relevant to briefly highlight applications of the CLIMEX model to several species (Sutherst et al. 2000). These generally focus on areas which may become bioclimatically suitable, as opposed to areas from which the exotic species may be lost (although exceptions do occur, such as the models for the New Zealand flatworm, *Arthurdendyus triangulatus*; Evans and Boag 1996), and therefore provide a skewed impression of future range change of invasive and pest species. In addition, some models may not be based on the total suitable current climate space (e.g. Thuiller et al. 2004). In the UK, nevertheless, increases amounting to 102 % of suitable climate space by 2060–2070 are predicted for the Colorado potato beetle (*Leptinotarsa decemlineata*; Baker et al. 2000), and a substantial increase in the risk of the Southern pine beetle (*Dendroctonus frontalis*) has also been predicted (Evans and Boag 1996). Similarly, simulations at global scales are likely to shed light on the potential new areas susceptible to be invaded by specific species (Roura-Pascual et al. 2004) or even species from specific biomes (Thuiller et al. 2005b). For instance, the Argentine ant (*Linepithema humile*), native to central South America, is now found in many Mediterranean and subtropical regions around the world. Projections using niche-based models onto four general circulation model scenarios of future (2050s) climates predicted the species to retract its range in tropical regions but to expand in higher-latitude areas, tropical coastal Africa and southeast Asia. Although niche-based models lack the predictive rigour of more mechanistic models – the inherent correlative approach relies indeed on many assumptions – they nevertheless offer rapid and useful tools for screening purposes (Panetta and Lawes 2005).

Alternatively, more process-based models, principally based on the description of plant-functional types, have shown that worldwide ecosystems are likely to experience change which can be likened to a sustained and intensifying disturbance, in that ongoing plant-functional type distribution and dynamic changes will occur in combination with (and possibly accelerated by) changes in structure and function (Cramer et al. 2001). This will alter the availability of resources, and create the physical and niche space to favour species or plant-functional types with opportunistic responses to increasing resource availability.

Examples can be found across the spectrum of plant-functional types: aggressive shrub species such as sagebrush, for example, require only minor disturbance and space creation through the removal of herbaceous species to establish in the Sierra Nevada of California (Berlow et al. 2002). Increasing minimum temperatures reduce productivity in indigenous short grass prairie dominants, favouring invasive herbaceous species (Alward et al. 1999). In the forests of Panama, a slight lengthening of the dry season is projected to cause extinction of 25 % of indigenous species, and favour drought-tolerant invasive species (Condit et al. 1996).

Greater niche availability may also occur for insect herbivores. This may be due to a combination of increased levels of disturbance, and a potential increase in resources. This may be affected through an increase in plant height and changes in plant architecture, which may provide additional feeding and sheltering sites, as well as through an increase in the growing season and, thus, a lengthening in the temporal availability of resources. Additionally, departures (extinctions) of native species as their climatic tolerances are exceeded may also provide vacant niches.

12.2.4 Other Factors

Besides elevated CO_2 levels, warming, and changes in rainfall patterns, climate change is also expected to increase the frequency of droughts, floods, storms, and extreme events such as hurricanes and wildfires (Chaps. 13–17). Not only are these likely to cause large-scale ecosystem disturbances but can also affect the composition and structure of ecosystems, and these factors may provide opportunities for invasion (due to niche availability) and increase in abundance of alien species. Variability in climate is predicted to change, with models indicating that there may be greater extremes in dry and wet seasons, and also in temperature, and this may allow non-native species to become problematical, even in areas where the average climate is unsuitable. Reservoirs of such species may occur in urban areas or in protected environments such as greenhouses. In other cases, propagule pressure will be high, and establishment and spread may take place during these windows of opportunity. For example, the CLIMEX model run for the Colorado potato beetle (*L. decemlineata*) in Norway suggests that both in the last decade and in the future, there will be periods during which it is possible for establishment to take place over a substantial area, although average climate is unsuitable in most locations (Rafoss and Sæthre 2003). Climatic variability may also lead to disruptions in the synchrony between natural enemies and their hosts, and may alter the effectiveness of biocontrol agents which have been released against non-native species. Under intensifying climate change conditions, it is conceivable that novel niches will become available, mainly to species with high rates of fecundity and dispersal – typical of exotic species and attributes of an invasive.

12.2.5 Increased Fire Frequency

Overall predictions of climate change on fire frequency are strongly limited by the lack of a globally applicable model of fire in ecosystems. A substantial fraction of the world's natural ecosystems are strongly influenced by fire, especially but not exclusively in the southern Hemisphere (Bond et al. 2005). Changing fire regimes have the strong potential to radically alter ecosystems, leading to switches in vegetation dominance and structure with substantial implications for management strategies and biodiversity (Briggs et al. 2002).

Invasive C_4 grasses are causing accelerated fire cycles (Brooks et al. 2004), reduced nutrient availability, and leading to forest loss in Asia, Africa and the Americas (Sage and Kubien 2003), and on oceanic islands (Cabin et al. 2002). Fecundity and dispersal potential may be key attributes in such invasions – the invasion of *Hyparrhenia* grass species in South America is due to seed availability at post-fire sites, fire-stimulated seed germination, and rapid seedling growth (Baruch and Bilbao 1999). C_4 grass invasions may be further promoted by even warmer and drier conditions (Sage and Kubien 2003), which may have ever stronger negative impacts on indigenous species and fire regimes.

Overall, it seems feasible that increasing temperatures and dry spells, combined with the positive effects of rising atmospheric CO_2 on plant productivity, may facilitate an increased fire frequency in fire-prone ecosystems. However, an even more powerful interaction may arise where invasive species themselves generate sufficient biomass to fuel accelerated fire regimes, such as in the Cape Floristic Region, or even introduce fire as a novel disturbance (Brooks et al. 2004). Even in the Cape Floristic Region, where extensive work on invasive alien species has been done, little is known about how fire and invasive alien species will interact in the future (van Wilgen and Richardson 1985). However, it is clear that the interaction has the potential to significantly increase extinction risk for many species (Mooney and Hobbs 2000).

In summary, the possible synergy between warmer conditions and productive exotic species has the potential to transform host ecosystems, with major negative implications for biodiversity.

12.3 Perspectives

The above sections have outlined and explored key facets of the complexity challenging our developing understanding of how climate change and CO_2 increase could potentially affect the dynamics of alien plant invasions around the globe. We show that it is difficult to identify clear determinants of future invasibility of ecosystems and invasion potential of introduced species, due to the complexity of both main drivers of change, to interactions with disturbance, and to species interactions. Some already-established alien plant species are likely to be strongly favoured whereas others will probably show little response or even a negative response, where their bioclimatic requirements closely match those of the invaded ecosystem. Given the multiple linkages and complex feedback and feed-forward loops implicated, we need to draw heavily on natural experiments and ongoing observations to inform us of the nature and magnitude of effects which are likely or possible. Manipulative experiments are currently underway – these will shed light on crucial aspects which will fine-tune our understanding and our ability to model the spread of alien plant invasions within the context of global environmental change. Finally, dynamic modelling tools must continue to guide new research questions and identify key unknowns to accelerate our understanding of this critical issue.

New paradigms for conservation are needed to accommodate potential changes in the status of biological invasions worldwide. For example, alien species, even those known to be highly invasive, may well be better than *no* species in some ecosystems, e.g. where such species provide essential cover/binding or other key ecosystem services, or act as nurse plants for native species. There is a clear need to give urgent attention to building such scenarios into frameworks for conservation planning.

Acknowledgements D.M.R. acknowledges support from the DST-NRF Centre of Excellence for Invasion Biology. W.T., G.F.M. and D.M.R. received support from the International Research Network (GDRI) project "France South Africa – Dynamics of biodiversity in Southern African ecosystems and sustainable use in the context of global change: processes and mechanisms involved"

References

Ainsworth EA, Long SP (2005) What have we learned from 15 years of free-air CO_2 enrichment (FACE)? A meta-analytic review of the responses of photosynthesis, canopy properties and plant production to rising CO_2. New Phytol 165:351–372

Alward RD, Detling JK, Milchunas DG (1999) Grassland vegetation changes and nocturnal global warming. Science 283:229–231

Baker RHA, Sansford CE, Jarvis CH, Cannon RJC, MacLeod A, Walters KFA (2000) The role of climatic mapping in predicting the potential geographical distribution of non-indigenous pests under current and future climates. Agric Ecosyst Environ 82:57–71

Baruch Z, Bilbao B (1999) Effects of fire and defoliation on the life history of native and invader C-4 grasses in a Neotropical savanna. Oecologia 119:510–520

Berlow EL, D'Antonio CM, Reynolds SA (2002) Shrub expansion in montane meadows: the interaction of local-scale disturbance and site aridity. Ecol Appl 12:1103–1118

Bond WJ, Richardson DM (1990) What can we learn from extinctions and invasions about the effects of climate change? S Afr J Sci 86:429–433

Bond WJ, Midgley GF, Woodward FI (2003) The importance of low atmospheric CO_2 and fire in promoting the spread of grasslands and savannas. Global Change Biol 9:973–982

Bond WJ, Woodward FI, Midgley GF (2005) The global distribution of ecosystems in a world without fire. New Phytol 165:525–538

Briggs JM, Knapp AK, Brock BL (2002) Expansion of woody plants in tallgrass prairie: a fifteen-year study of fire and fire-grazing interactions. Am Midl Nat 147:287–294

Brooks ML, D'Antonio CM, Richardson DM, Grace J, Keeley JE, DiTomasso JM, Hobbs RJ, Pellant M, Pyke D (2004) Effects of invasive alien plants on fire regimes. BioScience 54:677–688

Cabin RJ, Weller SG, Lorence DH, Cordell S, Hadway LJ, Montgomery R, Goo D, Urakami A (2002) Effects of light, alien grass, and native species additions on Hawaiian dry forest restoration. Ecol Appl 12:1595–1610

Cannon RJC (1998) The implications of predicting climate change for insects pests in the UK, with emphasis on non-indigenous species. Global Change Biol 4:785–796

Chapin FS III, Körner C (1996) Artic and alpine biodiversity: its patterns, causes and ecosystem consequences. In: Mooney HA, Cushman JH, Medina E, Sala OE, Schulze E-D (eds) Functional role of biodiversity: a global perspective. Wiley, Chichester, pp 7–32

Condit R, Hubbell SP, Foster RB (1996) Assessing the response of plant functional types to climatic change in tropical forests. J Veg Sci 7:405–416

Cramer W, Bondeau A, Woodward FI, Prentice IC, Betts RA, Brovkin V, Cox PM, Fisher V, Foley JA, Friend AD, Kucharik C, Lomas MR, Ramankutty N, Sitch S, Smith B, White A, Young-Molling C (2001) Global response of terrestrial ecosystem structure and function to CO_2 and climate change: results from six dynamic global vegetation models. Global Change Biol 7:357–373

Crawley MJ (1989) Chance and timing in biological invasions. In: Drake JA, Mooney HA, di Castri F, Groves RH, Kruger FJ, Rejmánek M, Williamson M (eds) Biological invasions: a global perspective. Wiley, Chichester, pp 407–424

Danner BT, Knapp AK (2003) Abiotic constraints on the establishment of *Quercus* seedlings in grassland. Global Change Biol 9:266–275

Davis AJ, Lawton JH, Shorrocks B, Jenkinson LS (1998) Individualistic species responses invalidate simple physiological models of community dynamics under global environmental change. J Anim Ecol 67:600–612

Drake BG, Gonzalez-Meler MA, Long SP (1997) More efficient plants: a consequence of rising atmospheric CO_2? Annu Rev Plant Physiol Plant Mol Biol 48:609–639

Dukes JS (2000) Will the increasing atmospheric CO_2 concentration affect the success of invasive species? In: Mooney HA, Hobbs HA (eds) Invasive species in a changing world. Island Press, Washington, DC, pp 95–113

Dukes JS, Mooney HA (1999) Does global change increase the success of biological invaders. Trends Ecol Ecol 14:135–139

Dukes JS, Mooney HA (2004) Disruption of ecosystem processes in western North America by invasive species. Rev Chil Hist Nat 77:411–437

Evans KA, Boag B (1996) The New Zealand flatworm: will climate change make it a potentially European problem? Aspects Appl Biol 45:335–338

Frenot Y, Chown SL, Whinam J, Selkirk PM, Convey P, Skotnicki M, Bergstrom DM (2005) Biological invasions in the Antarctic: extent, impacts and implications. Biol Rev 80:45–72

Grabherr G, Gottfried M, Pauli H (1994) Climate effects on mountain plants. Nature 369:448–450

Grigulis K, Lavorel S, Davies ID, Dossantos A, Lloret F, Vilà M (2005) Landscape-scale positive feedbacks between fire and expansion of the large tussock grass, *Ampelodesmos mauritanica* in Catalan shrublands. Global Change Biol 11:1042–1053

Grime JP (1998) Benefits of plant diversity to ecosystems: immediate, filter and founder effects. J Ecol 86:902–910

Gritti E, Smith B, Sykes MT (2006) Vulnerability of Mediterranean basin ecosystems to climate change and invasion by exotic plant species. J Biogeogr 33:145–157

Guisan A, Theurillat J-P (2000) Assessing alpine plant vulnerability to climate change: a modelling perspective. Integr Assessment 1:307–320

Guisan A, Thuiller W (2005) Predicting species distribution: offering more than simple habitat models. Ecol Lett 8:993–1009

Harrison SP, Prentice IC (2003) Climate and CO_2 controls on global vegetation distribution at the last glacial maximum: analysis based on paleovegetation data, biome modelling and paleoclimate simulations. Global Change Biol 9:983–1004

Hattenschwiler S, Körner C (2003) Does elevated CO_2 facilitate naturalization of the non-indigenous *Prunus laurocerasus* in Swiss temperate forests? Funct Ecol 17:778–785

Hobbs RJ, Arico A, Aronson J, Baron JS, Bridgewater P, Cramer VK, Epstein PR, Ewel JJ, Klink CA, Lugo AE, Norton D, Ojima D, Richardson DM, Sanderson EW, Valladares F, Vilà M, Zamora R, Zobel M (2006) Emerging ecosystems: theoretical and management aspects of the new ecological world order. Global Ecol Biogeogr 15:1–7

Kriticos DJ, Brown JR, Maywald GF, Radford ID, Nicholas DM, Sutherst RW, Adkins SW (2003) SPAnDX: a process-based population dynamics model to explore management and climate impacts on an invasive alien plant, *Acacia nilotica*. Ecol Modell 163:187–208

Lonsdale WM (1999) Global patterns of plant invasions and the concept of invasibility. Ecology 80:1522–1536

Midgley GF, Hannah L, Millar D, Thuiller W, Booth A (2003) Developing regional and species-level assessments of climate change impacts on biodiversity in the Cape Floristic Region. Biol Conserv 112:87–97

Mooney HA, Hobbs RJ (2000) Invasive species in a changing world. Island Press, Washington, DC

Nagel JM, Huxman TE, Griffin KL, Smith SD (2004) CO_2 enrichment reduces the energetic cost of biomass construction in an invasive desert grass. Ecology 85:100–106

Panetta FD, Lawes R (2005) Evaluation of weed eradication programs: the delimitation of extent. Diversity Distrib 11:435–442

Parmesan C, Yohe G (2003) A globally coherent fingerprint of climate change impacts across natural systems. Nature 421:37–42

Polley HW, Tischler CR, Johnson HB, Derner JD (2002) Growth rate and survivorship of drought: CO_2 effects on the presumed tradeoff in seedlings of five woody legumes. Tree Physiol 22:383–391

Prentice IC (1986) Vegetation response to past climatic variation. Vegetatio 67:131–141

Rafoss T, Sæthre M-G (2003) Spatial and temporal distribution of bioclimatic potential for the codling moth and Colorado potato beetle in Norway: model predictions versus climate and field data from the 1990s. Agric Forest Entomol 5:75–85

Rejmánek M (1999) Invasive plant species and invasible ecosystems. In: Sandlund OT, Schei PJ, Vilken A (eds) Invasive species and biodiversity management. Kluwer, Dordrecht, pp 79–102

Rejmánek M, Richardson DM (1996) What attributes make some plant species more invasive? Ecology 77:1655–1661

Rejmánek M, Richardson DM, Higgins SI, Pitcairn MJ, Grotkopp E (2005) Ecology of invasive plants: state of the art. In: Mooney HA, Mack RN, McNeely JA, Neville L, Schei PJ, Waage J (eds) Invasive alien species: a new synthesis, Island Press, Washington, DC, pp 104–161

Richardson DM, Bond WJ, Dean WRJ, Higgins SI, Midgley GF, Milton SJ, Powrie LW, Rutherford MC, Samways MJ, Schulze RE (2000) Invasive alien organisms and global change: a South African perspective. In: Mooney HA, Hobbs RJ (eds) Invasive species in a changing world. Island Press, Washington, DC, pp 303–349

Roura-Pascual N, Suarez AV, Gómez C, Pons P, Touyama Y, Wild AL, Peterson AT (2004) Geographical potential of Argentine ants (*Linepithema humile* Mayr) in the face of global climate change. Proc R Soc Lond B 271:2527–2534

Sage RF, Kubien DS (2003) Quo vadis C-4? An ecophysiological perspective on global change and the future of C4 plants. Photosynth Res 77:209–225

Smith SD, Huxman TE, Zitzer SF, Charlet TN, Housman DC, Coleman JS, Fenstermaker LK, Seemann JR, Nowak RS (2000) Elevated CO_2 increases productivity and invasive species success in an arid ecosystem. Nature 408:79–82

Sutherst RW, Maywald HF, Russel BL (2000) Estimating vulnerability under global change: modular modelling of pests. Agric Ecosyst Environ 82:303–319

Thuiller W, Brotons L, Araújo MB, Lavorel S (2004) Effects of restricting environmental range of data to project current and future species distributions. Ecography 27:165–172

Thuiller W, Lavorel S, Araújo MB, Sykes MT, Prentice IC (2005a) Climate change threats to plant diversity in Europe. Proc Natl Acad Sci USA 102:8245–8250

Thuiller W, Richardson DM, Pyšek P, Midgley GF, Hughes GO, Rouget M (2005b) Niche-based modelling as a tool for predicting the risk of alien plant invasions at a global scale. Global Change Biol 11:2234–2250

Thuiller W, Midgley GF, Hughes GO, Bomhard B, Drew G, Rutherford MC, Woodward FI (2006a) Endemic species and ecosystem vulnerability to climate change in Namibia. Global Change Biol 12:759–776

Thuiller W, Richardson DM, Rouget M, Proches S, Wilson JRU (2006b) Interactions between environment, species traits and human uses describe patterns of plant invasions. Ecology 87:1755–1769

van Wilgen BW, Richardson DM (1985) The effects of alien shrub invasion on vegetation structure and fire behavior in South African fynbos shrublands: a simulation study. J Appl Ecol 22:955–966

Walther G-R, Post E, Convey P, Menzel A, Parmesan C, Beebee TJC, Fromentin J-M, Hoegh-Gudberg O, Bairlein F (2002) Ecological responses to recent climate change. Nature 416:389–395

Welk E, Schubert K, Hoffman MH (2002) Present and potential distribution of invasive garlic mustard (*Alliaria petiolata*) in North America. Diversity Distrib 8:219–233

Zavaleta ES, Hulvey KB (2004) Realistic species loss disproportionately reduce grassland resistance to biological invaders. Science 306:1175–1177

Ziska LH (2003) Evaluation of the growth response of six invasive species to past, present and future atmospheric carbon dioxide. J Exp Bot 54:395–404

Section IV

Ecological Impact of Biological Invasions

Short Introduction

Wolfgang Nentwig

It is well known that some invasive alien species have a very strong influence on ecological interactions with other species. Such case studies are numerous and range from mere anecdotal reports to very well-analysed quantifications. Here, we present two of these in detail. The one dealing with marine jellyfish belongs to a group of hardly credible stories in ecology, where a few individuals of a few alien species transferred into a supposedly stable and well-buffered ecosystem have changed all major ecological patterns within a remarkably short time period (Chap. 14).

Our other case study concerns the river Rhine, subjected to heavy invasions by a whole armada of alien species, arriving by their own means or by ship via artificial waterways overcoming biogeographical barriers. The impact has been tremendous: the existing species assemblage has been completely replaced by aliens, causing a dramatic loss of biodiversity (Chap. 15).

Such a loss of species is deplorable, but it is feared that this and other such stories are not yet over – we can apprehend even more the next cascade of effects caused by invaders, i.e. possible alterations of ecosystem services, such as energy, nutrient and water cycling, and also of many biotic community interactions. Strong effects on ecosystem services, usually not recognized by our society as being important, urgently need to be counteracted because of tremendous potential economic consequences (Chap. 13).

A hidden danger within the context of alien species is hybridization. Genetically, a foreign species may completely absorb a native species or it may introduce parts of its own genome into a native species. The results are similar: the native species disappears and biodiversity is reduced. Hybridization or introgression has long been underestimated but today is considered to represent one of the most dangerous aspects of biological invasion (Chap. 16).

In this book on biological invasions, it may at first glance seem surprising to find a chapter dealing with genetically modified organisms. Though these are not alien species in a strict sense, they differ from their ancestors in at least one newly modified property, and this may suffice to already cause an ecologically modified behaviour. We still have very limited experience with genetically modified organisms under field conditions. Thus, it seems justified, within the framework of our broader knowledge on invasive alien species, to take a closer look at such organisms (Chap. 17).

13 Impacts of Invasive Species on Ecosystem Services

Heather Charles and Jeffrey S. Dukes

13.1 Introduction

The impacts of invasive species on ecosystem services have attracted worldwide attention. Despite the overwhelming evidence of these impacts and a growing appreciation for ecosystem services, however, researchers and policymakers rarely directly address the connection between invasions and ecosystem services. Various attempts have been made to address the ecosystem processes that are affected by invasive species (e.g., Levine et al. 2003; Dukes and Mooney 2004), but the links between these mechanisms and ecosystem services are largely lacking in the literature. Assessments of the economic impacts of invasive species cover costs beyond those associated with ecosystem services (e.g., control costs), and generally do not differentiate by ecosystem service type. Additionally, while advances have been made in quantifying non-market-based ecosystem services, their loss or alteration by invasive species is often overlooked or underappreciated.

Ecosystem services are the benefits provided to human society by natural ecosystems, or more broadly put, the ecosystem processes by which human life is maintained. The concept of ecosystem services is not new, and there have been multiple attempts to list and/or categorize these services, especially as the existence of additional services has been recognized (e.g., Daily 1997; NRC 2005). For the purposes of this chapter, we address ecosystem services in the framework put forward by the Millennium Ecosystem Assessment (2005). The services we list are primarily those enumerated in the Millennium Ecosystem Assessment (2005), with minimal variation in wording, and inclusion of several additional services not explicitly stated in this assessment. This framework places services into four categories (in italics). *Provisioning services* are products obtained from ecosystems, and include food (crops, livestock, fisheries, etc.), freshwater, fiber (timber, cotton, silk, etc.), fuel, genetic resources, biochemicals/pharmaceuticals/natural medicines, and ornamental resources. *Regulating services* are obtained from the regulation of ecosystem

Ecological Studies, Vol. 193
W. Nentwig (Ed.)
Biological Invasions

processes, and include air quality regulation, climate regulation, water regulation (timing and extent of flooding, runoff, etc.), water purification, waste treatment, disease regulation, natural pest control, pollination, erosion control, and coastal storm protection. *Cultural services* are non-material benefits, and include aesthetic values, recreation/tourism, spiritual/religious values, educational/scientific values, cultural heritage values, inspiration, and sense of place. *Supporting services* are overarching, indirect, and occur on large temporal scales, but are necessary for the maintenance of other services. They include photosynthesis, primary production, nutrient cycling, water cycling, soil formation and maintenance of fertility, as well as atmospheric composition. This framework includes both goods, which have direct market values, and services that in turn maintain the production of goods and biodiversity, and directly or indirectly benefit humans (Daily 1997).

In this chapter, we introduce concepts associated with the valuation of ecosystem services, and discuss how costs generated by invasions relate to impacts on ecosystem services. We link the effects of invasive species on community dynamics and ecosystem processes to effects on ecosystem services. Risks for specific ecosystem types and the organism types most likely to change particular services are discussed. Finally, we present examples of invasive species that alter each of these services. While the majority of these species negatively affect ecosystem services, several exceptions exist. We conclude by assessing the overall vulnerability of each category of ecosystem service to alteration by invasive species, suggesting future research needs, and discussing educational and collaborative opportunities in this field.

13.2 Relating Costs of Invasives to Valuation of Ecosystem Services

13.2.1 Valuing Ecosystem Services

In order to understand how invasive species affect ecosystem services, one must first understand how ecosystem services are valued, and how these values relate to the costs of invasive species. Economic valuation of ecosystem services (and goods) typically involves several components. All goods and services are categorized within a framework of total economic value (Fig. 13.1), and subsequently assigned monetary value (Costanza et al. 1997).

The framework initially differentiates between use and non-use values. Use values further divide into direct and indirect use values. Direct use values involve human interaction with nature, and include both consumptive and non-consumptive uses. Consumptive use refers to products consumed locally or sold in markets, whereas non-consumptive use typically refers to cultural

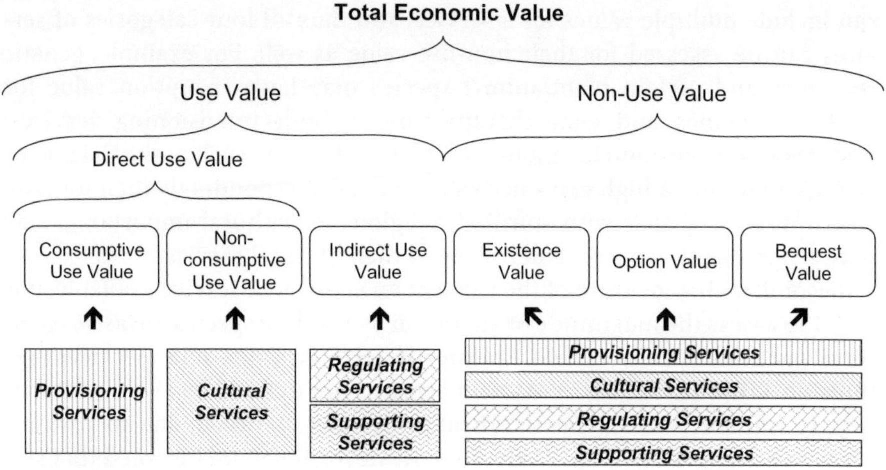

Fig. 13.1 Framework for economic valuation of ecosystem services (see text for further explanation)

services such as recreation and tourism. Indirect use values encompass species that humans rely on indirectly through trophic and other interactions (e.g., natural pest control), and services that are closely tied to ecosystem processes. Examples are productive inputs such as soil fertility, pollination, water purification, and flood control, all of which are extremely important in agriculture. Non-use values, while less tangible, are critical to a comprehensive assessment of economic valuation. They derive from the continued existence and intrinsic value of a service, good, species, habitat, etc., and include existence, option and bequest values.

These three values are succinctly explained by an example taken from Daily (1997), where non-use values for a hypothetical freshwater site include the value of knowing the site exists, irrespective of whether or not an individual visits the site (existence value); the value of preserving the option of enjoying the site in the future (option or future use value); and the value of ensuring that one's descendants will be able to enjoy the site (bequest value). While the literature on ecological economics includes several variations of this framework, all versions include the same basic principles (e.g., Daily 1997; NRC 2005; Born et al. 2005).

We link the total economic value framework to our discussion of ecosystem services and invasive species in two ways. First, the categories of ecosystem services can be connected to the categories of valuation in a generalized manner (Fig. 13.1). Provisioning services, which include all goods, fall into the consumptive use category. Most cultural services are considered to have non-consumptive use values. Regulating and supporting services are typically classified as having indirect use values. As mentioned above, the framework can include multiple values for a service, and thus all four categories of services can be assessed for their non-use value as well. For example, genetic resources and certain plant/animal species may have an option value for future medicines and gene therapy targets, both provisioning services. Endangered species and locations with high endemism, such as the Galapagos Islands, may have a high existence value and a correspondingly high tourism value. Sites or species with spiritual, religious, or cultural importance may have a significant bequest value, owing to their cultural services.

Second, with a measure of the value of an ecosystem service available, it is easier to assess the magnitude of alteration by invasive species. Invasives pose threats to human society that are proportional to the value of the services they threaten. Overall, because ecosystem services are defined by their contribution to human society, the significance of any alteration due to invasive species is dependent on their valuation. However, it should be noted that services may be undervalued if they are poorly understood or underappreciated.

13.2.2 Interpreting Invasive Impacts

Invasive impacts or costs are often classified as economic, environmental, or social in nature. Economic impacts are those of direct consequence to humans, typically leading to monetary losses. Environmental impacts are those that affect ecosystem structure and function, often referring to loss of biodiversity or unique habitats. Social impacts focus predominantly on human health and safety, but can also cover quality of life, recreational opportunities, cultural heritage, and other aspects of social structure. Where do ecosystem services fit into this classification? A unique facet of the concept of ecosystem services is the conjoining of ecological integrity and human benefit. As such, impacts will fall into all three categories with a good deal of overlap. Thus, all three types of impacts are useful in determining which services are affected by invasive species, and the magnitude of these effects.

Economic impact assessments give clues to some of the most significant impacts to humans by way of ecosystem services, but two caveats exist. First, economic assessments include control and management costs that are critical in determining control vs. prevention strategies, but do not address ecosystem services. Second, and more pertinent, economic assessments do not fully assess the alteration of certain ecosystem services, due to their subjective nature and the difficulty of assigning value. This includes almost all supporting services, and many regulating and cultural services. Since market values are easier to assign, and changes to these values are felt sooner and more acutely, economic assessments are necessarily biased toward provisioning services. Environmental impact assessments cover many of these remaining services, but often indirectly (e.g., biodiversity itself is not an ecosystem service per se), and without connections made to human benefits lost or gained.

Social impact assessments cover a smaller range of services, and some are not tied to ecosystem services (e.g., invasive insects that bite humans).

Nevertheless, we can make a few generalizations from impact assessments. Impacts of invasive species on ecosystem services related to agriculture, industry, and human health are substantial, well quantified, and typically negative (Chap. 18). These impacts affect the delivery of food, freshwater, and fiber, as well as water purification, pollination, natural pest control, disease regulation, soil fertility, and nutrient and water cycling. Invasives are having substantial, if not fully quantified, impacts on cultural services including aesthetic values, recreation, and tourism, in both riparian and upland areas (Eiswerth et al. 2005). Decreased biodiversity and species extinctions linked to invasive species threaten the continued delivery and quality of many ecosystem services. Finally, negative alterations of ecosystem services far outweigh positive alterations. Chapter 19 provides further discussion of economic and social impacts, as well as methods of impact assessment. Table 13.1 lists several studies that have quantified invasive species' impacts on specific ecosystem services, and includes both positive and negative impacts.

Table 13.1 Monetary impacts to ecosystem services associated with various invasive species

Invasive species	Geographic location	Ecosystem services altered	Monetary impact[a]	Reference
Acacia melanoxylon (blackwood), *Acacia cyclops* (rooikrans), *Eucalyptus* spp. (gum trees) and other woody shrubs and trees	Cape Floristic Region, South Africa (fynbos)	Food (sour figs, honeybush tea), fiber (thatching reed, timber), ornamental resources (flowers, greens, ferns), medicine, essential oils (buchu)	–2,852,984[b]	Turpie et al. (2003)
		Water (mountain catchments)	–67,836,059[b]	
		Pollination (bee keeping)	–27,783,728[b]	
		Ecotourism	–830,683[b]	
		Fuel (*Acacia cyclops* as firewood)	+2,799,492[b]	
Bemisia tabaci (whitefly)- and *B. tabaci*-transmitted viruses	Mexico	Food (melon, sesame), fiber (cotton)	–33 million	Oliveira et al. (2001)
	Brazil	Food (beans, tomatoes, melon, okra, cabbage)	–5 billion (for 5–6 years)	
	Florida, USA	Food (tomato, due to *Tomato mottle virus*)	–140 million	
	North America, Mediterranean Basin, Middle East	Food (lettuce, sugar beets, melon, due to *Lettuce infectious yellow virus*)	–20 million	

Table 13.1 (*Continued*)

Invasive species	Geographic location	Ecosystem services altered	Monetary impact[a]	Reference
Melaleuca quinquenervia	South Florida, USA (wetlands, open-canopied forests)	Recreation (park use)	–168 to 250 million	Serbesoff-King (2003)
		Tourism (Everglades National Park and rest of south Florida)	–250 million to 1 billion	
		Natural hazard regulation (increased fires)	–250 million	
		Various cultural services (endangered species loss)	–10 million	
		Ornamental resources (nurseries)	–1 million	
		Food (honey production)	+15 million	
Myriophyllum spicatum (Eurasian watermilfoil)	Western Nevada and northeast California; Truckee River watershed, USA	Recreation (swimming, boating, fishing, etc.)	–30 to 45 million	Eiswerth et al. (2000)
		Water quality, water supplies, non-use value	Unquantified negative costs	
Pomacea canaliculata (golden apple snail)	Philippines (rice systems)	Productivity losses (rice output)	–12.5 to 17.8 million	Naylor (1996)
Sus scrofa domestica (feral pig)	Florida, USA (three state parks; forest and wetland)	Habitat degradation (with implications for recreation, tourism, aesthetics, endangered species loss, erosion control, water quality)	–5,331 to 43,257 ha^{-1}, depending on park, season, and ecosystem type	Engeman et al. (2003)
Tamarix spp. (tamarisk)	Western United States, especially Colorado River	Irrigation water	–38.6 to 121 million	Zavaleta (2000)
		Municipal water	–26.3 to 67.8 million	
		Hydropower	–15.9 to 43.7 million	
		Natural hazard regulation (flood control)	–52 million	

[a] Costs are indicated with a negative sign (–) and benefits with a positive sign (+). Values are in US $ and represent annual losses, unless otherwise indicated

[b] Values were converted from year 2,000 Rands (R) to US $; 7 R=1 $

13.3 Mechanisms of Alteration

Ecosystems are characterized by their structure (composition and biological/physical organization) and functions or processes, which lead to the production and maintenance of ecosystem services. Invasive species alter the production, maintenance, and quality of services by a variety of mechanisms. As understanding of invasion biology has increased, so too has recognition and comprehension of these mechanisms. The mechanisms are interrelated, since they all affect aspects of the defining characteristics of ecosystem structure and function. However, they can be grouped into three categories to enhance ease in understanding (Fig. 13.2).

13.3.1 Species Extinctions and Community Structure

Invasive effects on native biodiversity and community structure are well known, but few studies have examined the mechanisms that lead to these effects (Levine et al. 2003). Invasive species may alter community structure through exploitation competition (indirect interactions such as resource use), and interference competition (direct interactions such as allelopathy in plants; Callaway and Ridenour 2004). Invasive impacts on other species interactions, including predation, herbivory, parasitism, and mutualisms, can change the abundance of species with certain key traits that influence ecosys-

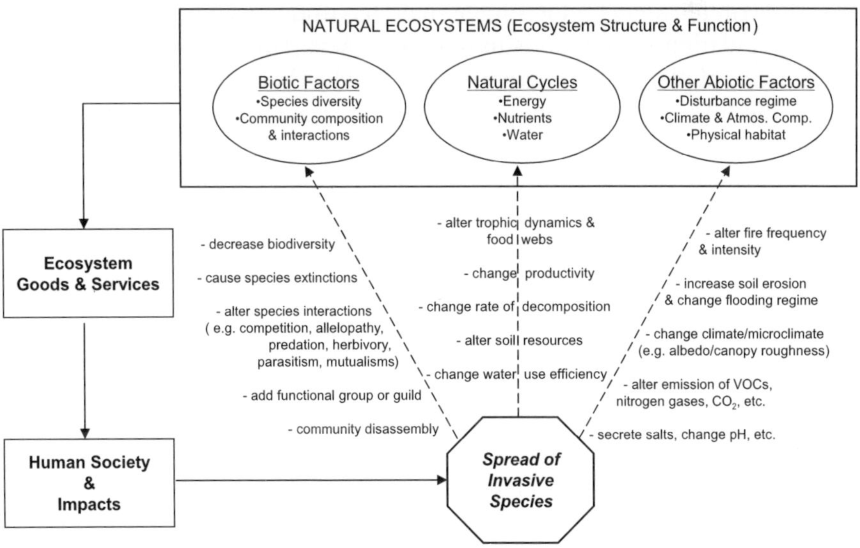

Fig. 13.2 Mechanisms of ecosystem service alteration by invasive species

tem processes (Chapin et al. 2000). A handful of nonnative animals, plants, and pathogens have also been implicated in extinctions of native species, in particular invasive animals on islands.

Changes in species and community structure can affect ecosystem services both directly and indirectly. Direct effects include the decline in abundance of economically valuable species, in particular those used for food, forage, fiber, fuel, or medicine. Aesthetic values are commonly lost with the arrival of "nuisance species" such as invasive vines or aquatic floating plants. Invasives that disrupt mutualisms pose risks particularly for pollination and natural pest control services. Decreased genetic diversity and species extinctions also lead to loss of option value. For example, the brown tree snake (*Boiga irregularis*) is blamed for the extinction of multiple bird and other species in Guam, with negative impacts on tourism, and unknown costs in genetic resources (Fritts and Rodda 1998). Indirect effects include a potential decrease in ecosystem resistance and resilience to change, due to the hypothesized link between stability and changes in biodiversity (Hooper et al. 2005). Finally, positive feedbacks due to interactions of invasive species may lead to increased vulnerability to further invasion, and potential degradation of ecosystem services (Simberloff and Von Holle 1999).

13.3.2 Energy, Nutrient, and Water Cycling

Invasive species' impacts also operate at the ecosystem level through the alteration of natural cycles. Energy flows can be altered by changes in trophic interactions, food webs and keystone species. For example, the herbivore *Pomacea canaliculata* (golden apple snail) has dramatically decreased aquatic plant populations in wetlands in Southeast Asia. This in turn has led to the dominance of planktonic algae, high nutrient levels, high phytoplankton biomass, and turbid waters, with implications for water quality and purification (Carlsson et al. 2004). Productivity can be altered by invasive species that use resources more efficiently, or that eliminate a prominent life form (Dukes and Mooney 2004). Since primary productivity is itself an ecosystem service, this shift could be detrimental to humans. Changes in decomposition rate, such as might occur if an invasive species altered litter chemistry, can affect nutrient cycling as well.

Nutrient cycling can also be altered by invasive plants that fix nitrogen, leach chemicals that inhibit nitrogen fixation by other species, release compounds that alter nutrient availability or retention, including nitrogen and phosphorus, and alter topsoil erosion or fire frequency (Dukes and Mooney 2004). The best studied of these mechanisms is the introduction of leguminous species with mutualistic nitrogen-fixing microorganisms, largely due to the dramatic effects of the invaders *Myrica faya* (fire tree) in Hawaii, New Zealand and Australia, and *Acacia mearnsii* (black wattle) in South Africa

(Levine et al. 2003). Ehrenfeld (2003) has shown that invasive plant impacts on nutrient cycling can vary in magnitude and direction across both invader types and sites, indicating that patterns are not universal, and that effects on ecosystem services can be either positive or negative. Alteration of nutrient cycling has additional implications for maintenance of soil fertility and primary production.

Invasive plant species have been shown to alter hydrological cycles by changing evapotranspiration rates and timing, runoff, and water table levels. Impacts are greatest when the invaders differ from natives in traits such as transpiration rate, leaf area index, photosynthetic tissue biomass, rooting depth, and phenology (Levine et al. 2003). Changes to water cycles may affect both water supply and regulation. Well-studied examples of invasive plants using more water than do native plants, and thus decreasing the water supply for humans, include *Tamarisk* spp. (salt cedar) in riparian zones of the southwestern United States, and pines in the Cape region of South Africa.

13.3.3 Disturbance Regime, Climate, and Physical Habitat

Several invasive species alter disturbance regimes (including fire, erosion, and flooding), or act as agents of disturbance themselves, particularly in soil disturbance (Mack and D'Antonio 1998). Fire enhancement can occur when grasses invade shrublands and increase fire frequency, extent, or intensity, whereas fire suppression is more likely to occur when trees invade grasslands and decrease fine fuel load and fire spread (Mack and D'Antonio 1998). These impacts are significant since they can cause a shift in ecosystem type and related species – for example, from shrublands to grasslands. Affected ecosystem services might include air purification or quality, atmospheric composition (e.g., through increased nitrogen volatilization), forage quality for cattle, and primary production. Mammalian invaders often increase erosion and soil disturbance, whereas woody plant invaders are more likely to affect water regulation by causing flooding and sedimentation in aquatic settings.

Maintenance of climate and atmospheric composition, both ecosystem services, are two of the least-studied mechanisms, perhaps because changes can occur over large temporal and spatial scales. Hoffmann and Jackson (2000) used modeling simulations to show that conversion of tropical savanna to grassland could both reduce precipitation and increase mean temperatures. However, the impetus for this study was land use change, not invasive species per se. On a smaller scale, experiments have shown that even a handful of invasive plants can alter a given microclimate. Finally, invasive species may alter atmospheric composition by changing rates of carbon dioxide sequestration, or the emission of volatile organic compounds and other biologically important gases (Dukes and Mooney 2004). Huxman et al. (2004) note that CO_2 and water flux to the atmosphere will be affected by the species-specific soil microclimate, and show differences in these fluxes between native and invasive grasses.

Invasive species can also alter the physical habitat. Both plant and animal invaders are capable of outcompeting natives and taking over habitat, and certain invaders additionally make the habitat less suitable for other species. Invasive plants may decrease the suitability of soil for other species by secreting salts (e.g., *Tamarisk*, Zavaleta 2000; the iceplant *Mesembryanthemum crystallinum*, Vivrette and Muller 1977), by acidifying the soil, or by releasing novel chemical compounds, as in allelopathy (Callaway and Ridenour 2004).

13.4 Which Ecosystems Are at Risk and Which Invasives Have the Greatest Impact?

Predicting which invasive species will have the greatest impact on ecosystem services would have both economic and societal benefits, and allow us to improve our prevention and management strategies. Unfortunately, the relationships between ecosystem impacts and ecosystem service impacts are difficult to characterize. We expect that species with the greatest ecological impacts will also have the greatest impacts on ecosystem services, but this has not been tested. Likewise, the relationship between community invasibility and the intensity of impacts is also debatable (Levine et al. 2003). Some generalizations can be made regarding the species most likely to alter ecosystem processes. Invasives that add a new function or trait have the potential to significantly impact ecosystem processes as their ranges expand, often by the addition of a new functional type based on traits related to resource use (e.g., nitrogen fixers), phenology, feeding habits, habitat preference, etc. (Chapin et al. 1996). Even without the addition of a new function or trait, an invader that comprises a large proportion of the biomass at a given trophic level may measurably alter ecosystem structure and function (Dukes and Mooney 2004). Invasive species of all taxa are capable of altering ecosystem services.

Which invasive species might pose the greatest threat to a given ecosystem service in a given system? This question is difficult to answer; few concrete patterns exist, and we currently rely on a handful of species-specific examples. We can broadly say that specific ecosystem types are susceptible to alteration of particular ecosystem services (Table 13.2). For simplicity, we use the six ecosystem types delineated by The State of the Nation's Ecosystems (The H. John Heinz III Center for Science Economics and the Environment 2002): coasts and oceans, farmlands, forests, fresh waters, grasslands and shrublands, and urban and suburban areas. These generalizations are necessarily subjective, based on our review of the literature. One notable source of information on a broad range of invader taxa and habitat and ecosystem types is the Global Invasive Species Database (http://www.issg.org/database).

Table 13.2 Ecosystem types differ in ecosystem services most at risk and prevalent invasive species types

Ecosystem type	Ecosystem services most at risk	Prevalent invader types	Invader examples and impacts	Other
Coasts and oceans	– Commercial fisheries – Shellfish beds – Water purification – Waste treatment – Disease regulation – Recreation, tourism	– Alga, seaweeds – Mollusks – Crustaceans – Fish	Caulerpa seaweed (*Caulerpa taxifolia*) – Forms dense mats in Mediterranean Sea – Negative impacts on aquaculture/fishing (Verlaque 1994) Green crab (*Carcinus maenus*) – Consumes native commercially important clams in Tasmania (Walton et al. 2002)	– Isolated areas more susceptible (e.g., Mediterranean and Black seas) – Long-distance dispersal makes eradication difficult
Farmlands and croplands	– Natural pest control – Pollination – Nutrient cycling – Primary production	– Insects – Pathogens – Grasses – Forbs – Birds	Sweet potato whitefly (*Bemisia tabaci*) – Consumes crops, transmits plant viruses and fungi; affects crops and ornamentals (Oliveira et al. 2001) Banana bunchy top virus – Invaded tropical Asia, Africa, Australia by vector aphid; damages fruit (Dale 1987)	– Large economic losses can result from introduced pests and crop-specific pathogens
Forests	– Timber – Nonwood products – Genetic resources – Ornamental resources – Aesthetic value	– Fungal pathogens – Forbs – Shrubs and vines – Insects – Mammals	Chestnut blight (*Cryphonectria parasitica*) Dutch elm disease (*Ophiostoma ulmi*) White pine blister rust (*Cronartium ribicola*) – Species-specific fungal pathogens with negative aesthetic and genetic impacts	– Subsistence economies at risk due to dependence on forest products (Daily 1997)

Table 13.2 (*Continued*)

Ecosystem type	Ecosystem services most at risk	Prevalent invader types	Invader examples and impacts	Other
Fresh waters (rivers, streams, lakes, ponds, wetlands, riparian areas)	- Water purification - Water regulation - Erosion control - Disease regulation - Recreation, tourism	- Aquatic plants - Fish - Mollusks - Amphibians	Zebra mussel (*Dreissena polymorpha*) - Threatens water supply, quality, and native clams following rapid dispersal through Great Lakes (Griffiths et al. 1991) Whirling disease (*Myxobolus cerebralis*) - Threatens trout in rivers in the USA, with impacts on recreation (Koel et al. 2005)	- Isolated lakes very susceptible - Rivers and riparian areas difficult to control; can easily transport propagules
Grasslands and shrublands (including desert and tundra)	- Livestock forage - Genetic resources - Air quality regulation - Nutrient cycling - Cultural heritage	- Grasses - Forbs - Shrubs - Trees - Mammals	Starthistle (*Centaurea solstitialis*) - Decreases livestock forage yield and quality, and depletes soil moisture (Gerlach 2004) Mesquite (*Prosopsis glandulosa*), *Acacia* spp. - Alter nitrogen and carbon cycling in arid lands worldwide (Geesing et al. 2000)	- Invasive species have decreased rangeland quality in many regions of the world
Urban and suburban	- Disease regulation - Aesthetic value - Cultural heritage	- Weedy plants - Small mammals - Birds - Pathogens	House mouse (*Mus musculus*) Norway rat (*Rattus norvegicus*) Grey squirrel (*Sciurus carolinensis*) - Can spread disease, and decrease aesthetic value by invading fragmented landscapes	- Close proximity of humans adds to adverse impacts on disease regulation

13.5 Case Studies and Examples

13.5.1 Provisioning Ecosystem Services

We have identified a range of examples of invasive species that covers a substantial breadth of services, species, and locations. Provisioning services are perhaps the easiest to assess, since impacts occur on a shorter time scale and are often felt more acutely, at least initially, than for other services. Crops are negatively impacted by invasives eating them, such as the European starling (*Sturnus vulgaris*) feeding on grain and fruit crops such as grapes (Somers and Morris 2002), and by decreases in land productivity and agricultural yields. Livestock are impacted indirectly by invasives that decrease forage quality or quantity, such as the unpalatable leafy spurge (*Euphorbia esula*) avoided by cattle in the mid-western United States (Kronberg et al. 1993), or directly by pathogens such as rinderpest, which is fatal to cattle and has led to famines in many parts of the world. Although many economically important crop and livestock species are invasive, they are typically under human management.

Marine food resources can be impacted by invasive predators such as the European green crab (*Carcinus maenus*; Table 13.2), and by competition with invasives such as the comb jelly (*Mnemiopsis leidyi*), which has devastated fisheries in the Black Sea as well as other seas (Shiganova et al. 2001). Impacts of invasives on water resources are among the best studied, particularly in the South African fynbos. Water is a critical resource in this semiarid region, and multiple invasive species, including *Melia azedarach*, pines, wattle (*Acacia mearnsii*), mesquite (*Prosopis* spp.) and *Lantana camara*, have substantially decreased available surface water and streamflow through their high evapotranspiration rates (Gorgens and van Wilgen 2004).

Timber and other structural support materials are particularly susceptible to termite (*Coptotermes* spp.) damage in South America (Constantino 2002) and other parts of the world. Fuel resources such as wood presumably share the same threats. Cotton and other fiber crops are susceptible to various invasive agricultural pests such as the red imported fire ant (*Solenopsis invicta*), which consumes beneficial arthropods (Eubanks 2001). Ornamental resources, especially trees, are susceptible to attack, and even death from the aphid *Cinara cupressi* throughout Europe and Africa (Watson et al. 1999), as well as from pathogens such as *Phytophthora* spp. It is important to note that many invasive plants have been introduced because they have ornamental value, despite negative impacts they may now have caused. Finally, due to their high option value, genetic resources, biochemicals, pharmaceuticals, and the like are at risk whenever there is a loss of biodiversity. Invasives that lead to species extinctions, such as the small Indian mongoose (*Herpestes javanicus*) or the rosy wolf snail (*Euglandina rosea*), may irretrievably alter these services. In addition, invasions into hotspots of biodiversity such as the tropics and aridlands pose significant risks to current and future sources of these provisioning services.

13.5.2 Regulating Ecosystem Services

Invasive species also alter regulating services, with far-reaching effects on human society. Fires release particulates, carbon monoxide and dioxide, and nitrogen oxides, leading to decreased air quality. Thus, invasives such as cheat grass (*Bromus tectorum*) that increase fire frequency will enhance these emissions. In addition, several invasive plants, including kudzu (*Pueraria montana*) and eucalyptus, emit large amounts of isoprene, which is highly reactive in the atmosphere and enhances the production of air pollutants (Wolfertz et al. 2003). Emission of isoprene and other volatile organic compounds also leads to the production of ozone and greenhouse gases such as carbon monoxide and methane, thereby altering climate regulation. On a smaller scale, invasives may alter microclimates. For example, smooth cordgrass (*Spartina alterniflora*) reduces light levels in salt marsh plant canopies, potentially decreasing estuarine algal productivity (Callaway and Josselyn 1992).

Invasives generally have a negative effect on water regulation. Salt cedar (*Tamarix* spp.) forms thickets along riparian corridors enhancing sediment capture and channel narrowing. This has decreased the water holding capacity of many waterways in the southwestern United States, leading to more frequent and extensive flooding and associated flood control costs (Zavaleta 2000). Water purification occurs in multiple types of ecosystems, but most notably in wetlands. The common carp (*Cyprinus carpio*) has been shown to decrease water quality in a degraded wetland in Spain by increasing turbidity and nutrient concentrations (Angeler et al. 2002). Aquatic invasive plants and mollusks may also impact waste treatment by clogging water pipes.

Disease regulation is altered by the invasion of human disease pathogens themselves (e.g., *Vibrio cholerae*, cholera-causing bacteria), or the invasion of disease vectors, particularly invasive mosquitoes such as *Aedes aegypti*, native to Africa, which enhanced the spread of yellow fever in the Americas and of dengue in tropical Asia (Juliano and Lounibos 2005). Natural pest control and pollination are well studied, due to wide recognition of their high economic value. Pest control is altered directly by invasives that consume or compete with either beneficial or detrimental insects, and indirectly by invasives that harbor additional pests. This complicated role is illustrated by the red imported fire ant (*Solenopsis invicta*), an intraguild predator that consumes both insect pests of soybeans and native biological control agents (Eubanks 2001). Impacts on pollination are equally complex. Honey bees (*Apis mellifera*) have been introduced worldwide for pollination services, but research suggests they may competitively displace native bee faunas, which are typically better pollinators (Spira 2001). Invasive plants may also threaten pollination services by luring pollinators from native species, as was shown with *Impatiens glandulifera* in central Europe (Chittka and Schurkens 2001).

Alteration of erosion control is linked to a large number of invasives. Despite the fact that many invasives were originally introduced to dampen erosion, many in fact increase erosion. Examples range from large mammals such as feral pigs (*Sus scrofa domestica*), which uproot plants, disturb soil, and are particularly damaging on islands (Mack and D'Antonio 1998), to small invertebrates such as the isopod *Sphaeroma quoyanum*, which has increased marsh erosion in California due to its burrowing activities (Talley et al. 2001). Since marshes also protect coasts from natural hazards, including hurricanes and strong waves, this loss of sediment is likely to decrease this service as well.

13.5.3 Cultural Ecosystem Services

Alteration of cultural services is far more difficult to assess, given the subjective nature of these services. For example, purple loosestrife (*Lythrum salicaria*) may actually increase the aesthetic value of wetlands for some observers, due to its brightly colored profusion of flowers, whereas others might find the sight distasteful, given their concerns about the species' effects on water quality and wildlife habitat provision. By the same token, the ability of natural ecosystems to provide inspiration is very personal and has the potential to change over time, even for one individual. In addition, the specific cultural, spiritual, religious, or other values held by an individual or group may be unknown. Nevertheless, the impacts of many invasives can be assumed to apply to a majority of individuals. For example, aesthetic values are lost during intense Asian gypsy moth (*Lymantria dispar*) invasions into forests in the northeastern United States, due to defoliation and corresponding high tree mortality (Hollenhorst et al. 1993). Invaders also cause substantial losses to recreation and tourism, particularly ecotourism. Aquatic macrophytes that form dense layers or beds are a notorious nuisance for boating, swimming, and diving. Examples are found worldwide in both fresh and salt water, and include *Caulerpa taxifolia*, *Hydrilla verticillata*, and *Sargassum muticum* (cf. Global Invasive Species Database). Terrestrial invasive plants may also form dense stands crowding out native species, and impacting recreation and tourism by making natural areas less accessible and by potentially reducing wildlife and rare-plant viewing. Examples include *Melaleuca quinquenervia*, *Mimosa pigra*, Japanese knotweed (*Fallopia japonica*), and the cactus *Opuntia stricta* (cf. Global Invasive Species Database).

Several invasives have provided positive recreation and tourism opportunities, especially in the area of fishing. These include large mouth bass

(*Micropterus salmoides*), brown trout (*Salmo trutta*) and rainbow trout (*Oncorhynchus mykiss*; Global Invasive Species Database). To put this in perspective, however, most of these invasives cause damage to other ecosystem services. Educational values are certainly lost whenever species become extinct, particularly in areas with high endemism such as the Galapagos Islands, considered a natural laboratory for evolutionary studies. Several endemic plants are considered to have disappeared from these islands due to *Lantana camara* invasion (Mauchamp et al. 1998). Overall, we conclude that all cultural services are altered by invasive species, with some positive effects, but predominantly negative effects. Despite the challenge in placing monetary values on these services, it is critical to recognize their widespread influence.

13.5.4 Supporting Ecosystem Services

Invasive species also directly alter supporting services. These impacts can be elusive, since they occur on large temporal and spatial scales to services not used directly by humans (i.e., they have non-use value). However, supporting services are necessary for the maintenance of all other services – when invasive species alter these, they often alter other, supported services. Thus, most of the examples given in Sects. 13.3.2 and 13.3.3 are not only mechanisms of alteration by invasive species, but also impacts on supporting services. A few additional examples are presented here. Studies of direct alteration to photosynthesis are limited in number. Aquatic plants that form floating mats, such as water hyacinth (*Eichhornia crassipes*), can decrease macroinvertebrate abundance by blocking light transmission and decreasing photosynthesis by phytoplankton and other plants, leading to anoxic conditions (Masifwa et al. 2001). Primary production may increase or decrease if an invasion leads to a shift in the major vegetation type of an area. In many cases, invasive plants increase net primary productivity, as is the case with giant reed (*Arundo donax*) and *Phragmites* in marshes (Ehrenfeld 2003). However, a recent study of buffelgrass (*Pennisetum ciliare*), which has been introduced to the Sonoran desert in Mexico to serve as cattle forage, shows that converted areas have lower net primary productivity than areas with native desert vegetation (Franklin et al. 2006).

Soil formation may be indirectly affected by changes in decomposition rates, soil carbon mineralization, and geomorphological disturbance processes (e.g., erosion), as well as succession (Mack and D'Antonio 1998). Maintenance of soil fertility is directly connected to nutrient cycling. Japanese barberry (*Berberis thunbergii*) and Japanese stilt grass (*Microstegium vimineum*), which have invaded forests in the eastern United States, can significantly alter microbial communities, leading to changes in nitrification and increased soil nutrient concentrations (Ehrenfeld 2003). Finally, atmospheric composition can be altered by changes in net ecosystem carbon exchange.

Reduced carbon sequestration rates in sagebrush communities invaded by annual grasses (Prater et al. 2006) will contribute to climate warming, illustrating the linkages among these global changes (Chap. 12).

13.6 Conclusions

Across invader taxa, ecosystem types, and geographic locations, invasive species are capable of altering ecosystem services by affecting populations, community interactions, ecosystem processes, and abiotic variables. Virtually all ecosystem services can be negatively impacted by invasive species, although positive impacts do exist. Many invasive species cause cascading effects in communities and/or affect both biotic and abiotic components of ecosystems. This usually leads to an influence on multiple ecosystem services. Different ecosystem types are susceptible to the alteration of specific services. Option values illustrate how invasive species may impact future ecosystem services by threatening native species and communities.

Our assessment found a general lack of work in the area of invasive species and their alteration of ecosystem services. To date, scientific research has focused largely on predicting invasibility, comparing invader and native traits, and assessing environmental impacts, particularly on biodiversity. Ecological economics has generally addressed a limited number of ecosystem services, namely, those with direct market valuation. More recently, several papers have examined the causal mechanisms underlying invasive species' impacts. These studies have begun to link invasive species, ecosystem structure and function, and ecosystem goods and services. Several studies also hint at impacts to ecosystem services, but do not directly address these services. Research in this area is critical for several reasons. First, impact assessments for invasive species are not complete without considering implications for human society. Comprehensive assessments allow us to better predict impacts, particularly for species in similar taxa. Second, this research has the potential to increase our understanding of invasive impacts on ecosystem structure and function outside the domain of ecosystem services. Because invasive species' impacts on ecosystem services overlap with environmental impacts (e.g., altered biodiversity), scientists will gain knowledge relating to impacts on all native species. This may also lead to advances in understanding invasibility and community interactions. Third, increased awareness of invasive species' impacts could inform decisions on allocating resources for the control of invasives, and for the protection of ecosystem services and "natural" ecosystems. Finally, increased research efforts will be critical in predicting the effects of invasive species in conjunction with other global changes, including climate and land use, which have been shown to affect ecosystem service supply (Schroter et al. 2005). Dialogue between ecologists, economists, and policymakers is critical to moving this research agenda forward.

The four categories of ecosystem services provide a useful framework for assessing our overall knowledge of invasive species' impacts on ecosystem services. Table 13.3 gives a qualitative assessment of several aspects of these four types of services, and suggests a path forward by identifying areas currently lacking research. In particular, supporting and regulating services both have a high value, but a low level of research. Given that their susceptibility to invasive impacts is uncertain and high, respectively, this is evidently an area where research is needed. Recognition of the value of ecosystem services, and the many examples and mechanisms by which invasive species affect ecosystem services, leads to several additional opportunities. The general public is still largely unaware of the extent of invasive species' impacts. In addition, society does not often appreciate the extent of its dependence on natural ecosystems (Daily 1997). This creates an opportunity to educate the general public about both issues in tandem, leading to better understanding and appreciation for both. Specific examples of alteration to ecosystem services will also allow policymakers and land managers to prioritize eradication and control campaigns. As with many unquantified threats to human society attributable to global changes, it would be prudent to err on the side of caution in estimating and managing the threats posed by invasive species (i.e., the precautionary principle). As our understanding of the links between invasive species, ecosystem structure and function, and provision of ecosystem goods and services increases, so too will our ability to recognize invasive species' impacts on ecosystem services, and to better manage these impacts.

Table 13.3 Qualitative assessment of the value of ecosystem services and current knowledge of their susceptibility to, and the amount of research focused on, invasive species' impacts

Services	Provisioning	Regulating	Cultural	Supporting
Value	High	High	Medium	Very high
Susceptibility to to alteration by invasive species	High	High	Medium to low	Uncertain
Amount of research on invasive impacts	Medium	Low	Medium	Very low

References

Angeler DG, Alvarez-Cobelas M, Sanchez-Carrillo S, Rodrigo MA (2002) Assessment of exotic fish impacts on water quality and zooplankton in a degraded semi-arid floodplain wetland. Aquat Sci 64:76–86

Born W, Rauschmayer F, Brauer I (2005) Economic evaluation of biological invasions – a survey. Ecol Econ 55:321–336

Callaway JC, Josselyn MN (1992) The introduction and spread of smooth cordgrass (*Spartina alterniflora*) in South San-Francisco Bay. Estuaries 15:218–226

Callaway RM, Ridenour WM (2004) Novel weapons: invasive success and the evolution of increased competitive ability. Frontiers Ecol Environ 2:436–443

Carlsson NOL, Bronmark C, Hansson LA (2004) Invading herbivory: the golden apple snail alters ecosystem functioning in Asian wetlands. Ecology 85:1575–1580

Chapin FS, Reynolds HL, D'Antonio CM, Eckhart VM (1996) The functional role of species in terrestrial ecosystems. In: Walker B, Steffen W (eds) Global change and terrestrial ecosystems. Cambridge University Press, Cambridge, pp 403–428

Chapin FS, Zavaleta ES, Eviner VT, Naylor RL, Vitousek PM, Reynolds HL, Hooper DU, Lavorel S, Sala OE, Hobbie SE, Mack MC, Diaz S (2000) Consequences of changing biodiversity. Nature 405:234–242

Chittka L, Schurkens S (2001) Successful invasion of a floral market – an exotic Asian plant has moved in on Europe's river-banks by bribing pollinators. Nature 411:653

Constantino R (2002) The pest termites of South America: taxonomy, distribution and status. J Appl Entomol 126:355–365

Costanza R, d'Arge R, de Groot R, Farber S, Grasso M, Hannon B, Limburg K, Naeem S, O'Neill RV, Paruelo J, Raskin RG, Sutton P, van den Belt M (1997) The value of the world's ecosystem services and natural capital. Nature 387:253–260

Daily GC (ed) (1997) Nature's services: societal dependence on natural ecosystems. Island Press, Washington, DC

Dale JL (1987) Banana bunchy top - an economically important tropical plant-virus disease. Adv Virus Res 33:301–325

Dukes JS, Mooney HA (2004) Disruption of ecosystem processes in western North America by invasive species. Rev Chil Hist Nat 77:411–437

Ehrenfeld JG (2003) Effects of exotic plant invasions on soil nutrient cycling processes. Ecosystems 6:503–523

Eiswerth ME, Donaldson SG, Johnson WS (2000) Potential environmental impacts and economic damages of Eurasian watermilfoil (*Myriophyllum spicatum*) in western Nevada and northeastern California. Weed Technol 14:511–518

Eiswerth ME, Darden TD, Johnson WS, Agapoff J, Harris TR (2005) Input-output modelling, outdoor recreation, and the economic impacts of weeds. Weed Sci 53:130–137

Engeman RM, Smith HT, Shwiff SA, Constantin B, Woolard J, Nelson M, Griffin D (2003) Prevalence and economic value of feral swine damage to native habitat in three Florida state parks. Environ Conserv 30:319–324

Eubanks MD (2001) Estimates of the direct and indirect effects of red imported fire ants on biological control in field crops. Biol Control 21:35–43

Franklin KA, Lyons K, Nagler PL, Lampkin D, Glenn EP, Molina-Freaner F, Markow T, Huete AR (2006) Buffelgrass (*Pennisetum ciliare*) land conversion and productivity in the plains of Sonora, Mexico. Biol Conserv 127:62–71

Fritts TH, Rodda GH (1998) The role of introduced species in the degradation of island ecosystems: a case history of Guam. Annu Rev Ecol Syst 29:113–140

Geesing D, Felker P, Bingham RL (2000) Influence of mesquite (*Prosopis glandulosa*) on soil nitrogen and carbon development: implications for global carbon sequestration. J Arid Environ 46:157–180

Gerlach JD (2004) The impacts of serial land-use changes and biological invasions on soil water resources in California, USA. J Arid Environ 57:365–379

Global Invasive Species Database (http://www.issg.org/database)

Gorgens AHM, van Wilgen BW (2004) Invasive alien plants and water resources in South Africa: current understanding, predictive ability and research challenges. S Afr J Sci 100:27–33

Griffiths RW, Schloesser DW, Leach JH, Kovalak WP (1991) Distribution and dispersal of the zebra mussel (*Dreissena polymorpha*) in the Great-Lakes Region. Can J Fish Aquat Sci 48:1381–1388

Hoffmann WA, Jackson RB (2000) Vegetation-climate feedbacks in the conversion of tropical savanna to grassland. J Climate 13:1593–1602

Hollenhorst SJ, Brock SM, Freimund WA, Twery MJ (1993) Predicting the effects of gypsy-moth on near-view aesthetic preferences and recreation appeal. Forest Sci 39:28–40

Hooper DU, Chapin FS, Ewel JJ, Hector A, Inchausti P, Lavorel S, Lawton JH, Lodge DM, Loreau M, Naeem S, Schmid B, Setala H, Symstad AJ, Vandermeer J, Wardle DA (2005) Effects of biodiversity on ecosystem functioning: a consensus of current knowledge. Ecol Monogr 75:3–35

Huxman TE, Cable JM, Ignace DD, Eilts JA, English NB, Weltzin J, Williams DG (2004) Response of net ecosystem gas exchange to a simulated precipitation pulse in a semiarid grassland: the role of native versus non-native grasses and soil texture. Oecologia 141:295–305

Juliano SA, Lounibos LP (2005) Ecology of invasive mosquitoes: effects on resident species and on human health. Ecol Lett 8:558–574

Koel TM, Bigelow PE, Doepke PD, Ertel BD, Mahony DL (2005) Nonnative lake trout result in Yellowstone cutthroat trout decline and impacts to bears and anglers. Fisheries 30:10–19

Kronberg SL, Muntifering RB, Ayers EL, Marlow CB (1993) Cattle avoidance of leafy spurge – a case of conditioned aversion. J Range Manage 46:364–366

Levine JM, Vila M, D'Antonio CM, Dukes JS, Grigulis K, Lavorel S (2003) Mechanisms underlying the impacts of exotic plant invasions. Proc Roy Soc Lond B Biol 270:775–781

Mack MC, D'Antonio CM (1998) Impacts of biological invasions on disturbance regimes. Trends Ecol Evol 13:195–198

Masifwa WF, Twongo T, Denny P (2001) The impact of water hyacinth, *Eichhornia crassipes* (Mart) Solms on the abundance and diversity of aquatic macroinvertebrates along the shores of northern Lake Victoria, Uganda. Hydrobiologia 452:79–88

Mauchamp A, Aldaz I, Ortiz E, Valdebenito H (1998) Threatened species, a re-evaluation of the status of eight endemic plants of the Galapagos. Biodiversity Conserv 7:97–107

Millennium Ecosystem Assessment (2005) Ecosystems and human well-being: synthesis. Island Press, Washington, DC

NRC (2005) Valuing ecosystem services: toward better environmental decision-making. National Research Council, National Academy Press, Washington, DC

Naylor R (1996) Invasions in agriculture: assessing the cost of the golden apple snail in Asia. Ambio 25:443–448

Oliveira MRV, Henneberry TJ, Anderson P (2001) History, current status, and collaborative research projects for *Bemisia tabaci*. Crop Protect 20:709–723

Prater MR, Obrist D, Arnone JA, DeLucia EH (2006) Net carbon exchange and evapotranspiration in postfire and intact sagebrush communities in the Great Basin. Oecologia 146:595–607

Schroter D, Cramer W, Leemans R, Prentice IC, Araujo MB, Arnell NW, Bondeau A, Bugmann H, Carter TR, Gracia CA, de la Vega-Leinert AC, Erhard M, Ewert F, Glendining M, House JI, Kankaanpaa S, Klein RJT, Lavorel S, Lindner M, Metzger MJ, Meyer J, Mitchell TD, Reginster I, Rounsevell M, Sabate S, Sitch S, Smith B, Smith J, Smith P, Sykes MT, Thonicke K, Thuiller W, Tuck G, Zaehle S, Zierl B (2005) Ecosystem service supply and vulnerability to global change in Europe. Science 310:1333–1337

Serbesoff-King K (2003) *Melaleuca* in Florida: a literature review on the taxonomy, distribution, biology, ecology, economic importance and control measures. J Aquat Plant Manage 41:98–112

Shiganova TA, Mirzoyan ZA, Studenikina EA, Volovik SP, Siokou-Frangou I, Zervoudaki S, Christou ED, Skirta AY, Dumont HJ (2001) Population development of the invader ctenophore *Mnemiopsis leidyi*, in the Black Sea and in other seas of the Mediterranean basin. Mar Biol 139:431–445

Simberloff D, Von Holle B (1999) Positive interactions of nonindigenous species: invasional meltdown? Biol Invasions 1:21–32

Somers CM, Morris RD (2002) Birds and wine grapes: foraging activity causes small-scale damage patterns in single vineyards. J Appl Ecol 39:511–523

Spira TP (2001) Plant-pollinator interactions: a threatened mutualism with implications for the ecology and management of rare plants. Nat Area J 21:78–88

Talley TS, Crooks JA, Levin LA (2001) Habitat utilization and alteration by the invasive burrowing isopod, *Sphaeroma quoyanum*, in California salt marshes. Mar Biol 138:561–573

The H. John Heinz III Center for Science Economics and the Environment (2002) The state of the nation's ecosystems. Cambridge University Press, Cambridge

Turpie JK, Heydenrych BJ, Lamberth SJ (2003) Economic value of terrestrial and marine biodiversity in the Cape Floristic Region: implications for defining effective and socially optimal conservation strategies. Biol Conserv 112:233–251

Verlaque M (1994) Checklist of introduced plants in the Mediterranean – origins and impact on the environment and human activities. Oceanol Acta 17:1–23

Vivrette NJ, Muller CH (1977) Mechanism of invasion and dominance of coastal grassland by *Mesembryanthemum crystallinum*. Ecol Monogr 47:301–318

Walton WC, MacKinnon C, Rodriguez LF, Proctor C, Ruiz GA (2002) Effect of an invasive crab upon a marine fishery: green crab, *Carcinus maenas*, predation upon a venerid clam, *Katelysia scalarina*, in Tasmania (Australia). J Exp Mar Biol Ecol 272:171–189

Watson GW, Voegtlin DJ, Murphy ST, Foottit RG (1999) Biogeography of the *Cinara cupressi* complex (Hemiptera: Aphididae) on Cupressaceae, with description of a pest species introduced into Africa. Bull Entomol Res 89:271–283

Wolfertz M, Sharkey TD, Boland W, Kuhnemann F, Yeh S, Weise SE (2003) Biochemical regulation of isoprene emission. Plant Cell Environ 26:1357–1364

Zavaleta E (2000) The economic value of controlling an invasive shrub. Ambio 29:462–467

14 Biological Invasions by Marine Jellyfish

William M. Graham and Keith M. Bayha

14.1 Introduction

Comparatively little research has been conducted on the ecology of invasive organisms in marine ecosystems when balanced against their terrestrial counterparts (Carlton and Geller 1993). Perhaps rates of invasions in marine systems are simply lower than in other systems, but more likely lack of scrutiny, difficulty with taxonomic resolution, and unusual life-history characters of marine organisms cause the vast majority of invasions to go unreported. Regardless, those few well-studied marine invasions have resulted in tremendous ecological, economic, social, and health problems (e.g., Carlton et al. 1990; Hallegraeff and Bolch 1992; Kideys 1994; Grosholz and Ruiz 1995; Chaps. 4 and 5). Among marine communities that have been extensively studied (e.g., the Chesapeake Bay, San Francisco Bay, and the Black Sea), non-indigenous species are extremely common, and encompass a broad range of taxonomic and trophic groups (Ruiz et al. 1997). Moreover, many marine communities contain remarkably large numbers of 'cryptogenic' species (i.e., species that have unknown origins) that are, in fact, likely to have been introduced (Carlton 1996; Dawson et al. 2005).

Gelatinous zooplankton, broadly referred to here as jellyfish, extend systematically across several phyla including Cnidaria (medusae, siphonophores, and a variety of other forms), Ctenophora (comb jellies), and Chordata (pelagic tunicates including the sea gooseberries and close allies). Not unexpectedly, jellyfish present an unusual and difficult set of issues when considering their potential as bioinvaders. The first, and perhaps the most critical, is the probability that far more invasions have actually occurred than are being described in the literature, due to incomplete historical systematic treatment, generally poor taxonomic appreciation by non-specialists, and species crypsis. Contributing to the latter is the ability of jellyfish to exhibit morphological plasticity when introduced into a new abiotic or trophic environment. Such plasticity creates great confusion over diagnostic traits used in species identification. A

Ecological Studies, Vol. 193
W. Nentwig (Ed.)
Biological Invasions

second issue is that most jellyfish have the ability to appear or disappear suddenly, creating difficulty in tracking the progression or tracing the history of invasion. Jellyfish 'blooms' are also implicated in dramatic ecological and economic problems ranging from fisheries damage to the clogging of seawater intakes of power plants and ships. Third, life-histories of gelatinous animals often involve cryptic benthic stages, rapid asexual propagation, or self-fertilization. This also severely limits our ability to understand origins and fates of invasions. We present below details of a few important jellyfish invasions, including medusae and ctenophores (note that no documented reports of invasive planktonic tunicates exist), and then discuss further critical issues impeding a complete understanding of jellyfish invasions.

14.2 Ctenophores

14.2.1 *Mnemiopsis leidyi*

The premiere example of a marine bioinvasion is the introduction of the ctenophore *Mnemiopsis leidyi* into the Black Sea during the late 1980s. The chronological history of the *Mnemiopsis* invasion of the Black Sea ecosystem has been extensively reviewed elsewhere (GESAMP 1997; Bilio and Niermann 2004), but a brief synopsis in provided here. *M. leidyi* is native to the eastern seaboard of the Americas, from Massachusetts (USA) to the Patagonian coast of Argentina, including the Caribbean Sea (Agassiz 1860; Mayer 1912; Mianzan 1999). In the Black Sea region, *M. leidyi* was first reported from the coast of the Crimean peninsula in 1982, and had spread to the whole of the Black Sea and the Sea of Azov by 1988 (Vinogradov et al. 1989; GESAMP 1997). By this time, the ctenophore population had increased dramatically, with local biomass reaching 1.2 kg m^{-3} (Vinogradov et al. 1989) to 2.0 kg m^{-3} (Mutlu et al. 1994). Vinograov et al. (1989) estimated a total of 840,000,000 t over the entire sea.

By the early 1990s, ctenophore populations had spread through the Sea of Marmara to include the Aegean and the eastern Mediterranean (Kideys and Niermann 1993, 1994). While biomass declined modestly during the 1990s, occasional intense blooms, such as one in 1995, still occurred (Shiganova et al. 2001; Kideys and Romanova 2001). Major concern among scientists and resource managers grew when *Mnemiopsis* was found in the Caspian Sea in 1999. Spread was rapid, and the species occupied the entire sea by 2000 (Ivanov et al. 2000), with concentrations as high as 2,285 individuals m^{-2} (Kideys and Moghim 2003).

Coincident with increased ctenophore biomass was substantial reduction in fishery harvest, including the collapse of the planktivorous anchovy

Engraulis encrasiclolus fishery. In 1990, total harvest of this species was 66,000 t, which reflected a ~78 % reduction from 295,000 t harvested in 1988 (GFCM 1993, as referenced in Kideys 1994). It was hypothesized that *M. leidyi* triggered the massive decline in the harvest of *E. encrasicolus* both by direct feeding on eggs and larvae, and by competition with adults and larval anchovies for zooplankton food (Vinogradov and Shushkina 1992; Kideys 1994). In the Caspian Sea, reductions in zooplankton biomass and diversity followed *Mnemiopsis* blooms, to the extent that only the calanoid copepod *Acartia* could be found in measurable concentrations (Stone 2005). Similarly to harvest reductions in the Black Sea, catches of the zooplanktivorous fish *Clupeonella* sp. dropped precipitously in the Caspian Sea (Kideys 2001a, 2001b, as referenced in Kideys 2002 and Stone 2005).

It was previously assumed that invasions into both the Black and Caspian seas were mediated by ballast water discharge (Vinogradov et al. 1989; Ivanov et al. 2000). Although it has been both insinuated (Kideys 2002; Shiganova et al. 2003) and explicitly illustrated (http://www.zin.ru/projects/invasions/gaas/mnelei_d.htm) that the invaded *M. leidyi* populations originated from the Atlantic seaboard of the United States, we emphasize that there has been no definitive evidence to support this. A recent genetic study indicated a source region in the Southern US Atlantic (south of Cape Hatteras) or Gulf of Mexico, and could rule out South America but not the Caribbean Sea as source of the Black/Caspian Sea invaders (Bayha 2005). Given the widespread natural range of this ctenophore, as well as its subtle population genetic structure, the source region remains unresolved.

14.2.2 *Beroë ovata*

The invasion by *Mnemiopsis leidyi* was followed in 1997 by that of *Beroë ovata* (Vinogradov et al. 2000), a ctenophore that feeds almost exclusively on other ctenophores, especially *Mnemiopsis* (Kremer and Nixon 1976). With the invasion of *Beroë* into the Black Sea, *Mnemiopsis* concentrations decreased (Shiganova et al. 2001, 2003) and zooplankton concentrations, including fish eggs and larvae, increased (reviewed by Kideys 2002; Gordina et al. 2005). Given the perceived success the *Beroë* invasion has had in the Black Sea, there have been calls for the purposeful introduction of this ctenophore predator into the Caspian Sea. However, with the checkered past history of purposeful animal introductions (Chap. 23), many have advised caution for the Caspian Sea, until *Beroë*'s survival (Volovik and Korpakova 2004; Kideys et al. 2004a) and trophic ecology (Kideys et al. 2004b) are studied further.

14.3 Medusae (Cnidaria)

14.3.1 *Phyllorhiza punctata* (Scyphozoa)

The scyphomedusa *Phyllorhiza punctata* appeared suddenly and in spectacular numbers in the northern Gulf of Mexico during the summer of 2000 (Graham et al. 2003). *Phyllorhiza punctata* was first described from Pt. Jackson, Australia (von Lendenfeld 1884), and, presumably due to its conspicuous medusa stage, has a relatively well-documented history of invading tropical and subtropical environments around the globe over the past 200 years. The first recorded invasion of *P. punctata* was into the Swan-Canning estuary, Perth (Western Australia) between 1837–1838, presumably by early Australian shipping (Rippingale and Kelly 1995). *P. punctata* subsequently appeared in Pearl Harbor, Hawaii, USA (1941; Devaney and Eldridge 1977), Laguna Joyunda, Puerto Rico (1945 at the latest; Garcia-Sais and Durbin 1993), San Diego, California, USA (1990 at the latest; Colin and Arneson 1995), Danajon Bank, Bohol Island, The Philippines (Heeger et al. 1992), and Bahia de Todos os Santos, Brazil (between 1991–1999; da Silveira and Cornelius 2000). *P. punctata* also appeared in the eastern Mediterranean (in 1965, as referenced by Galil et al. 1990), but apparently did not persist there. This well-documented history of invasion over an extended period may make *P. punctata* a particularly instructive model toward understanding invasions of jellyfish in particular, and of invasive marine species in general.

The occurrence of the very large population estimated at 10×10^6 medusae across the north-central Gulf of Mexico (Alabama, Mississippi, and Louisiana) in the summer of 2000 was unexpected, as the species had never been documented north of the Caribbean Sea. The timing of the occurrence was coincident with the incursion of tropical water from the Caribbean into the Gulf of Mexico. Using a modeling approach to assess transport mechanisms via currents measured during the 2000 bloom, Johnson et al. (2005) were able to suggest a plausible advection of medusae out of the Caribbean Sea by the northward-flowing Loop Current in the Gulf of Mexico. However, in the absence of direct evidence of transport of medusae from the Caribbean to the northern Gulf via this current (Johnson et al. 2005), we must also consider that cryptic populations have existed for some time in the Gulf of Mexico. This is supported by anecdotal reports by fishermen of this species occurring as small populations around coastal Louisiana for at least several years prior to 2000. *P. punctata* has been documented in the coastal waters of south-central Louisiana in each summer since 2000 (unpublished data).

The potential ecological and economic impacts of *P. punctata* were judged as high, and it was feared that the ecology of the northern Gulf of Mexico would be altered permanently, along with the valuable fishing industry that depends on it (Graham et al. 2003). These fears were fueled by the costs of *P.*

punctata to the shrimp industry – for Mississippi alone, these have been estimated to be US$ 10 million for 2000 (Graham et al. 2003, and references therein).

14.3.2 *Cassiopea andromeda* (Scyphozoa)

The 'upside-down' jellyfishes (*Cassiopea* spp.) are common in tropical to subtropical shallow water ecosystems such as mangroves, in both the Atlantic and Pacific, and including the Indo-Pacific. While the medusa stage is capable of swimming, it is best described as sedentary, with the algal symbiont-containing oral mass oriented upward (i.e., bell pointing downward). Large dispersal distances by this medusa are not realistic, lending further credence to the probability that invasive scyphozoan populations are spread during the polypoid stage. While human health, fisheries, or other commercial impacts are not documented in assessments of either indigenous or invasive populations (Spanier 1989; Holland et al. 2004), the sedentary nature of *Cassiopea* spp. medusae allows us to gain some insight into possible human-mediated spread of other problematic jellyfish.

The systematics of *Cassiopea*, similarly to that of other jellyfish, is complicated by historical introductions. Holland et al. (2004) derived a global molecular phylogeny of *Cassiopea* spp., in an attempt to gain historical clarity on the putative *C. andromeda* invasion of the Hawaiian islands. Using mitochondrial sequence information (cytochrome c oxidase I), they resolved six species based upon reasonable genetic divergence. Moreover, they proposed that the invasive population of Hawaiian *C. andromeda* represents two separate invasion events, one from the Indo-Pacific, the other from the Atlantic. Holland et al. (2004) suspected that the Indo-Pacific invasion was due to US naval ships during World War II. However, ship activity between the Hawaiian islands and the western Indo-Pacific was also heavy in the decades prior to the war.

Another invasion of *C. andromeda* into the Mediterranean Sea is perhaps the only instance of a 'smoking gun' where the actual invasion event has been observed. Galil et al. (1990) reviewed several early publications not readily available, but descriptive nonetheless of a sequence of interesting events. The first was the observation of 'large numbers' of *C. andromeda* within the Suez Canal itself near Toussuom south of Lake Timsah in 1886, only 17 years after completion of the canal (Keller 1888, as referenced in Galil et al. 1990). This was followed 15 years later by the first documented report of *C. andromeda* in the Mediterranean along the coast of Cyprus (Maas 1903, as referenced in Galil et al. 1990).

14.3.3 *Rhopilema nomadica* (Scyphozoa)

Another example of Lessepsian invasion (Chap. 5) is the tropical medusa, *Rhopilema nomadica*. Recognized by sea bathers as a painful stinger, seasonal blooms of *R. nomadica* create problems along recreational beaches of the eastern Mediterranean (Lotan et al. 1993, 1994; Gusmani et al. 1997). Medusae concentrations reported by Lotan and colleagues during the late 1980s were on the order of '600,000 per nautical mile' (Lotan et al. 1992, 1993), sufficient to create local problems such as clogging of fishing nets. Kideys and Gucu (1995) suggest the first appearance of this species into the Mediterranean to be in the mid-1970s. Subsequent to the initial period of colonization, populations were qualitatively noted to increase (Galil et al. 1990). The species appears to be limited to the eastern Mediterranean Sea, with reports of it occurring along the coasts of Egypt, Israel, Lebanon, Syria, and Turkey (Galil et al. 1990; Kideys and Gucu 1995).

14.3.4 *Aurelia* spp. (Scyphozoa)

The genus *Aurelia*, known as moon jellyfish, is a conspicuous member of coastal ecosystems from polar to tropical seas (Kramp 1961, 1970). As perhaps the most studied of all the scyphomedusae (literally 100s of publications, ranging from fundamental ecology to microgravitational impacts on sensory development), the genus is known to school children and scientists alike. Until recently, three generally accepted species had been recorded within the genus: the polar *A. limbata*, the north Pacific *A. labiata*, and the cosmopolitan *A. aurita* (Dawson and Jacobs 2001). However, recent molecular genetic work by Dawson and colleagues has described a far more diverse genus (Dawson and Jacobs 2001; Dawson and Martin 2001; Dawson 2003; Dawson et al. 2005), with most species considered 'cryptic', since general morphological characters alone are not sufficient to differentiate between these (sensu Mayr and Ashlock 1993). Surprising to many, the common moon jellyfish was not a single species, *A. aurita*, but perhaps as many as 12 species (Dawson 2003) with enough morphological similarity to confuse traditional taxonomists (Sect. 14.6).

Since medusae of *Aurelia* spp., similarly to those of most scyphomedusae, persist for weeks to months, diffusion processes, ocean currents, and active swimming could potentially disperse this stage over 1,000s of kilometers (e.g., Johnson et al. 2005). However, Dawson et al. (2005) showed that, despite the high dispersal potential of *Aurelia* spp., a molecular phylogeny of the genus exhibits substantial biogeographic regionalization, indicating that genetic isolation is more common than previously recognized.

Dawson et al. (2005) did note, however, a key exception where one cryptic species (*Aurelia* sp. 1) showed global distribution likely related to historical shipping activity. They concluded, based on this species' limited ability to traverse the Pacific Ocean, that its global distribution was invasive and mediated (possibly multiple times) by shipping (Dawson 2003; Dawson et al. 2005). Another of their species (*Aurelia* sp. 4) was also identified as invasive in Hawaii from an Indo-Pacific origin (Dawson et al. 2005). In sum, these findings illustrate a realization that many of our highly recognizable 'cosmopolitan' species are, in fact, probably historically invasive.

14.3.5 *Maeotias marginata*, *Blackfordia virginica*, and *Moerisia lyonsii* (Hydrozoa)

Three species of invasive hydromedusae are noteworthy: *Maeotias marginata*, *Blackfordia virginica*, and *Moerisia lyonsii*. All three are believed native to the Black Sea/Sea of Azov region: *M. marginata*: Borcea (1928) and Ostroumoff (1896), both as discussed in Mills and Sommer (1995); *B. virginica*: Thiel (1935), as discussed in Mills and Sommer (1995); *M. lyonsii*: Kramp (1959), as discussed in Calder and Burrell (1967). However, all three have successfully invaded regions of the Atlantic, Pacific, and Indian (except *M. marginata*) oceans (Calder and Burrell 1967; Mills and Sommer 1995; Mills and Rees 2000; Väinölä and Oulasvirta 2001). Interestingly, all have been found in the Chesapeake Bay (Mayer 1910; Calder and Burrell 1967, 1969) and San Francisco Bay (Mills and Sommer 1995; Mills and Rees 2000). While no major negative impacts have been described for any of these three invasive hydromedusae, *M. lyonsii* was noted by several authors for its ability to foul experimental or culturing mesocosms (Sandifer et al. 1974; Petersen et al. 1998; Purcell et al. 1999).

14.4 Jellyfish Invasions: Blooms and Ecosystem Controls

Gelatinous zooplankton blooms exert tremendous pressure on marine planktonic food webs, including in regions of important commercial fisheries (Purcell 1985, 1989; Purcell and Arai 2001; Purcell et al. 2001). In its native range *M. leidyi* seasonally forms large blooms, especially in enclosed bays and estuaries, exerting significant predation pressure on zooplankton species (Feigenbaum and Kelly 1984; Purcell and Decker 2005), as well as on the eggs and larvae of economically important fish and shellfish (Purcell et al. 1991, 1994). This is especially true for the Chesapeake Bay (USA), where *M. leidyi*, along with the scyphozoan jellyfish *Chrysaora quinquecirrha*, has been termed a keystone predator (Purcell and Decker 2005). Because *M. leidyi* has broad environmental tolerances to both salinity and temperature, inhabiting both estuarine and coastal regions (Kremer 1994; GESAMP 1997), its 'invasiveness' is perceived to be particularly high.

Interestingly, though, definitive negative relationships between fish populations and gelatinous zooplankton predation or competition are difficult to resolve. Still, if a common relationship between jellyfish and fisheries does exist, it could be best exemplified by the ctenophore *Mnemiopsis leidyi* invasion into the Black Sea and Sea of Azov. The arrival of this ctenophore into the Black Sea in the early 1980s (Kideys 1994; Shiganova 1998; Shiganova and Bulgakova 2000) resulted in heavy predation on the eggs and larvae of anchovies, as well as on a shared zooplankton prey resource of anchovies, and likely contributed to the collapse of the regionally important anchovy fishery (Sect. 14.2.1).

The connection between increased ctenophore biomass and the collapse of fisheries is a compelling one, and the reduction in the Black Sea anchovy harvest has repeatedly been linked to food competition and predation by large ctenophore populations (Kideys 1994; GESAMP 1997). Recent studies, however, triggered a reexamination of this generally accepted interpretation (reviewed by Bilio and Niermann 2004), and indicated that the answer is far more complex than the simple top-down pressures of predation and competition. For instance, modeling work (Gucu and Oguz 1998; Wiesse et al. 2002) has indicated the Black Sea would be incapable of sustaining the 840,000,000 t of *Mnemiopsis* estimated by Vinogradov et al. (1989), and cited errors in interpolation. Other studies have indicated that over-harvesting of anchovy may have had a greater role in the fishery collapse than have had either ctenophore predation or food competition (Gucu 2002). Additionally, a reanalysis of the economic impact of the *Mnemiopsis* invasion by Knowler (2005) indicated that earlier assessments may have overestimated this impact as much as tenfold.

Bilio and Niermann (2004) concluded that the Black Sea anchovy collapse was probably due to numerous factors, including over-harvesting of the anchovy stock exacerbated by predation and food competition by *Mnemiopsis*, as well as a long-term regime shift in the composition of the Black Sea plankton that may have favored *Mnemiopsis*. The significance of the ctenophore's invasion impact in explaining the anchovy collapse is not entirely clear (although likely substantial), but that the story is more complicated than originally thought is not surprising, given the many factors negatively influencing anchovy populations in a water body as anthropogenically impacted as the Black Sea.

14.5 The Role of Life-Histories

Studies into the ecology of invasions by hydromedusae and scyphomedusae are hampered by their complex bipartite life-history: a typically cryptic sessile, asexually reproducing polypoid stage is followed by a pelagic, sexually reproducing medusa stage. Depending upon food availability and other environmental variables, the polypoid stages of scyphozoans can asexually produce large numbers of planktonic young medusae, or remain dormant for extended periods, perhaps even years or decades. Such a life-history trait is not only important as a potential vector (Sect. 14.7), but makes monitoring for invasions all the more difficult, since polyps are typically inconspicuous members of a larger fouling community. In addition, similarly to the *P. punctata* invasion of the Gulf of Mexico in 2000, little warning would likely precede the sudden onset of an invasive jellyfish bloom.

Lotan et al. (1992) conducted a study involving the culture and growth of the asexual polyp stage of *Rhopilema nomadica*. They determined that the period from planula settlement, through polyp generation, to the release of new medusae was on the order of 3–4 months. This would suggest that several cohorts of medusae could be produced from newly recruited planula larvae 2–3 times per year. Likewise, *Phyllorhiza punctata* also has a bipartite life-history involving both medusa and polyp. Rippingale and Kelly (1995) provided laboratory measured growth rates of polyps, and they report fully grown polyps after only 2–3 days post-settlement of planula larvae. This may be of little value to the Gulf of Mexico invasive population, as this *P. punctata* population is entirely male (Graham et al. 2003; Bolton and Graham 2004). [Interestingly, a purported invasive population of *Cassiopea andromeda* sampled along fish farm canals on Oahu (Hawaii) by Hofmann and Hadfield (2002) was also entirely male, though the ecological implications of this observation were not discussed.]

While the ctenophore *Mnemiopsis* is holoplanktonic (i.e., no sessile stage), it has its own life-history peculiarity. *Mnemiopsis* is a simultaneous hermaphrodite capable of an extremely high degree of fecundity (Mayer 1912; Reeve et al. 1989). This ctenophore is capable of reproduction within days of hatching (Martindale 1987), producing thousands of eggs with very little energy investment (Reeve et al. 1989). As a result, *Mnemiopsis* is able to very rapidly increase its population size in response to higher food concentrations (Feigenbaum and Kelly 1984; Kremer 1994). Accordingly, an initial invading population, say contained within ballast water, could be very small (theoretically, only one), but still lead to a successful invasion.

14.6 Taxonomic Confusion, Species Crypsis, and Morphological Plasticity

Historical taxonomic confusion has often complicated initial efforts to study jellyfish invasions, and nowhere is that more true than in the case of the Black Sea ctenophore invasions. Initially, the Black Sea *Mnemiopsis* was identified alternatively as *M. mccradyi* (Zaika and Sergeeva 1990) or *M. leidyi* (Vinogradov et al. 1989), the two most recently recognized *Mnemiopsis* species. However, many doubted these terminologies (Seravin 1994; Harbison and Volovik 1994), and the validity of the two *Mnemiopsis* species has been questioned based on morphological grounds (Seravin 1994; Harbison and Volovik 1994; Oliveira and Migotto 2006). A recent molecular study has indicated that one species exists worldwide, i.e., *M. leidyi* (Bayha 2005). Another notable example of such confusion is that of the ctenophore *Beroë*. The same name (*B. ovata*) had historically been used for two morphologically different animals (one from the western Atlantic and Caribbean, the other from the Mediterranean), and it was initially believed that *Beroë* may have invaded from the Mediterranean. However, both genetic (Bayha et al. 2004) and morphological (Seravin et al. 2002) evidence indicated that the invasive animal has a western Atlantic origin (eastern seaboard of the Americas and the Caribbean). Consequently, Bayha et al. (2004) proposed that, pending a thorough systematic revision of the genus *Beroë*, the species be termed *Beroë ovata* sensu Mayer (as opposed to the incorrect *Beroë ovata* Mayer 1912).

Species crypsis and morphological plasticity can also hinder efforts to study invasive animals. Invasive patterns of *Aurelia* and *Cassiopea* are excellent examples of the problems associated with species crypsis. Neither invasion was fully appreciated until genetic techniques were employed (Greenberg et al. 1996; Holland et al. 2004; Dawson et al. 2005), because the invaders were historically confused with morphologically similar natives. Where species crypsis confuses multiple species for one, morphological plasticity may confound species invasion patterns by having only one invading species recognized as multiple others. This phenomenon has hindered studies of the invasion of *Phyllorhiza punctata* into the Gulf of Mexico and elsewhere (Bolton and Graham 2004). Again, only molecular genetic approaches will allow us to fully unravel interrelationships between invasive jellyfish exhibiting morphological plasticity.

14.7 Transport of Invasive Marine Jellyfish

The primary vector for the introduction of non-indigenous species into marine ecosystems is widely perceived to be shipping traffic, which enables organisms to cross natural oceanic barriers (Carlton and Geller 1993; Holland 2000; Chap. 4). Over the past 20–30 years, increases in the speed, size, and volume of shipping traffic on a global scale, along with stricter regulation of the use of anti-fouling paints, have resulted in concomitantly accelerating rates of non-indigenous species introductions into marine ecosystems (Ruiz et al. 1997, and references therein). In addition, oil and gas drilling platforms are routinely moved over large distances, sometimes between ocean basins, as new fields or markets emerge for exploration. In terms of the total number of fixed platforms in the sea, individual drill platform movements are extremely small. Still, the relocation of even a single platform via ocean towing will transport not only individuals, but perhaps entire exotic populations or mature communities.

Both ballast water and hull fouling are plausible mechanisms of jellyfish introductions. While ships carry large volumes of ballast water containing an enormous variety of potentially invasive organisms (Carlton and Geller 1993), only the holoplanktonic ctenophores appear suited to ballast water-mediated introduction. By contrast, hull fouling transport, either by ships or on drill platforms, is more likely for the hydrozoans and scyphozoans, owing to their bipartite life-histories that include a sessile polypoid stage. Despite the early observations described by Galil et al. (1990) of *C. andromeda* medusae swimming into the Mediterranean through the Suez Canal, it is more likely that translocation occurs not by the medusa stage, but rather by the sessile polyps, a counterintuitive argument supported by recent molecular genetic research (Dawson et al. 2005).

The aquarium trade has also received extensive attention as an important mode of marine introductions of ornamental fish and invertebrates. Recently, Bolton and Graham (2006) reported on incipient introductions resulting from the transport of 'live rock' materials associated with the aquarium trade. 'Live rock', either naturally collected or artificially cultured, refers to rock coated with algae, invertebrates, and microorganisms that is used for the purpose of increasing the aesthetics of home aquaria. Live rock became apparent as a potential vector of invasive organisms when material collected in the Indo-Pacific (likely Indonesia or Fiji) was imported to a local pet store in Florida (USA). Several weeks after purchase, a home aquarium enthusiast reported the appearance of numerous small jellyfish that ultimately were identified as 'upside-down' jellyfish *Cassiopea* spp., a recognized globally invasive genus (Sect. 14.3.2). The fact that the live rock trade in the United States receives little attention, much less appropriate quarantine measures, makes it a concern for aquatic invasive research (Bolton and Graham 2006).

14.8 Conclusions

Due to a few noteworthy cases described in this chapter, jellyfish have gained notoriety as potentially invasive animals. In fact, when one considers the physiological, ecological, and life-history traits of jellyfish (i.e., rapid growth, asexual propagation, intensive predators, crypsis, and morphological plasticity), this make them almost perfectly suited as invasive organisms. Yet, the limited examples presented here reflect the paucity of information we have regarding rates of jellyfish invasions in marine ecosystems. The likelihood remains that many invasions of jellyfish have gone undetected. Perhaps the greater concern should be that, until appropriate techniques and taxonomic appreciation are further developed, many more invasions by jellyfish will occur in the future.

Acknowledgements. While we owe the body of this work to numerous researchers, we especially appreciate communications and collaborations with T.F. Bolton, M.N. Dawson, G.R. Harbison, and A.E. Kideys. Funds in support of the authors' own research were provided by the Mississippi-Alabama Sea Grant Consortium (WMG), and Sigma Xi GIAR (KMB).

References

Agassiz L (1860) Contributions to the Natural History of the United States of America, vol 3. Little, Brown & Co, Boston

Bayha KM (2005) The molecular systematics and population genetics of four coastal ctenophores and scyphozoan jellyfish of the U.S. Atlantic and Gulf of Mexico. PhD Dissertation, The University of Delaware, Newark

Bayha KM, Harbison GR, McDonald JH, Gaffney PM (2004) Preliminary investigation on the molecular systematics of the invasive ctenophore *Beroe ovata*. In: Dumont H, Shiganova TA, Niermann U (eds) Aquatic invasions in the Black, Caspian and Mediterranean Seas. Kluwer, Dordrecht, pp 167–176

Bilio M, Niermann U (2004) Is the comb jelly really to blame for it all? *Mnemiopsis leidyi* and the ecological concerns about the Caspian Sea. Mar Ecol Prog Ser 269:173–183

Bolton TF, Graham WM (2004) Morphological variation among populations of an invasive jellyfish. Mar Ecol Prog Ser 278:125–139

Bolton TF, Graham WM (2006) Jellyfish on the rocks: bioinvasion threat of the international trade in aquarium live rock. Biol Invasions 8:651–653

Calder DR, Burrell VG Jr (1967) Occurrence of *Moerisia lyonsi* (Limnomedusae, Moerisiidae) in North America. Am Midl Nat 78:540–541

Calder DR, Burrell VG Jr (1969) Brackish water hydromedusa *Maeotias inexpectata* in North America. Nature 222:694–695

Carlton JT (1996) Biological invasions and cryptogenic species. Ecology. 77:1652–1655

Carlton JT, Geller JB (1993) Ecological roulette: the global transport of nonindigenous marine organisms. Science 261:78–82

Carlton JT, Thompson JK, Schemel, LE, Nichols FH (1990) Remarkable invasion of San Francisco Bay (California USA) by the Asian clam *Potamocorbula amurensis*. I. Introduction and dispersal. Mar Ecol Prog Ser 66:81–94

Colin PL, Arneson C (1995) Tropical Pacific invertebrates. Coral Reef Foundation, Coral Reef Press, Beverly Hills

Cowan JH, Houde ED (1992) Size-dependent predation on marine fish and larvae by ctenophores, scyphomedusae, and planktivorous fish. Fisheries Oceanogr 1:113–126

da Silveira FL, Cornelius PFS (2000) New observations on medusae (Cnidaria, Scyphozoa, Rhizostomeae) from northeast and south Brazil. Acta Biol Leopold 22:9–18

Dawson MN (2003) Macro-morphological variation among cryptic species of the moon jellyfish, *Aurelia* (Cnidaria: Scyphozoa). Mar Biol 143:369–379

Dawson MN, Jacobs DK (2001) Molecular evidence for cryptic species of *Aurelia aurita* (Cnidara, Scyphozoa). Biol Bull 200:92–96

Dawson MN, Martin LE (2001) Geographic variation and ecological adaptation in *Aurelia* (Scyphozoa, Semaeostomeae): some implications from molecular phylogenetics. Hydrobiologia 451:259–273

Dawson MN, Sen Gupta A, England MH (2005) Coupled biophysical global ocean model and molecular genetic analyses identify multiple introductions of cryptogenic species. Proc Natl Acad Sci USA 102:11968–11973

Devaney DM, Eldridge LG (1977) Reef and shore fauna of Hawaii. Section 1. Protozoa through Ctenophora. Bishop Mus Spec Publ 64:1–277

Elton CS (1958) The ecology of invasion by plants and animals. Methuen, London

Feigenbaum D, Kelly M (1984) Changes in the lower Chesapeake Bay food chain in the presence of the sea nettle *Chrysaora quinquecirrha* (Scyphomedusa). Mar Ecol Prog Ser 19:39–47

Galil BS, Spanier E, Ferguson WW (1990) The scyphomedusae of the Mediterranean coast of Israel, including two Lessepsian migrants new to the Mediterranean. Zool Mediterr 64:95–105

Garcia-Sais JR, Durbin E (1993) Zooplanktivorous predation by large scyphomedusae *Phyllorhiza punctata* (Cnidaria: Scyphozoa) in Laguna Joyuda. J Exp Mar Biol Ecol 173:71–93

GESAMP (1997) Opportunistic settlers and the problem of the ctenophore *Mnemiopsis leidyi* invasion in the Black Sea. International Maritime Organization, London, GESAMP Reports and Studies 58

Gordina AD, Zagorodnyaya JA, Kideys AE, Bat L, Satilmis HH (2005) Summer ichthyoplankton, food supply of fish larvae and impact of invasive ctenophores on the nutrition of fish larvae in the Black Sea during 2000 and 2001. J Mar Biol Assoc UK 85:537–548

Graham WM, Martin DL, Felder DL, Asper VL, Perry HM (2003) Ecological and economic implications of a tropical jellyfish invader in the Gulf of Mexico. Biol Invasions 5:53–69

Greenberg N, Garthwaite RL, Potts DC (1996) Allozyme and morphological evidence for a newly introduced species of *Aurelia* in San Francisco Bay, California. Mar Biol 125:401–410

Grosholz ED, Ruiz GM (1995) The spread and potential impact of the recently introduced European green crab, *Carcincus maenas*, in central California. Mar Biol 122:239–247

Gucu AC (2002) Can overfishing be responsible for the successful establishment of *Mnemiopsis leidyi* in the Black Sea? Estuar Coastal Shelf Sci 54:439–451

Gucu AC, Oguz T (1998) Modeling trophic interrelationships in the Black Sea. In: Ivanov LI, Oguz T (eds) Ecosystem modeling as a management tool for the Black Sea. In: NATO-TU Black Sea Project, vol 2. Kluwer, Dordrecht, pp 359–371

Gusmani L, Avian M, Galil B, Patriarca P, Rottini G (1997) Biologically active polypeptides in the venom of the jellyfish *Rhopilema nomadica*. Toxicon 35:637–648

Hallegraeff GM (1993) A review of harmful algal blooms and their apparent global increase. Phycologia 32:79–99

Hallegraeff GM, Bolch CJ (1992) Transport of diatom and dinoflagellate resting spores in ships' ballast water: implications for plankton biogeography and aquaculture. J Plankton Res 14:1067–1084

Harbison GR, Volovik SP (1994) The ctenophore, *Mnemiopsis leidyi*, in the Black Sea: a holoplanktonic organism transported in the ballast of ships. In: Proc Conf Workshop Non-Indigenous Estuarine & Marine Organisms (NEMO) and Introduced Marine Species. NOAA Tech Rep, US Dept Commerce, US Govt Print Office, Washington, DC, pp 25–36

Heeger T, Piatkowski U, Moeller H (1992) Predation on jellyfish by the cephalopod *Argonauta argo*. Mar Ecol Prog Ser 88:293–296

Hofmann DK, Hadfield MG (2002) Hermaphroditism, gonochorism, and asexual reproduction in *Cassiopea* sp. – an immigrant in the islands of Hawai'i. Invert Reprod Develop 41:215–221

Holland BS (2000) Genetics of marine bioinvasions. Hydrobiologia 420:63–71

Holland BS, Dawson MN, Crow GL, Hofmann DK (2004) Global phylogeny of *Cassiopea* (Scyphozoa: Rhizostomeae): molecular evidence for cryptic species and multiple invasions of the Hawaiian Islands. Mar Biol 145:1119–1128

Ivanov VP, Kamakin AM, Ushivtez VB, Shiganova TA, Zhukova O, Aladin N, Wilson SI, Harbinson GR, Durmont HJ (2000) Invasion of the Caspian Sea by comb jellyfish *Mnemiopsis leidyi* (Ctenophora). Biol Invasions 2:255–258

Johnson DR, Perry HM, Graham WM (2005) Using nowcast model currents to explore transport of non-indigenous jellyfish into the Gulf of Mexico. Mar Ecol Prog Ser 305:139–146

Kideys AE (1994) Recent dramatic changes in the Black Sea ecosystem: the reason for the sharp decline in Turkish anchovy fisheries. J Mar Syst 5:171–181

Kideys AE (2002) Fall and rise of the Black Sea ecosystem. Science 297:1482–1484

Kideys AE, Gucu AC (1995) *Rhopilema nomadica*: a Lessepsian scyphomedusan new to the Mediterranean coast of Turkey. Israel J Zool 41:615–617

Kideys AE, Moghim M (2003) Distribution of the alien ctenophore *Mnemiopsis leidyi* in the Caspian Sea in August 2001. Mar Biol 143:163–171

Kideys AE, Niermann U (1993) Intrusion of *Mnemiopsis mccradyi* (Ctenophora; Lobata) into the Mediterranean Sea. Senckenbergiana marit 23:43–47

Kideys AE, Niermann U (1994) Occurrence of *Mnemiopsis* along the Turkish coasts (from northeastern Mediterranean to Istanbul). ICES J Mar Sci 51:423–427

Kideys AE, Romanova ZA (2001) Distribution of gelatinous macrozooplankton in the southern Black Sea during 1996-1999. Mar Biol 139:535–547

Kideys AE, Finenko GA, Anninsky BE, Shiganova TA, Roohi A, Tabari MR, Youseffyan M, Rostamian MT, Rostami H, Negarestan H (2004a) Physiological characteristics of the ctenophore *Beroe ovata* in Caspian Sea water. Mar Ecol Prog Ser 266:111–121

Kideys AE, Finenko GA, Anninsky BA, Shiganova TA, Roohi A, Tabari MR, Youseffan M, Rostamiya MT (2004b) Feeding, reproduction and growth of ctenophore *Beroe* cf *ovata* in the low salinity conditions of the Caspian Sea. In: Dumont H, Shiganova TA, Niermann U (eds) Aquatic invasions in the Black, Caspian and Mediterranean Seas. Kluwer, Dordrecht, pp 193–200

Knowler D (2005) Reassessing the costs of biological invasion: *Mnemiopsis leidyi* in the Black Sea. Ecol Econ 52:197–199

Kramp PL (1961) Synopsis of the medusae of the world. J Mar Biol Assoc UK 40:1–469

Kramp PL (1970) Zoogeographical studies on Rhizostomeae (Scyphozoa). Vid Medd Dansk Naturhist For 133:7–30

Kremer P (1994) Patterns of abundance for *Mnemiopsis* in US coastal waters – a comparative review. ICES J Mar Sci 51:347–354

Kremer P, Nixon S (1976) Distribution and abundance of the ctenophore, *Mnemiopsis leidyi* in Narragansett Bay. Estuar Coastal Mar Sci 4:627–639

Lotan A, Ben-Hillel R, Loya Y (1992) Life cycle of *Rhopilema nomadica*: a new immigrant scyphomedusan in the Mediterranean. Mar Biol 112:237–242

Lotan A, Ben-Hillen R, Loya Y (1993) Aggregation and dispersal of *R. nomadica*, a tropical immigrant medusa in the Mediterranean Sea. Israel J Zool 39:67

Lotan A, Fine M, Ben-Hillel R (1994) Synchronization of the life cycle and dispersal pattern of the tropical invader scyphomedusan *Rhopilema nomadica* is temperature dependent. Mar Ecol Prog Ser 109:59–65

Martindale MQ (1987) Larval reproduction in the ctenophore *Mnemiopsis mccradyi* (Order Lobata). Mar Biol 94:409–414

Mayer AG (1910) Medusae of the World. Hydromedusae, vol II. Carnegie Institute, Washington, DC

Mayer AG (1912) Ctenophores of the Atlantic Coast of North America. Carnegie Institute, Washington, DC

Mayr E, Ashlock PD (1993) Principles of systematic zoology. McGraw-Hill, New York

Mianzan HW (1999) Ctenophora. In: Boltovskoy D (ed) South Atlantic zooplankton. Backhuys, Leiden, pp 561–573

Mills CE, Rees JT (2000) New observations and corrections concerning the trio of invasive hydromedusae *Maeotias marginata (=M. inexpectata)*, *Blackfordia virginica* and *Moerisia* sp. in the San Francisco estuary. Scientia marina 64 suppl 1:151–155

Mills CE, Sommer F (1995) Invertebrate introductions in marine habitats: two species of hydromedusae (Cnidaria) native to the Black Sea, *Maeotias inexspectata* and *Blackfordia virginica*, invade San Francisco Bay. Mar Biol 122:279–288

Mooney HA, Hobbs RJ (eds) (2000) Invasive species in a changing world. Island Press, Washington, DC

Mutlu E, Bingel F, Gucu AC, Melnikov VV, Niermann U, Ostr NA, Zaika VE (1994) Distribution of the new invader *Mnemiopsis* sp. and the resident *Aurelia aurita* and *Pleurobrachia pileus* populations in the Black Sea in the years 1991-1993. ICES J Mar Sci 51:407–421

Oliveira OM, Migotto AE (2006) Pelagic ctenophores from the Sao Sebastiao Channel, southeastern Brazil. Zootaxa 1183:1–26

Petersen JE, Sanford LP, Kemp WM (1998) Coastal plankton responses to turbulent mixing in experimental ecosystems. Mar Ecol Prog Ser 171:23–41

Pimental D, Lach L, Zuniga, Morrison D (2000) Environmental and economic costs of nonindigenous species in the United States. Bioscience 50:53–65

Purcell JE (1985) Predation of fish eggs and larvae by pelagic cnidarians and ctenophores. Bull Mar Sci 37:739–755

Purcell JE (1989) Predation on fish larvae and eggs by hydromedusae *Aequorea victoria* at a herring spawning ground in British Columbia. Can J Fish Aquat Sci 46:1415–1427

Purcell JE, Arai MN (2001) Interactions of pelagic cnidarians and ctenophores with fish: a review. Hydrobiologia 451:27–44

Purcell JE, Decker MB (2005) Effects of climate on relative predation by scyphomedusae and ctenophores on copepods in Chesapeake Bay during 1987-2000. Limnol Oceanogr 50:376–387

Purcell JE, Cresswell FP, Cargo DG, Kennedy VS (1991) Differential ingestion of bivalve larvae by the scyphozoan *Chrysaora quinquecirrha* and the ctenophore *Mnemiopsis leidyi*. Biol Bull 180:103–111

Purcell JE, Nemazie DA, Dorsey SE, Houde ED, Gamble JC (1994) Predation mortality of bay anchovy *Anchoa mitchilli* eggs and larvae due to scyphomedusae and ctenophores in the Chesapeake Bay. Mar Ecol Prog Ser 114:47–58

Purcell JE, Båmstedt U, Båmstedt A (1999) Prey, feeding rates, and asexual reproduction rates of the introduced oligohaline hydrozoan *Moerisia lyonsi*. Mar Biol 134:317–325

Purcell JE, Shiganova TA, Decker MB, Houde ED (2001) The ctenophore *Mnemiopsis leidyi* in native and exotic habitats: US estuaries versus the Black Sea basin. Hydrobiologia 451:145–176

Reeve MR, Syms MA, Kremer P (1989) Growth dynamics of a ctenophore (*Mnemiopsis*) in relation to variable food supply. 1. Carbon biomass, feeding, egg production, growth and assimilation efficiency. J Plankton Res 11:535–552

Rippingale RJ, Kelly SJ (1995) Reproduction and survival of *Phyllorhiza punctata* (Cnidaria: Rhizostomeae) in a seasonally fluctuating salinity regime in Western Australia. Mar Freshwater Res 46:1145–1151

Ruiz GM, Carlton JT, Grosholz ED, Hines AH (1997) Global invasions of marine and estuarine habitats by non-indigenous species: mechanisms, extent, and consequences. Am Zool 37:621–632

Sandifer PA, Smith TLJ, Calder DR (1974) Hydrozoans as pests in closed-system culture of larval decapod crustaceans. Aquaculture 4:55–59

Seravin LN (1994) The systematic revision of the genus *Mnemiopsis* (Ctenophora, Lobata). 2. Species attribution on *Mnemiopsis* from the Black Sea and the species composition of the genus *Mnemiopsis* (in Russian). Zool Zhurnal 73:19–34

Seravin LN, Shiganova TA, Luppova NE (2002) History of studying the ctenophore *Beroe ovata* (Ctenophora, Atentaculata, Beroida) and some structural features of its representative from the Black Sea (in Russian). Zool Zhurnal 81:1193–1200

Shiganova TA (1998) Invasion of the Black Sea by the ctenophore *Mnemiopsis leidyi* and recent changes in pelagic structure. Fisheries Oceanogr 7:305–310

Shiganova TA, Bulgakova YV (2000) Effects of gelatinous zooplankton on Black Sea and Sea of Azov fish and their food resources. ICES J Mar Sci 57:641–648

Shiganova TA, Mirzoyan ZA, Studenika EA, Volovik SP, Siokou-Frangou I, Zervoudaki S, Christou ED, Skirta AY, Dumont HJ (2001) Population development of the invader *Mnemiopsis leidyi*, in the Black Sea and in other sea of the Mediterranean basin. Mar Biol 139:431–445

Shiganova TA, Musaeva EI, Bulgakova YuV, Mirzoyan ZA, Martynyuk ML (2003) Invaders *Mnemiopsis leidyi* (A. Agassiz) and *Beroe ovata* Mayer 1912, and their influence on the pelagic ecosystem of northeastern Black Sea. Biol Bull 30:180–190

Spanier E (1989) Swarming jellyfishes along the Mediterranean coast of Israel. Israel J Zool 36:55–56

Stone R (2005) Attack of the killer jellies. Science 309:1805–1806

Väinölä R, Oulasvirta P (2001) The first record of *Maeotias marginata* (Cnidaria, Hydrozoa) from the Baltic Sea: a Pontocaspian invader. Sarsia 86:401–404

Vinogradov ME, Shushkina EA (1992) Temporal changes in community structure in the open Black Sea. Oceanology 32:485–491

Vinogradov ME, Shushkina EA, Musayeva EI, Sorokin PY (1989) A newly acclimated species in the Black Sea: the ctenophore *Mnemiopsis leidyi* (Ctenophora: Lobata). Oceanology 29:220–224

Vinogradov ME, Shishkina EA, Anokhina LL, Vostokov SV, Kucheruk NV, Lukashova TA (2000) Dense aggregations of the ctenophore *Beroe ovata* (Eschscholtz) near the north-east shore of the Black Sea. Oceanology 40:46–49

Volovik SP, Korpakova IG (2004) Introduction of *Beroe* cf *ovata* to the Caspian Sea needed to control *Mnemiopsis leidyi*. In: Dumont H, Shiganova TA, Niermann U (eds)

Aquatic invasions in the Black, Caspian and Mediterranean Seas. Kluwer, Dordrecht, pp 177–192

von Lendenfeld R (1884) Über eine Übergangsform zwischen Semostomen und Rhizostomen. Zool Anz 5:380–383

Wiesse T, Gomoiu MT, Scheffel U, Bordrecht F (2002) Biomass and size composition of the comb jelly *Mnemiopsis* sp. in the north-western Black Sea during spring 1997 and summer 1995. Estuar Coastal Shelf Sci 54:423–437

Wilcove DS, Rothstein D, Dubow D, Phillips, Losos E (1998) Quantifying threats to imperiled species in the United States. Bioscience 48:607–615

Zaika VE, Sergeeva NG (1990) Morphology and development of *Mnemiopsis mccradyi* (Ctenophora, Lobata) in the Black Sea. Hydrobiol J 26:1–6

15 Effects of Invasive Non-Native Species on the Native Biodiversity in the River Rhine

Bruno Baur and Stephanie Schmidlin

15.1 Introduction

Besides habitat degradation, the impacts of non-native invasive species are a major cause of extinction of native species (Groombridge 1992; Sala et al. 2000; Cox 2004). Invading species may interact with the native biota in a variety of ways, for example, by competition, predation, parasitism, disease and hybridization. Some non-native species may enter an ecosystem and remain at low densities for many years or disappear gradually whereas others might have a profound impact on the existing community by changing species abundance, food webs and energy fluxes. Linking invasion patterns with interspecific processes is often difficult but such information is crucial to predict the impacts of non-native species on the biodiversity of newly invaded locations (Moyle and Light 1996; Williamson 1996, 1999).

The Convention on Biodiversity exhorts the contracting parties to "prevent the introduction, control or eradicate those alien species which threaten ecosystems, habitats or species" (Glowka et al. 1994). To implement these directives, detailed knowledge on native biodiversity, and on potential interactions between invading non-native species and native species is required. Compared to the attention paid to extinctions in terrestrial habitats, much less focus has been given to species loss in freshwater ecosystems, and this despite several studies demonstrating a growing number of extinctions in freshwater animal species (fishes, molluscs, crayfishes; e.g. Kaufman 1992; Strayer 1999; Ricciardi and Rasmussen 1999).

This chapter examines the impact of invasive non-native species on the biodiversity in the river Rhine. The occurrence and spread of non-native species are relatively well documented in the Rhine (e.g. Tittizer et al. 2000; Geitler et al. 2002; Rey et al. 2004). Quantitative studies on changes in abundance of non-native species and on species composition of native communities complement these reports (e.g. Van den Brink et al. 1990, 1996; Haas et

Ecological Studies, Vol. 193
W. Nentwig (Ed.)
Biological Invasions

al. 2002). We review major changes in the biota of the river Rhine, focusing on mechanisms underlying changes in species abundance following the invasion of non-native species. Our emphasis is on benthic macroinvertebrates but interactions with other animals are also considered. Along the way, we identify important gaps in knowledge and suggest areas for further research.

15.2 The River Rhine

With a length of 1,320 km and a catchment area of 185,000 km^2, the river Rhine is one of the largest rivers in central Europe (Van Urk 1984; Friedrich and Müller 1984). It originates in the Eastern Swiss Alps, flows north to form the frontier with Liechtenstein and Austria (Alpenrhein), and empties into Lake Constance (Fig. 15.1). The Rhine (High Rhine) then re-emerges and flows west, mainly on the border between Switzerland and Germany. In Basel, it turns to the north and forms the southern part of the border between France and Germany (Upper Rhine) in a wide valley, before entering Germany exclusively (Middle Rhine). Here, the Rhine encounters some of its main tributaries (the Neckar, the Main and then the Moselle). Between Bingen and Bonn, the Rhine flows through the Rhine gorge, a formation created by erosion (this gorge is a UNESCO World Heritage Site since 2002). After passing the Ruhr area, the Rhine (Lower Rhine) turns west into The Netherlands. After crossing the border, it splits into three main distributaries, the Waal, the IJssel and the Nederrijn/Lek, before discharging into the North Sea.

The flow regime can be characterized as rain-fed/snow-fed, the highest water levels usually being attained in March–May and the lowest in August–November. The mean annual river discharge of the Rhine is 1,032 m^3 s^{-1} in Basel and 2,260 m^{3}s^{-1} (range 800–12,000 m^3 s^{-1}) at the Dutch border. This results in the minimum and maximum water levels differing by up to 8 m in The Netherlands (Van Geest et al. 2005).

The deterioration of the Rhine started in the Middle Ages, with the deforestation of large areas on the floodplains (Nienhuis and Leuven 1998). By the early 18th century, almost all beech and oak forests had been replaced by grassland. The river morphology became increasingly degraded because of straightening, reduction of channel networks to a single channel, and disconnection from the floodplain. In the 19th century, major river regulations in the Upper and Lower Rhine modified the river bed. For example, in the so-called Tulla-correction carried out between 1817 and 1874 and also in subsequent channelisations, the Upper Rhine north of Basel was transformed from a river system up to 6 km wide, with numerous branches, slow-flowing meanders, islands, and sand and gravel flats, into a 130-m-wide, fast-flowing sealed canal

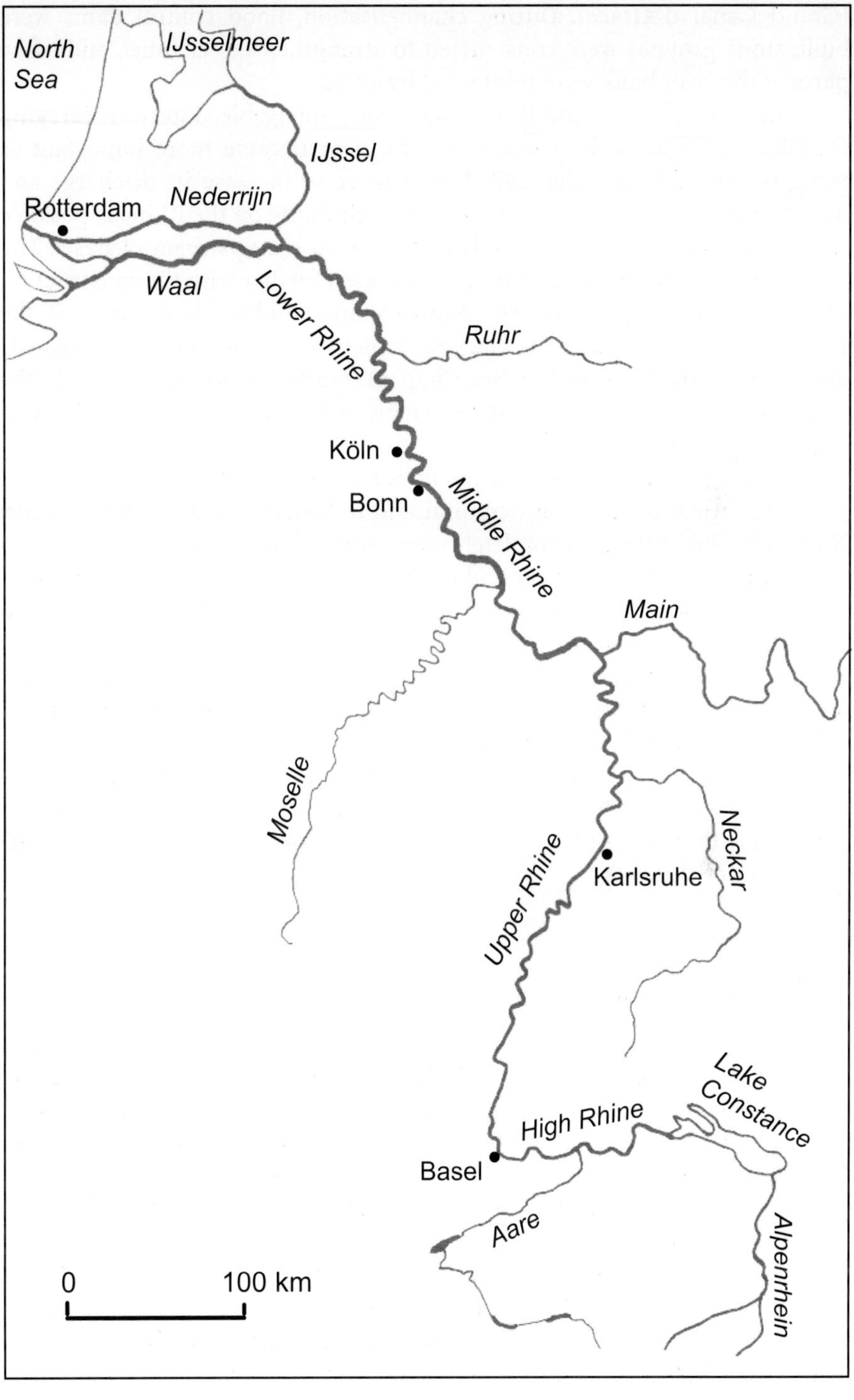

Fig. 15.1 Map of the river Rhine, with most of the locations mentioned in the text

(Grand Canal d'Alsace). During channelisation, flood control dams were built, stone groynes were constructed to strengthen the channel, and some parts of the river bank were reinforced by stones.

Since Roman times, the Rhine has been a navigable waterway, carrying travellers and goods deep inland. As the river became more important as transport route, it was channelled even more to increase its discharge and maintain its depth. In modern times, cargo shipping on the Rhine is possible from Rotterdam (North Sea) to Rheinfelden, 20 km upstream of Basel. The importance of international shipping increased further with the construction of river-connecting canals. The Rhine-Main-Danube Canal connects the Rhine via the Danube with the Black Sea which, in turn, is connected by canals and rivers to the Caspian Sea (see Chap. 5). Another navigation route to the Black Sea and Caspian Sea is the German Mittelland-Elbe-Vistula-Pripyat-Bug-Dnieper canal system.

Parallel to the channelisation, the floodplain of the Rhine has been extensively modified to extend agricultural and industrial areas and settlements. Nowadays, the Rhine is a completely man-manipulated river, more intensively used than ever before (Tittizer and Krebs 1996). Besides its function as transportation route, it provides water for communities and industry, is used to generate hydroelectric power, provides cooling water and a means of effluent transport, and is increasingly a focus for recreation. Despite profound alterations of river characteristics, the river still has a large (albeit not unlimited) self-cleaning capacity, and natural and semi-natural banks and areas of the floodplain, with abandoned meanders, brooks, sand and gravel pits, and remnants of riparian forest still harbour an extraordinarily high diversity of plants and animals, and are therefore of high conservation value (e.g. LfU 2000; Baur et al. 2002).

15.3 Native Biodiversity and Invasion History

Faunal diversity decreased dramatically in the river Rhine between 1900 and 1970 (Kinzelbach 1972; Van den Brink et al. 1990; Streit 1992). For example, species richness of selected groups of macroinvertebrates in the Dutch section of the Rhine declined from 83 species in 1900 to 43 species in 1940 and to 41 species in 1981/1987 (Van den Brink et al. 1990; Den Hartog et al. 1992). Omitting the non-native species arriving in the 20th century, however, the total number of species for 1940 would be 40, and only 27 for 1981/1987. Schöll (2002) presented a list of 21 typical riverine macroinvertebrate species (seven mayflies, 10 stoneflies and four caddis flies) occurring in the German part of the Rhine in 1900 – none were found in the river between 1960 and 2000. Most probably, these specialized benthic species went extinct in the river Rhine. However, the actual causes of extinction are unknown. In the Rhine near

Basel, the number of stonefly species declined from 13 to four between 1910 and 1990, and those of mayflies from 19 to 13 (Küry 1994).

The decline of the freshwater fauna in the river Rhine is linked to extensive habitat deterioration caused by channelisation and flow regulation by weirs, stream fragmentation, organic pollution from land-use activities, toxic contaminants from municipal and industrial sources, and interactions with an increasing number of non-native species (Streit 1992; Baur and Ringeis 2002; Van der Velde et al. 2002; Nehring 2003). Since the industrial revolution and the construction of sewage systems, domestic and industrial pollution have led to a gradual deterioration in water quality, and this from the second half of the 19th century to the end of the 1960s. Water quality was very poor during the period 1950–1970, with low oxygen levels, serious eutrophication, high chemical and organic pollution loads, salination caused by French potassium mines and mining water from brown coal mines in Germany, and thermal pollution (Rhine river water temperature has risen by approximately 2 °C above its natural value; Admiraal et al. 1993).

Faunal diversity in the river Rhine was lowest in the late 1960s, when levels of toxicants were highest and oxygen levels extremely low (Kinzelbach 1972; Streit 1992). During the period 1970–1986, waste water treatment plants were constructed along the river, resulting in improvements of water quality including an increase in oxygen levels and a reduction of some heavy metals and organic pesticides. Also, faunal diversity began to recover (Admiraal et al. 1993). Driven partly by the toxic spill following the Sandoz accident (see below), ministers from riparian countries decided in 1986 to establish the Rhine Action Programme. One of its aims is the restoration of the river ecosystem.

Haas et al. (2002) described three successional phases in the development of benthic communities in the German section of the Rhine, following the extreme toxic and organic contamination which the river has known in earlier times.

1. From 1970 to 1986, the aquatic community was species-poor and still in an early recovery. Because of the remaining organic pollution, only sewage-resistant taxa such as the leech *Erpobdella octoculata*, the isopod *Asellus aquaticus*, the snail *Radix ovata*, sponges, chironomids and oligochaetes occurred. The non-native zebra mussel *Dreissena polymorpha* started to colonise hard substrates. However, the major Sandoz industrial accident near Basel in 1986, when runoff from water used in firefighting carried nearly 30 t of toxic chemicals (insecticides, fungicides and herbicides) into the Rhine, caused serious damage to the flora and fauna over hundreds of kilometres, resetting the recovery process. In 1987, benthic faunal densities were still close to zero (Den Hartog et al. 1992). Yet, *D. polymorpha* was able to quickly recolonise the Rhine following the Sandoz spill because of the immigration of pelagic larvae from unaffected sites.
2. In 1987 and 1988, the non-native amphipod *Corophium curvispinum* (=*Chelicorophium curvispinum*), and the Asiatic clams *Corbicula fluminea*

and *C. fluminalis* invaded the Rhine (Sect. 15.4). Already in 1989, the population density of *C. curvispinum* in the Middle and Lower Rhine was so high that the hard substrate of the channel bottom had been essentially completely overgrown due to the species' engineering activity. The *D. polymorpha* population collapsed because adult shells were rapidly overgrown by *C. curvispinum*, and their muddy tubes inhibited the development of new *D. polymorpha* patches – the planktonic larvae can settle only on hard surfaces (Van der Velde et al. 1994; Tittizer and Krebs 1996; Haas et al. 2002).

3. A new phase started with the invasion of the amphipod *Dikerogammarus villosus* in 1995 (Sect. 15.4). In 2000, maximum densities of 3,000 individuals m^{-2} were recorded. Since 1996, the population densities of *C. curvispinum* have decreased whereas *D. polymorpha* has recovered and again reached high densities. Subsequent to the appearance of *D. villosus*, two other amphipods, *Gammarus tigrinus* and *Echinogammarus ischnus*, have declined in the Upper Rhine; *G. tigrinus* finally disappeared in 1999. In 1997 and 1998, three new non-native invertebrates reached the river Rhine, originating from the Danube and the Ponto-Caspic region: the isopod *Jaera istri*, the turbellarian worm *Dendrocoelum romanodanubiale* and the polychaete *Hypania invalida* (Haas et al. 2002).

There is an accelerating colonisation rate of non-native macroinvertebrate species in the Rhine (Fig. 15.2). The shape of the cumulative colonisation curve shows that 55% of the total number of colonisations were recorded after 1970. Thus, more than half of all colonisations in the 175-year record have been reported these last 35 years. The average rate of colonisation has increased from 0.15 new species established per year in the period 1831–1970 to 0.74 new species per year for the period 1971–2005. Considering exclusively the period 1991–2005, the current rate of colonisation averages 1.27 new species per year.

Similarly to macroinvertebrates, fish species composition in the river Rhine has altered in the past century. There is ample evidence that the river engineering works have had deleterious effects on the species number and abundance of fish (Lelek and Köhler 1989). Associated river modifications have led to the disappearance of specific spawning grounds, feeding biotopes and nursery areas, and to the obstruction of migration routes. The construction of fish passes at almost every weir along the main stream section seems to have been insufficient to prevent the decline of the migrating fish populations. Low oxygen concentration and the massive discharge of toxic materials contributed substantially to this decline. Since the water quality of the Rhine began to improve in the 1970s, however, the fish community has been recovering (Cazemier 1988; Lelek and Köhler 1989).

Lelek (1996) presented a list of 27 non-native fish species for the German part of the Rhine. Eighteen of the 27 species (67%) were intentionally intro-

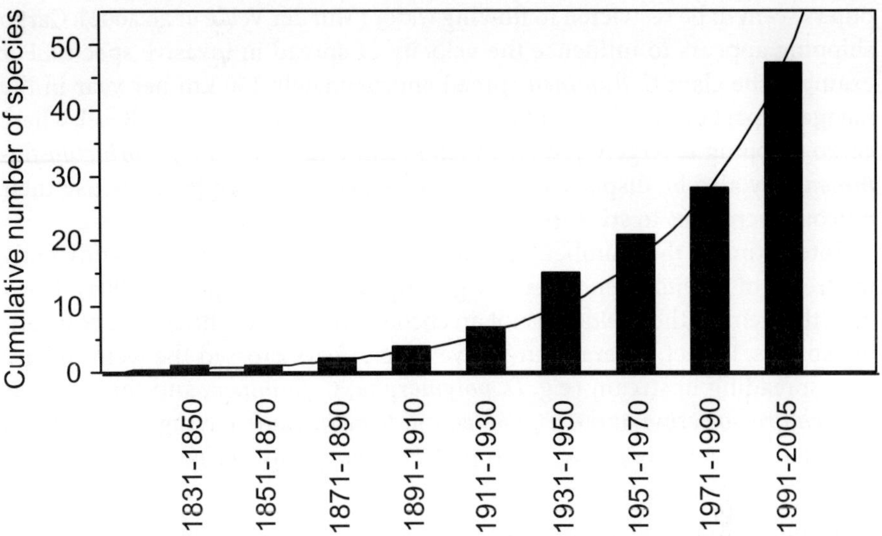

Fig. 15.2 Increasing number of non-native macroinvertebrate species colonising the river Rhine. Cumulative data are shown for periods of 20 years (note: the *last bar* includes data for only 15 years). The exponential model was fitted by least-squares regression ($y=5.5936 \times 10^{-22}\ e^{0.0265t}$, $n=9$, $R^2=0.98$, t indicates the year). Data were obtained from Tittizer et al. (2000), Geitler et al. (2002) and Rey et al. (2004)

duced by fishermen, another seven species (26 %) having been inadvertently introduced by the aquarium trade. Interestingly, among the phytoplankton, an ecologically important group, no non-native species have yet been observed in the Rhine (Nehring 2005).

About one of two non-native aquatic species entering German rivers could spread over large areas, and about one of five non-native species have become invasive (Nehring 2003). In the Rhine delta in The Netherlands, the proportion of non-native species in the biodiversity of river channels and floodplain lakes ranges from 7–10 % among macrophytes to 5–12 % among macroinvertebrates and 17–19 % among fish (Van den Brink et al. 1996). In the Middle and Upper Rhine, non-native species represent 10–15 % of total species richness (Haas et al. 2002). Non-native species dominate in terms of total abundance and biomass, however, the values exceeding 80 % (Tittizer et al. 2000; Haas et al. 2002).

Thus, species composition in the river Rhine has changed remarkably in the past four decades. Replacing characteristic riverine species, large numbers of euryoecious and non-native species, in particular macroinvertebrates and fish, have invaded this river system (e.g. Van den Brink et al. 1988, 1990). Some of the species entered the river via ports and estuaries, and then moved upstream whereas others moved downstream after entering via canals. Several of these species have penetrated into the larger, still-water expanses but others seem to be restricted to flowing water (Van der Velde et al. 2002). Cargo shipping appears to influence the velocity of spread in invasive species. For example, the clam *C. fluminea* spread approximately 150 km per year in the navigable part of the Rhine but only 2.4 km per year upstream of Basel, where cargo shipping is largely reduced (Schmidlin and Baur 2006). *Corbicula fluminea* may also be displaced by waterfowl, because juvenile clams use their mucous secretions to stick to ducks' feet.

Interestingly, the number of non-native species decreases significantly upstream of Rheinfelden where cargo shipping ends (Rey et al. 2004). However, the weir in Rheinfelden is not an absolute barrier for the spread of invading species. In fact, several non-native species have crossed the weir and are now spreading upstream (e.g. *D. polymorpha*, *C. fluminea*, and the annelids *Branchiura sowerbyi* and *Caspiobdella fadejewi*), some having even entered the tributary Aare (e.g. the gastropod *Potamopyrgus antipodarum* and the flatworm *Dugesia tigrina*; Rey et al. 2004).

15.4 Species Interactions and Mechanisms of Replacement

15.4.1 Amphipods

The amphipod *Corophium curvispinum*, originating from the Ponto-Caspic region, was first observed in the Middle and Lower Rhine in 1987 (Schöll 1990). A few years later, *C. curvispinum* was found to be by far the most numerous macroinvertebrate species in the Lower Rhine (Van den Brink et al. 1991). Its density increased up to 200,000 specimens m^{-2} on groynes (Van den Brink et al. 1993). It has been claimed that *C. curvispinum* had filled an 'empty niche' because it was the first tubiculous amphipod to colonise the Rhine (Den Hartog et al. 1992). The animals produced extensive mats of dense silty tubes which covered all available hard surface. As a consequence, other epilithic invertebrates were negatively affected by this muddy layer. Significant declines in population densities were recorded for the amphipod *Gammarus tigrinus*, the zebra mussel *Dreissena polymorpha*, the gastropod *Potamopyrgus antipodarum*, the caddis fly *Hydropsyche contubernalis*, and several species of Chironomidae (Van den Brink et al. 1993). The former three are non-native species whereas *H. contubernalis* is native. It has been suggested that these changes in abundance were at least partly the result of competition for food – *C. curvispinum*, *D. polymorpha* and *H. contubernalis* are all filter-feeders (Rajagopal et al. 1999). In fact, the exponential increase in the density of *C. curvispinum* during 1989–1991 coincided with a decrease in the concentrations of total organic carbon and total suspended matter in the Lower Rhine, which may be related to an increase in filtration capacity in the river.

Stable isotope analysis showed very similar values for carbon and nitrogen sources in the stone-dwelling *C. curvispinum*, *D. polymorpha* and the sand-dwelling Asiatic clams *Corbicula fluminea* and *C. fluminalis*, indicating a common source of phytoplankton and particulate organic matter for these filter-feeding animals (Marguillier et al. 1998).

Besides competition for food, there might also have been competition for space among benthic macroinvertebrates. For example, specimens of *D. polymorpha* were observed to be completely overgrown by the tubes of *C. curvispinum*. Moreover, in building its muddy tubes, the amphipod modifies the substrate, thereby preventing the settlement of larvae of *D. polymorpha*. However, relatively little is known about the fundamental features of tube building activity and filtration rate in *C. curvispinum*.

The impact of the population explosion of *C. curvispinum* on the density of other macroinvertebrates has also resulted in a shift in the diet of the European eel *Anguilla anguilla*. In 1989, prior to the population explosion of *C. curvispinum*, *Gammarus tigrinus* and *Dreissena polymorpha* dominated the diet of the eel (Van der Velde et al. 1998). In 1994, however, *C. curvispinum* occurred in 80 % of the eels sampled whereas *G. tigrinus* decreased in percentage occurrence from 32 to 4 %. Similarly, *D. polymorpha* was eaten to a far lesser extent in 1994 than in 1989 (Van der Velde et al. 1998). The perch *Perca fluviatilis* showed a similar shift in diet (Kelleher et al. 1998).

The amphipods *C. curvispinum* and *Dikerogammarus villosus* and the isopod *Jaera istri* act as intermediate hosts for a variety of parasites of the eel. In the German part of the Rhine, nine metazoan species were found to infest eels (Sures et al. 1999). Among-site differences in eel parasite diversity was related to the presence and abundance of invading crustacean species (Sures and Streit 2001).

Since 1984, there has also been a significant increase in the distribution and abundance of the amphipod *Gammarus tigrinus*, which originated from North America. In many sections of the Rhine, *G. tigrinus* has displaced *Gammarus duebeni*, a native and originally widespread species in Western Europe (Tittizer et al. 2000). In the late 1990s, however, the abundance of *G. tigrinus* declined sharply, coincidental with the invasion of the amphipod *Dikerogammarus villosus*. This species is native to the Ponto-Caspian region and has invaded Western Europe via the Main-Danube canal, appearing in the river Rhine at the German-Dutch border in 1994–1995 (Tittizer et al. 2000). *D. villosus* has wide environmental tolerances in terms of temperature and salinity, and thus is able to colonise various microhabitats.

Stable isotope analyses have shown that *D. villosus* is a predatory species whereas *G. duebeni* is a detrivorous/herbivorous amphipod. It was hypothesized that the rapid expansion of *D. villosus*, and its devastating impact on *G. duebeni* and related species may involve intraguild predation, rather than interspecific competition. In laboratory experiments, survival of female *G. duebeni* was lower when male *D. villosus* – rather than male *G. duebeni* – were present (Dick and Platvoet 2000). Similarly, *D. villosus* preyed upon *G. tigrinus*. *D. villosus* killed and consumed recently moulted and, less frequently, intermoult victims. Thus, the predatory impact of *D. villosus* is not restricted to the short (approximately 12 h) period of post-moult vulnerability, facilitating rapid eliminations of all stages of reproducing females (Dick and Platvoet 2000). No male *G. duebeni* was killed during the experiment, indicating that the larger males are more able to fend off any predatory attack, although this may not be the case at moult (Dick 1996). Compared to other freshwater amphipods, the large size of *D. villosus* might partly explain its successful predatory behaviour.

Gut content analyses showed that *D. villosus* preyed also on *C. curvispinum* in the wild, thereby interfering in the interspecific competition for space between the two filter-feeders *C. curvispinum* and *D. polymorpha*. In amphipods and many other arthropods, intraguild predation has been increasingly recognized as an important mechanism in structuring communities (e.g. Polis et al. 1989). In many cases, intraguild predation may override interspecific competition. In the present example, intraguild predation also appears to be the mechanism for the exclusion of both non-native and native species.

15.4.2 Molluscs

The zebra mussel *Dreissena polymorpha*, originating from the Caspian and Black Sea region, was first recorded in the Lower Rhine near Rotterdam in 1826. In the following decades, it expanded upstream and reached large densities (Kinzelbach 1972). However, water pollution in the mid-20th century, and subsequent competitive interactions with *C. curvispinum* strongly reduced *D. polymorpha* populations in the Rhine. Continuous improvement in water quality these past decades and reduced *C. curvispinum* densities have allowed the *D. polymorpha* populations to recover; nowadays, they have again attained densities of up to 40,000 individuals m^{-2}.

The zebra mussel attaches to solid surfaces using adhesive byssal fibres, and possesses a planktonic larval (veliger) stage which can remain in the water column for several weeks before settlement. Native unionid mussels have a complex life cycle in which the larvae are obligate parasites of fish, with survivorship dependent on the availability of appropriate fish hosts and accessibility to favourable habitats. Adult unionid mussels live partially buried in the sediments of lakes and rivers, with their posterior shell exposed to the water column, providing a suitable surface for colonisation by *D. polymorpha*. Infestation by *D. polymorpha* is considered to impair the metabolic activity (feeding, respiration, excretion) and locomotion of unionid mussels, thereby depleting their energy reserves and effectively starving them to death (Haag et al. 1993). Moreover, data from North America demonstrate that *D.*

polymorpha can also harm other suspension-feeding bivalves by depleting food resources (phytoplankton) through massive filtration (Caraco et al. 1997).

Dreissena polymorpha has virtually eliminated the native unionid fauna in many parts of the lower Great Lakes in North America (Ricciardi et al. 1998; Strayer 1999). In the Rhine, the decline of the highly specialized and endangered unionid mussels and other filter-feeding macroinvertebrates could also partly be due to competition with *D. polymorpha*. However, *D. polymorpha* is not harmful to all riverine species. In North America, the clam provides other benthic invertebrates with nourishment (in the form of faecal deposits) and shelter (interstitial spaces between clumped mussel shells), resulting in a local enhancement of abundance and diversity for these other species (Ricciardi 2005). Non-native deposit feeders may increase in abundance whereas native filter-feeders are out-competed by *D. polymorpha*. Among the invertebrates responding positively to zebra mussel colonisation are non-native oligochaetes, leeches, amphipods, gastropods, larval chironomids and aquatic weeds (Ricciardi et al. 1997; Karatayev et al. 2002). Thus, invading species may also have synergistic impacts which facilitate the establishment of other invaders.

The clams *Corbicula fluminea* and *C. fluminalis*, originating from Southeast Asia, were first recorded in the Lower Rhine in The Netherlands in 1985 (Bij de Vaate and Greijdanus-Klaas 1990). Six years later, the clams were found near Karlsruhe in the Upper Rhine and, in 1995, *C. fluminea* was reported near Basel in Switzerland (Rey et al. 2004). *C. fluminea* is restricted to the gravely–sandy river bottom because sticking structures are lacking. The clam reached densities of 1,800 individuals m^{-2} in the Rhine (Haas et al. 2002). Den Hartog et al. (1992) suspected that the spill of toxic waste from the Sandoz accident in 1986, affecting the Rhine over hundreds of kilometres, contributed to the clams' success because most macroinvertebrates were killed and, as a consequence, their niches were unoccupied.

Several mechanisms by which *Corbicula* may affect native bivalves have been proposed (Strayer 1999). Dense populations of *Corbicula* may deplete concentrations of phytoplankton and other edible suspended particles, thereby 'starving out' native bivalves. Modest to dramatic declines in phytoplankton or seston have been recorded in habitats with high *Corbicula* density in North America (Leff et al. 1990; Phelps 1994). Dense populations of *Corbicula* may ingest large numbers of unionid sperm, glochidia and newly metamorphosed juveniles (Strayer 1999). Because *Corbicula* pedal feeds on edible particles in the sediments, it may deplete also this food resource, affecting some sphaeriids and juvenile unionids which use benthic organic matter as food. *Corbicula* actively disturbs the sediments, so dense populations may reduce habitat quality and space for native bivalves.

Several studies show that the impact of *C. fluminea* on native benthic species depends on both site and community characteristics (Leff et al. 1990;

Strayer 1999). The clam severely affected native mollusc assemblages in some North American rivers but can coexist with other bivalves at other sites. Similar information on the impact of *Corbicula* on native macroinvertebrates in the river Rhine is not yet available.

15.5 Why Are There so many Non-Native Species in the Rhine?

The number of non-native animal species colonising the river Rhine is still increasing (Fig. 15.2). Furthermore, non-native plant species constitute a significant proportion of the vegetation of the river bank and floodplain (Schwabe and Kratochwil 1991). A variety of mutually non-exclusive hypotheses have been suggested to explain the success of invaders in the river Rhine: (1) vacant niches, (2) disturbances preventing strong interspecific competition, (3) the creation of new niches by invasive species, (4) ecosystem instability (invasional meltdown), (5) groups of co-adapted invaders, and (6) enemy-free space.

It has been argued that human alterations of habitat make a community vulnerable to invasions and that extreme natural disturbances facilitate the establishment of non-native species (Mack et al. 2000). Community vulnerability to invasions has been ascribed to a combination of several factors, such as the presence of vacant niches, habitat modification, and disturbance before and after invasion. Recent findings indicate that species-rich communities are less vulnerable to invasions (at least, in terrestrial habitats; Cox 2004). Moreover, invasibility is known to increase if a community lacks certain species present under normal conditions (Chap. 11).

The invasional meltdown model (Chap. 6) predicts that ecosystems subjected to a chronically high frequency of species introduction will become progressively unstable and easier to invade, as each introduced species has the potential to facilitate subsequent invaders (Simberloff and Von Holle 1999). Invasional meltdown may occur through one of two processes: frequent disturbance through species introductions progressively lowers community resistance to invasion, and increased introductions lead to a higher frequency of potential facilitations and synergies between invaders (Ricciardi 2005). Highly active invasion corridors (in the present case, canals) may introduce numerous species from one and the same region (e.g. the Ponto-Caspic region), and thus may reunite groups of co-adapted species, either in simultaneous introductions (e.g. a host arriving with its parasites) or in successive introductions, thereby assembling contiguous links of a non-native food web.

If co-adaptation reduces the intensity of predation and parasitism, then positive interactions probably dominate invasion 'groups', and successive introductions of co-adapted species might result in a higher success of invaders than would introductions of unacquainted species (Ricciardi 2005). This could be an alternative to the enemy release hypothesis, which relates the success of an invader to the absence of its natural predators and parasites in the invaded region (Chap. 6). Each of the examples presented in Sect. 15.4 could be explained by at least one of these six hypotheses. However, experimental tests of these hypotheses are lacking for the Rhine.

15.6 Conclusions

The river Rhine is a good example for how a combination of different factors structure benthic communities. River modification deteriorated certain habitats but also created new habitats. Prolonged pollution changed the original communities and caused the loss of certain species, creating open niches for pollution-tolerant invaders. Major disturbances, such as the Sandoz accident in 1986, subsequently enabled the invasion of many new species which reached unprecedented densities. The Rhine-Main-Danube Canal, opened in 1992–1993, provided additional opportunities for the immigration of non-native species from the Ponto-Caspian region, some of them being co-adapted. After reduction of the pollution in the Rhine, recolonisation seemed to favour invaders, rather than native species. These invaders suppressed the development of populations of native species. At the present day, the number of invaders is still increasing.

For the development of appropriate conservation strategies for the river Rhine, detailed knowledge of the ecological consequences of invasive non-native species for the native biota is required. The present review shows that, in most cases, negative impacts of invasive species on native species have been deduced from correlative evidence. Evidently, there is an urgent need for experimental studies on interactions between invasive and native species. Numerous rare native species in the Rhine are threatened with extinction by the combined impacts of environmental degradation and species invasions (e.g. by *D. polymorpha*). From a conservation perspective, the habitat requirements, population dynamics and persistence of rare native species deserve increased attention. Restoration to pristine conditions is not feasible in the Rhine. However, several promising ecological restoration projects are of vital importance to preserve those facets of the originally unique biodiversity of the river Rhine and its floodplain still present today.

Acknowledgements. We thank A. Baur and P. Stoll for constructive comments on the manuscript.

References

Admiraal W, Van der Velde G, Smit H, Cazemier WG (1993) The rivers Rhine and Meuse in the Netherlands – present state and signs of ecological recovery. Hydrobiologia 265:97–128

Baur B, Ringeis B (2002) Changes in gastropod assemblages in freshwater habitats in the vicinity of Basel (Switzerland) over 87 years. Hydrobiologia 479:1–10

Baur B, Zschokke S, Coray A, Schläpfer M, Erhardt A (2002) Habitat characteristics of the endangered flightless beetle *Dorcadion fuliginator* (Coleoptera: Cerambycidae): implications for conservation. Biol Conserv 105:133–142

Bij de Vaate A, Greijdanus-Klaas M (1990) The Asiatic clam, *Corbicula fluminea* Müller, 1774 (Pelecypoda, Corbiculidae), a new immigrant in the Netherlands. Bull Zool Mus Amsterdam 12:173–178

Caraco NF, Cole JJ, Raymond PA, Strayer DL, Pace ML, Findlay SEG, Fischer DT (1997) Zebra mussel invasion in a large, turbid river: phytoplankton response to increased grazing. Ecology 78:588–602

Cazemier WG (1988) Fish and their environment in large European river ecosystems. The Dutch part of the river Rhine. Science Eau 7:95–116

Cox GW (2004) Alien species and evolution. Island Press, Washington, DC

Den Hartog C, Van den Brink FWB, Van der Velde G (1992) Why was the invasion of the river Rhine by *Corophium curvispinum* and *Corbicula* species so successful? J Nat Hist 26:1121–1129

Dick JTA (1996) Post-invasion amphipod communities of Lough Neagh, Northern Ireland: influences of habitat selection and mutual predation. J Anim Ecol 65:756–767

Dick JTA, Platvoet D (2000) Invading predatory crustacean *Dikerogammarus villosus* eliminates both native and exotic species. Proc R Soc Lond B 267:977–983

Friedrich G, Müller D (1984) Rhine. In: Whitton BA (ed) Ecology of European rivers. Blackwell, Oxford, pp 265–315

Geitler O, Homma S, Kinzelbach R (2002) Bestandesaufnahme und Bewertung von Neozoen in Deutschland. Umweltbundesamt Berlin, UBA Texte 25/02:1–290

Glowka L, Burhenne-Guilmin F, Synge H (1994) A guide to the convention on biological diversity. IUCN, Gland

Groombridge B (ed) (1992) Global biodiversity: status of the earth's living resources. Chapman and Hall, London

Haag WR, Berg DJ, Garton DW, Farris JL (1993) Reduced survival and fitness in native bivalves in response to fouling by the introduced zebra mussel (*Dreissena polymorpha*) in western Lake Erie. Can J Fish Aquat Sci 50:13–19

Haas G, Brunke M, Streit B (2002) Fast turnover in dominance of exotic species in the Rhine river determines biodiversity and ecosystem function: an affair between amphipods and mussels. In: Leppäkoski E, Gollasch S, Olenin S (eds) Invasive aquatic species of Europe. Distribution, impacts and management. Kluwer, Dordrecht, pp 426–432

Karatayev AY, Burlakova LE, Padilla DK (2002) Impacts of zebra mussels on aquatic communities and their role as ecosystem engineers. In: Leppäkoski E, Gollasch S, Olenin S (eds) Invasive aquatic species of Europe. Distribution, impacts and management. Kluwer, Dordrecht, pp 433–446

Kaufman L (1992) Catastrophic change in species-rich freshwater ecosystems. BioScience 42:846–858

Kelleher B, Bergers PJM, Van den Brink FWB, Giller PS, Van der Velde G, Bij de Vaate A (1998) Effects of exotic amphipod invasions on fish diet in the Lower Rhine. Arch Hydrobiol 143:363–382

Kinzelbach R (1972) Einschleppung und Einwanderung von Wirbellosen in Ober- und Mittelrhein. Mainzer Naturwiss Arch 11:150–190

Küry D (1994) Die Wirbellosenfauna der Fliessgewässer in der Region Basel. Verh Naturf Ges Basel 104:19–44

Leff LG, Burch JL, McArthur JV (1990) Spatial distribution, seston removal, and potential competitive interactions of the bivalves *Corbicula fluminea* and *Elliptio complanata* in a coastal plain stream. Freshwater Biol 24:409–416

Lelek A (1996) Die allochthonen und die beheimateten Fischarten unserer grossen Flüsse – Neozoen der Fischfauna. In: Gebhardt H, Kinzelbach R, Schmidt-Fischer S (eds) Gebietsfremde Tierarten: Auswirkungen auf einheimische Arten, Lebensgemeinschaften und Biotope, Situationsanalyse. Ecomed, Landsberg, pp 197–215

Lelek A, Köhler C (1989) Zustandsanalyse der Fischartengemeinschaften im Rhein (1987–1988). Fischökologie 1:47–64

LfU (2000) Vom Wildstrom zur Trockenaue: Natur und Geschichte der Flusslandschaft am südlichen Oberrhein. Landesanstalt für Umweltschutz Baden-Württemberg, Verlag Regionalkultur, Ubstadt-Weiher

Mack RN, Simberloff D, Lonsdale WM, Evans H, Clout M, Bazzaz FA (2000) Biotic invasions: causes, epidemiology, global consequences, and control. Ecol Appl 10:689–710

Marguillier S, Dehairs F, Van der Velde G, Kelleher B, Rajagopal S (1998) Initial results on the trophic relationship based on *Corophium curvispinum* in the Rhine traced by stable isotopes. In: Nienhuis PH, Leuven RSEW, Ragas AMJ (eds) New concepts for sustainable management of river basins. Backhuys, Leiden, pp 171–177

Moyle PB, Light T (1996) Biological invasions of fresh water: empirical rules and assembly theory. Biol Conserv 78:149–161

Nehring S (2003) Gebietsfremde Arten in den deutschen Gewässern – ein Risiko für die Biodiversität. Schriftenr BMVEL Angew Wiss 498:40–52

Nehring S (2005) International shipping – a risk for aquatic biodiversity in Germany. In: Nentwig W, Bacher S, Cock MJW, Dietz H, Gigon A, Wittenberg R (eds) Biological invasions – from ecology to control. Neobiota 6:125–143

Nienhuis PH, Leuven RSEW (1998) Ecological concepts for the sustainable management of lowland river basins: a review. In: Nienhuis PH, Leuven RSEW, Ragas AMJ (eds) New concepts for sustainable management of river basins. Backhuys, Leiden, pp 7–33

Phelps HL (1994) The Asiatic clam (*Corbicula fluminea*) invasion and system-level ecological change in the Potomac River estuary near Washington DC. Estuaries 17:614–621

Polis GA, Myers CA, Holt RD (1989) The ecology and evolution of intraguild predation: potential competitors that eat each other. Annu Rev Ecol Syst 20:297–330

Rajagopal S, Van der Velde G, Paffen BGP, Van den Brink FWB, Bij de Vaate A (1999) Life history and reproductive biology of the invasive amphipod *Corophium curvispinum* (Crustacea: Amphipoda) in the Lower Rhine. Arch Hydrobiol 144:305–325

Rey P, Ortlepp J, Küry D (2004) Wirbellose Neozoen im Hochrhein. Ausbreitung und ökologische Bedeutung. BUWAL, Bern, Schriftenr Umwelt 380:1–88

Ricciardi A (2005) Facilitation and synergistic interactions between introduced aquatic species. In: Mooney HA, Mack RN, McNeely JA, Neville LE, Schei PJ, Waage JK (eds) Invasive alien species. Island Press, Washington, DC, pp 162–178

Ricciardi A, Rasmussen JB (1999) Extinction rates of North American freshwater fauna. Conserv Biol 13:1220–1222

Ricciardi A, Whoriskey FG, Rasmussen JB (1997) The role of the zebra mussel (*Dreissena polymorpha*) in structuring macroinvertebrate communities on hard substrata. Can J Fish Aquat Sci 54:2596–2608

Ricciardi A, Neves RJ, Rasmussen JB (1998) Impending extinctions of North American freshwater mussels (Unionoida) following the zebra mussel (*Dreissena polymorpha*) invasion. J Anim Ecol 67:613–619

Sala OE, Chapin FS III, Armesto JJ, Berlow E, Bloomfield J, Dirzo R, Huber-Sannwald E (2000) Global biodiversity scenarios for the year 2100. Science 287:1770–1774

Schmidlin S, Baur B (2006) Distribution and substrate preference of the invasive clam *Corbicula fluminea* in the river Rhine in the region of Basel (Switzerland, Germany, France). Aquat Sci (in press)

Schöll F (1990) Zur Bestandessituation von *Corophium curvispinum* Sars im Rheingebiet. Lauterbornia 5:67–70

Schöll F (2002) Das Makrozoobenthos des Rheins 2000. Kommission zum Schutz des Rheins, Koblenz, Bericht Nr 128

Schwabe A, Kratochwil A (1991) Gewässer-begleitende Neophyten und ihre Beurteilung aus Naturschutz-Sicht unter besonderer Berücksichtigung Südwestdeutschlands. NNA-Berichte 4/1:14–27

Simberloff D, Von Holle B (1999) Positive interactions of nonindigenous species: invasional meltdown? Biol Invasions 1:21–32

Strayer DL (1999) Effects of alien species on freshwater mollusks in North America. J N Am Benthol Soc 18:74–98

Streit B (1992) Zur Ökologie der Tierwelt im Rhein. Verh Naturforsch Ges Basel 102: 323–342

Sures B, Streit B (2001) Eel parasite diversity and intermediate host abundance in the River Rhine, Germany. Parasitology 123:185–191

Sures B, Knopf K, Wurtz J, Hirt J (1999) Richness and diversity of parasite communities in European eels *Anguilla anguilla* of the River Rhine, Germany, with special reference to helminth parasites. Parasitology 119:323–330

Tittizer T, Krebs F (eds) (1996) Ökosystemforschung: Der Rhein und seine Auen – eine Bilanz. Springer, Berlin Heidelberg New York

Tittizer T, Schöll F, Banning M, Haybach A, Schleuter M (2000) Aquatische Neozoen im Makrozoobenthos der Binnenwasserstrassen Deutschlands. Lauterbornia 39:1–172

Van den Brink FWB, Van der Velde G, Geelen JFM (1988) Life history parameters and temperature-related activity of an American crayfish, *Orconectes limosus* (Rafinesque, 1817) (Crustacea, Decapoda), in the area of the major rivers in The Netherlands. Arch Hydrobiol 114:275–289

Van den Brink FWB, Van der Velde G, Cazemier WG (1990) The faunistic composition of the freshwater section of the River Rhine in The Netherlands: present state and changes since 1900. In: Kinzelbach R, Friedrich G (eds) Biologie des Rheins. Limnol Aktuell 1:192–216

Van den Brink FWB, Van der Velde G, Bij de Vaate A (1991) Amphipod invasion on the Rhine. Nature 352:576

Van den Brink FWB, Van der Velde G, Bij de Vaate A (1993) Ecological aspects, explosive range extension and impact of a mass invader, *Corophium curvispinum* Sars, 1895 (Crustacea: Amphipoda), in the Lower Rhine (The Netherlands). Oecologia 93:224–232

Van den Brink FWB, Van der Velde G, Buijse AD, Klink AD (1996) Biodiversity in the Lower Rhine and Meuse river-floodplains: its significance for ecological river management. Neth J Aquat Ecol 30:129–149

Van der Velde G, Paffen BGP, Van den Brink FWB, Bij de Vaate A, Jenner HA (1994) Decline of zebra mussel populations in the Rhine. Competition between two mass invaders (*Dreissena polymorpha* and *Corophium curvispinum*). Naturwissenschaften 81:32–34

Van der Velde G, Rajagopal S, Van den Brink FWB, Kelleher B, Paffen BGP, Kempers AJ, Bij de Vaate A (1998) Ecological impact of an exotic amphipod invasion in the river Rhine. In: Nienhuis PH, Leuven RSEW, Ragas AMJ (eds) New concepts for sustainable management of river basins. Backhuys, Leiden, pp 159–169

Van der Velde G, Nagelkerken I, Rajagopal S, Bij de Vaate A (2002) Invasions by alien species in inland freshwater bodies in Western Europe: the Rhine delta. In: Leppäkoski E, Gollasch S, Olenin S (eds) Invasive aquatic species of Europe. Distribution, impacts and management. Kluwer, Dordrecht, pp 360–372

Van Geest GJ, Coops H, Roijackers RMM, Buijse AD, Scheffer M (2005) Succession of aquatic vegetation driven by reduced water-level fluctuation in floodplain lakes. J Appl Ecol 42:251–260

Van Urk G (1984) Lower Rhine-Meuse. In: Whitton BA (ed) Ecology of European rivers. Blackwell, Oxford, pp 437–468

Williamson M (1996) Biological invasions. Chapman and Hall, London

Williamson M (1999) Invasions. Ecography 22:5–12

16 Hybridization and Introgression Between Native and Alien Species

CARLO R. LARGIADÈR

16.1 Introduction

Human activities, such intentional and unintentional transplantations, and habitat alterations including the establishment of migration corridors, generate increasing opportunities for formerly allopatric taxa to meet and to hybridize. There is indeed increasing evidence that these introduced plant and animal taxa (including crop plants and domesticated animal taxa) frequently hybridize with native relatives and with other introduced taxa, leading to a growing concern that these hybridizations may compromise the genetic integrity of native taxa to the point of causing extinctions (Abbott 1992; Rhymer and Simberloff 1996; Levin et al. 1996; Ellstrand and Schierenbeck 2000; Vilà et al. 2000). A decade ago, Rhymer and Simberloff (1996) stated in their review on this topic that the known cases are probably just the tip of the iceberg. Using the search term 'hybridization and introgression', the Web of Science database yields a total of 1,178 research articles, of which 935 (or 80 %) have been published after 1995 (Fig. 16.1). Indeed, the evidence for natural and man-induced hybridization and introgression appears to have increased exponentially these last few years.

Presently, we still cannot answer the questions of how widespread and how important these processes are in the context of biological invasions. However, regarding the dramatic increase of evidence in the literature, we can already conclude that natural and anthropogenic hybridization and introgression are very frequent, and play an important role in animal and plant species, both at the inter- and intraspecific level. It is also very likely that this evidence will continue to grow – probably many cases of hybridization and introgression have so far remained unnoticed, because hybrids may be rare and because, in morphologically uniform groups, they are difficult to recognize. Furthermore, many taxa are still unknown, and many taxonomic groups and geographic areas are understudied. Finally, the manner in which hybridization

Ecological Studies, Vol. 193
W. Nentwig (Ed.)
Biological Invasions

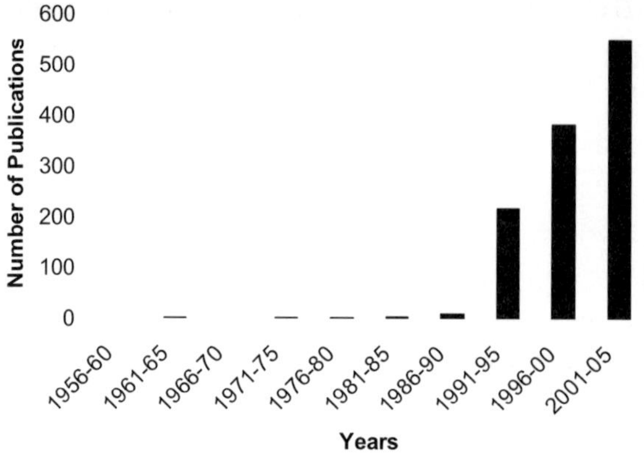

Fig. 16.1 Development of natural and anthropogenic hybridization and introgression as research topics. The literature search was carried out in the Web of Science (Thomson Scientific) citation database, using the search string 'hybridization and introgression'

and species are defined also strongly influences the perception of the problem.

Recently, Mallet (2005) reviewed studies of natural interspecific hybridization in plants and a variety of animals. He estimated that at least 25 % of plant species, and 10 % of animal species are involved in hybridization and potential introgression with other species. Plants seem more prone to hybridization than are animals. However, this difference may at least partly be due to the different historical attitudes of botanists and zoologists toward hybridization, which resulted possibly in greater attention being paid to this phenomenon by botanists (Dowling and Secor 1997). While botanists viewed hybridization and introgression as important processes in adaptive evolution (Rieseberg 1997), zoologists have tended to see them rather as problems being the converse of reproductive isolation, and thus challenging the 'reality' of biological species. However, there is increasing evidence that hybridization and introgression have also played an important role in the evolution of animals (Grant and Grant 1996; Dowling and Secor 1997). Taking into account the potential evolutionary importance of these processes changes considerably how we may perceive them in the context of biological invasions, e.g., that they may act as a stimulus for the evolution of invasiveness in transplanted species (Ellstrand and Schierenbeck 2000).

In this chapter, I intend to give a non-exhaustive overview on the empirical evidence of hybridization and introgression in the context of biological invasions. When any two taxa hybridize, the outcome of this event is difficult to generalize, and is modulated by many interacting external (e.g., habitat modifications) and evolutionary factors (e.g., mate preferences, heterosis etc.): the two taxa can merge completely, forming a 'hybrid swarm' leading to the extinction of the native taxon, which is replaced by a single mongrel species. Hybridization can be asymmetrical, i.e., genes introgress directionally from one taxon into the gene pool of the other, and hence lead to an acceleration of the replacement of a native gene pool. Even if hybridization ends with the formation of completely sterile first-generation (F1) hybrids, it may still accelerate the process of replacing native taxa through the wasting of reproductive efforts. Finally, in some cases, new genetic adaptations are generated in hybrids, leading to hybrid speciation. Thus, this complexity of possible outcomes, and the multitude of potential factors involved make a systematic treatment of the subject rather difficult. Here, I start with the definition of relevant terms, and review some technical aspects. I then summarize empirical studies of anthropogenic introgressive and non-introgressive hybridization. I continue by briefly discussing hybridization as a process capable of inducing invasiveness and leading to new biological diversity. By asking the question whether we can predict hybridization and its outcome, I examine some of the main factors that promote and affect the outcome of anthropogenic hybridization. Cases of gene flow from genetically modified organisms into wild populations are subject of Chap. 17, and will thus not be discussed here.

16.2 Definitions and Technical Aspects

16.2.1 Definition of Hybridization and Introgression

In this text, hybridization is defined as the interbreeding of genetically differentiated forms, regardless of their current taxonomic status. Accordingly, introgression is defined as the movement of genes between genetically differentiated forms (again regardless of their current taxonomic status) mediated by backcrossing (Avise 1994). Such broad definitions of hybridization and introgression are now widely used in the current literature. They account for the fact that inter- and intraspecific hybridization are both relevant in the context of biological invasions, and circumvent the difficulty and much debated issue of defining species. From a conservation perspective, we are concerned about loosing genetic diversity not only at the inter- but also at the intraspecific level. Firstly, intraspecific hybridization can homogenize the unique characteristics of geographically distinct populations and incipient species, which reduces the 'raw material' for future allopatric speciation, and thus reduces the source of future species diversity (Perry et al. 2002; Olden et al. 2004). Secondly, outbreeding depression, i.e., a reduction in fitness due to the mating of genetically divergent individuals, may occur in invaded populations. Outbreeding depression may arise through underdominance (heterozygote disadvantage) between alleles of the two parental populations, or through a breakup of coadapted gene complexes (Dobzhansky 1950; Lynch 1991). Similarly to the evolution of local adaptations by natural selection, coadapted gene complexes arise as intrinsic adaptations to the genetic environment of local populations, i.e., the evolution of 'local' complexes of genes that interact in a mutually favorable manner (Lynch 1996). An important feature of outbreeding depression through the breakup of coadapted gene complexes is that it theoretically may occur between populations that have adapted to similar extrinsic environments. An increasing number of studies presenting empirical evidence of outbreeding depression have been reported for several animal and plant taxa (e.g., Butcher and Williams 2002; Goldberg et al. 2005; Sagvik et al. 2005). Finally, the loss of locally adapted populations through introgressive hybridization leads to a general reduction of fitness of a species and its adaptive potential, and may thus make it even more vulnerable to invasions (Mallet 2005).

The gene exchange by introgression between native and invading species has been given many different names, most having pejorative connotations, e.g., 'genetic pollution' or 'contamination' (Rhymer and Simberloff 1996). So far, there has been no general agreement on a specific term to denote the genetic effects of man-induced introgressive hybridization. Also, most of the terms applied so far in this context have been used also in various other contexts. Here, I follow the suggestion of Rhymer and Simberloff (1996), and use the non-value-laden term 'mixing' to denote a mixing of gene pools associated, or not, with a decline in fitness.

16.2.2 Genetic and Statistical Tools

As was predicted by Rhymer and Simberloff (1996), this increase in evidence of genetic interactions between genetically distinct taxa is directly connected to the advent of various new molecular genetic tools, which have facilitated the detection of hybridization and introgression in cases where hybrid individuals were morphologically difficult to distinguish from their parental forms. Within this context, particularly important roles are played by the polymerase chain reaction (PCR; Mullis et al. 1986) for the in vitro amplification of specific DNA sequences, and the techniques and genetic markers based on PCR technology, such as rapid DNA sequencing methods (e.g., cycle sequencing) and microsatellite DNA markers (Tautz 1989). The latter allow a genetic resolution at the level of individuals (Estoup and Angers 1998). Besides that these techniques have become very much more efficient, affordable for small laboratories and easy to use even for non-molecular biologists, the major advantage of PCR-based technologies is their non-destructive and minuscule tissue requirements, enabling a non-invasive way to study rare or endangered species (Morin and Woodruff 1996).

Furthermore, recent statistical developments have facilitated the detection of hybridization and hybrid individuals in cases where no taxa-specific markers are available. For example, model-based Bayesian statistical techniques, which utilize the information of highly polymorphic markers such as microsatellites (Pritchard et al. 2000; Anderson and Thompson 2002), have already been widely applied to the study of hybridization and introgression in natural and man-induced cases (e.g., Beaumont et al. 2001; Barilani et al. 2005; Williams et al. 2005; Lecis et al. 2006).

The simultaneous use of cytoplasmic (mitochondrial and chloroplast DNA) and nuclear genetic markers has become an important standard in studies of introgressive hybridization. The combination of these two marker classes allow us to gain very detailed insight into the processes of hybridization and introgression (as reviewed, e.g., for aquatic organisms by Avise 2000).

These studies take advantage of the fact that these cytoplasmic genomes are usually maternally inherited, and thus show a pattern of inheritance different to that of recombining nuclear markers. A joint use of these marker classes provides information that cannot be obtained by using either marker class alone. For example, if only one sex of the invaded taxon hybridizes, then the maternally inherited markers will introgress asymmetrically in relation to nuclear makers, which are inherited by both sexes equally. The direction of the asymmetry will give us information on which combinations of mating occur in the interbreeding of the two parental taxa. If the interbreeding is restricted to males of the invading taxon with females of the native taxon, then we would observe that hybrid individuals always carry a mitochondrial genotype of the native taxon, while having a mixture of alleles of the two parental taxa at the nuclear markers. Thus, measures of association between specific alleles at nuclear markers and cytoplasmic genotypes can be used to formulate hypotheses of factors involved in hybrid formation, and the rate and direction of genetic introgression in hybridization events. This development of a cytonuclear theory and of statistical models provided an important framework for hypothesis testing using empirical data in hybridizing taxa (Asmussen et al. 1987; Scribner et al. 2000).

16.3 Basic Types of Anthropogenic Hybridization: Empirical Examples

16.3.1 Hybridization Without Introgression

Hybridization may contribute to the decline of native species even if F1-hybrids are all completely sterile. In this case, although hybridization is not a threat through genetic mixing, part of the reproductive effort will be wasted in the production of sterile hybrids (Rhymer and Simberloff 1996; Allendorf et al. 2001). A genetic model suggests that hybridization with sterile hybrids has little effect on the dynamics of the displacement (Huxel 1999), which primarily depends on the relative fitness of invaders and their immigration rate. However, the model assumed a constant population size, and that the two parental species produce (at their respective proportions) a sufficient number offspring to account for the loss of hybrid individuals in each generation. Thus, the accelerating effect in replacing the native species of hybridization without introgression may strongly depend on the fecundity of the native species, which may be further reduced indirectly by competition with hybrids for limited habitat resources.

So far, empirical evidence for such cases has been quite rare in animals. I also have not found a single well-documented case in plants, though there are several examples of crosses of plant species yielding sterile hybrid progeny (Ellstrand 1992). However, it may well be that many such cases have remained unnoticed, since computer models predict that such replacements of local populations may occur very rapidly within a few generations (Huxel 1999; Wolf et al. 2001). Furthermore, if sterile hybrid individuals have reduced vigor (e.g., die at very early life stages), then their presence may be overlooked. A case where the latter could apply concerns the formerly widespread and now threatened European mink (*Mustela lutreola*), whose populations have declined almost everywhere throughout its range. Several hypotheses have been put forward to explain the disappearance of this species. One important cause seems competition with the American mink (*M. vison*), which was introduced for fur farming or accidentally released in many parts of Europe in the 1920s–1930s. Based on crossing experiments with the two species, which showed that hybrid embryos are resorbed, it was suggested that, because the breeding season of *M. vison* starts earlier and because *M. vison* males are stronger than the European mink males, the *M. vison* males would mate with the native *M. lutreola* females, thereby preventing *M. lutreola* reproduction (Maran and Henttonen 1995).

Current genetic evidence for hybridization confined primarily to the F1-generation involves freshwater fish species. In a first example, genetic data collected over a period of 8 years from a stream in western Montana (northwestern North America) indicated a rapid displacement of native bull trout (*Salvelinus confluentus*) by introduced brook trout (*S. fontinalis*), with very low introgression (Leary et al. 1993). Only two of 75 hybrids detected throughout western Montana were not F1-individuals, and a comparison of the genetic data of mtDNA and ten diagnostic nuclear markers showed that both sexes of each species interbreed with the corresponding sex of the other.

A second, more recent example concerns two freshwater minnow species (*Pseudorasbora pumila* and *P. parva*) in Japan (Konishi and Takata 2004). *P. parva* native to western Japan has been accidentally introduced during the transplantations of other cyprinid fish species into eastern Japan. Over the last 30 years, *P. parva* has largely replaced *P. pumila* native to eastern Japan. In the contact zone of the two species, only F1-hybrids showing exclusively mtDNA haplotypes of *P. pumila* have been detected, even after following the genetic structure of hybridizing populations over a 5-year period. The data indicate that the F1-hybrids are sterile, and resulted from mating only between *P. pumila* females and *P. parva* males. The data also suggest that the rapid replacement of *P. pumila* by *P. parva* has been promoted by asymmetrical hybridization without introgression, but with *P. pumila* females wasting considerably greater reproductive efforts than did *P. parva* males.

16.3.2 Hybridization with Introgression

The primary genetic consequence of introgressive hybridization is that the genomes of the two hybridizing forms are recombined, i.e., genes from one taxon introgress into the gene pool of the other taxon, and thus lead to a mixing of the two gene pools. This introgression may be asymmetrical at the level of the taxa involved, as well as at the level of genes. For example, in the case of the brown trout (*Salmo trutta*) and its sister species the Atlantic salmon (*S. salar*), experimentally backcrossing F1-hybrids to Atlantic salmon yielded viable offspring, whereas all attempts of backcrosses to brown trout failed, suggesting that an asymmetric introgression of brown trout genes into the Atlantic salmon gene pool is possible (Garcia-Vazquez et al. 2004).

In natural hybrid zones, asymmetric introgression patterns may be observed for different genes, indicating that some genes introgress 'more easily' than others (Avise 1994). While some of these asymmetries arise through selective advantages conferred by particular alleles in the new genetic background of the introgressed species, others may arise from different behaviors or fitnesses between the sexes. For example, in plants, male and female gene flow is generally strongly decoupled through different dispersal modes of pollen and seeds (Arnold 1992), resulting in asymmetrical introgression patterns of nuclear and maternally inherited gene markers (Arnold et al. 1991). These observations of natural cases of hybridization have important implications for man-induced introgressive hybridization. Firstly, this is not a uniform process leading to a single predictable outcome. Secondly, it is modulated by various factors, such as behavioral components, which may be changed in disturbed habitats or in domesticated varieties, as will be discussed in more detail in the examples presented below.

Examples of man-induced introgressive hybridization have been reported for many taxonomic groups of plants and animals. However, invertebrates, although being the most diverse group of animals, are clearly underrepresented, which reflects the fact that research of genetic effects of biological invasions is strongly focused on birds, mammals, and fishes. Perry et al. (2002) have recently reviewed the importance of anthropogenic hybridization in the freshwater fauna as a threat to North American biodiversity, and suggested that hybridization with introduced species may represent an underestimated threat for crayfish species. This country harbors about 75% (ca. 390 species) of the world's known crayfish species, of which about 30 % are threatened or endangered (Lodge et al. 2000). Indeed, the rusty crayfish (*Orconectes rusticus*) has been shown to be hybridizing with, and displacing native *O. propinquus* (Perry et al. 2001). *O. rusticus* is native to southwestern Ohio, and has been introduced widely as fish bait throughout the United States, where it has become a serious pest over the last 35 years. A detailed study by Perry et al. (2001) revealed a quite interesting situation. Patterns of cytonuclear disequilibrium between allozymes and mtDNA suggested that the majority of F1-hybrids were offspring of matings between *O. rusticus* females and *O. propinquus* males, although *O. rusticus* males were expected to outcompete the native *O. propinquus* males, which are smaller in size than their introduced congeners. Compared to both parental forms, these authors also found no reduction of fecundity and early survivorship, but rather a competitive superiority in hybrids. They stated that these results, at first sight, seemed to be at odds with the expectation that introgressive hybridization would enhance the displacement of the native gene pools by the invading genes, since the asymmetrical gene flow of native nuclear genes into the invaders gene pool should operate rather in the opposite direction. However, by assuming that the higher competitive ability of early-generation hybrids translates into higher relative fitness, and using a simple one-locus model, they estimated that introgressive hybridization would accelerate the elimination of pure *O. propinquus* by 4.8–36.3 % above that due to the previously documented ecological interaction.

Among birds, several duck species have been intensively studied in the context of anthropogenic hybridization. These studies provide impressive examples on how habitat modifications and transplantations, followed by introgressive hybridization, have led to declines of several native taxa throughout the world (Kulikova et al. 2004; Williams et al. 2005). A nearly complete genetic mixing of the New Zealand grey duck (*Anas s. superciliosa*) with introduced mallards (*A. platyrhynchos*) threatens the native taxa to become extinct, and to be replaced by a new mongrel species (Rhymer et al. 1994). An especially noteworthy example concerning duck species involves the American black duck (*A. rubripes*) and *A. platyrhynchos*. These two species have been primarily allopatric prior to the settlement by Europeans of North America. Habitat alteration and game-farm mallard releases during the 20th century enabled mallards to extend their range, and to come into contact and interbreed with the morphologically and behaviorally similar black duck (Mank et al. 2004). Molecular data based on modern specimens suggested the black duck to be a recent evolutionary derivative of a more broadly distributed mallard–black duck ancestor (Avise et al. 1990). However, the genetic analysis of modern and museum specimens clearly showed a dramatic decrease in genetic differentiation between the two taxa, indicating that the present-day genetic similarity is the consequence of gene flow through introgressive hybridization (Mank et al. 2004). This study is a nice example on the usefulness of historical museum specimens to study such dynamic processes.

At the intraspecific level, an interesting case of anthropogenic introgressive hybridization in birds in the wild concerns common quails (*Coturnix c. conturnix*) and domesticated Japanese quails (*Conturnix c. japonica*). The common quail's breeding range extends from the Atlantic to Lake Baikal, and from the Arctic Circle to the tropics (Guyomarc'h et al. 1998). Its breeding range overlaps only in small areas with the natural distribution of the Japanese quail. The decline of the common quail at northern latitudes over the last few decades has stimulated the release of Japanese quail as game species. Wild common quails migrate toward North Africa, while such migratory behavior is absent or reduced in domesticated Japanese or hybrid quails (Deregnaucourt et al. 2005). The long-distance migration is thought to be an adaptation to avoid unfavorable winter conditions in Northern and Central Europe, where the common quail is generally not observed during this time (Guyomarc'h et al. 1998). Hybrids between common and Japanese quails have recently been detected in the wild in several European countries and in Africa, and also in some captive-reared stocks of Japanese quail (Barilani et al. 2005). Thus, this introgressive hybridization may potentially lead to a gradual decline of the migratory behavior of wild common quail populations (Deregnaucourt et al. 2005).

Among vertebrates, fishes seem to hybridize most frequently at the interspecific level, which has been explained by their generally external mode of fertilization, coupled with weakly developed reproductive isolating mechanisms (Hubbs 1955). Many freshwater fishes have been intensively transplanted or hybridized in aquaculture in order to increase yields. Canals and river diversions have removed isolating barriers that historically prevented the range overlap of allopatric species (Chap. 5). Finally, natural habitats of many freshwater fish have been altered, e.g., through canalizations of rivers, often resulting in a consolidation of spawning activities in reduced or altered habitats of decreasing complexity. Thus, it is not surprising that there should be numerous documented cases of anthropogenic introgressive hybridization for this group of species. Scribner et al. (2000) report that about 50 % (or 81) of a total of 163 reviewed cases of interspecific hybridization, encompassing 168 species from 19 freshwater fish families, could be attributed to the aforementioned types of human impact. There are also many studies reporting introgressive hybridization at the intraspecific level, encompassing mainly economically important salmonid species that have been extensively transplanted over a century for harvest enhancement, and to compensate for population declines due to habitat deterioration. This has involved often massive introductions from domesticated hatchery strains (Largiadèr and Scholl 1995). Utter (2000) reviewed the patterns of subspecific introgression in two salmonid genera of North America (*Oncorhynchus* spp.) and Europe (*Salmo* spp.). He found that freshwater resident populations were more susceptible than anadromous ones to introgression from genetically distinct lineages, including some lineages that had been isolated for more than a million years. By contrast, within major genetic lineages, anadromous populations appear to be more susceptible to introgression. In general, there is a great variability observed in the extent of introgression in different populations, ranging from hardly detectable introgression of exotic genes following several decades of intensive introductions (Largiadèr and Scholl 1996), to nearly complete displacement of the native gene pools (Largiadèr and Scholl 1995).

Regarding the literature on anthropogenic introgressive hybridization in mammals, evidence of gene flow between domesticated varieties and wild conspecifics or closely related species seems to have particularly increased in the last few years. This involves, for example, the European wildcat (*Felis silvestris silvestris*) and domestic cats (*F. s. catus*; Beaumont et al. 2001; Lecis et al. 2006), domestic ferrets (*Mustela furo*) and the European polecat (*M. putorius*; Davison et al. 1999), wolves (*Canis lupus*) and domestic dogs (*C. l. familiaris*; Randi and Lucchini 2002; Vilà et al. 2003), coyotes (*C. latrans*) and dogs (Adams et al. 2003), bison (*Bison bison*) and domestic cattle (*Bos taurus*; Halbert et al. 2005), and other bovine species (Nijman et al. 2003).

Although there is no systematic survey dealing specifically with man-induced inter- and intraspecific introgressive hybridization in plant species to assess the extent of these phenomena as a threat to native biological diversity, plants still represent probably the best-studied group in this context. Vilà et al. (2000) have recently compiled and reviewed a large number of human-mediated plant hybridizations, which encompass a wide range of taxonomic groups and the full range of potential outcomes, as described above for animal species. Also, the gene flow between domesticated and wild plant species has received increased attention since the advent of transgenics. A recent review reports that 12 of the world's 13 most important crop species hybridize with wild relatives (Ellstrand et al. 2002), and substantial evidence has now been compiled that at least 48 cultivated plant taxa hybridize with one or more wild relatives somewhere in the world (Ellstrand 2003a, 2003b). Two further reviews focused on the extinction of rare species through hybridization with numerically superior invading species (Ellstrand and Elam 1993; Levin et al. 1996). Yet, there are also cases of small invading populations threatening larger populations of native species. For example, smooth cord grass (*Spartina alterniflora*), which was introduced into the salt marshes of San Francisco Bay in the mid-1970s, shows a much higher male fitness than does the native California cord grass (*S. foliosa*). This fitness difference seems to reverse the direction of gene flow, which would normally be expected to occur from the more abundant to the rarer taxon, and thus may ultimately lead to the extinction of the more widespread, abundant native species (*S. foliosa*) through progressive introgression of genes of the rare invader (*S. alterniflora*) into the native gene pool (Anttila et al. 1998).

16.4 Hybridization as a Stimulus for the Evolution of Invasiveness and the Emergence of Anthropogenic Hybrid Taxa

The effect of anthropogenic hybridization and introgression is not always restricted to accelerating the replacement of native taxa and locally adapted populations, leading to a loss of genetic diversity at the species and subspecies level. In some cases, hybridization, often combined with habitat modifications, may generate new invasive hybrid taxa that are better adapted than their parental species, and thus induce invasiveness in situ (Abbott 1992). Hybridization may directly produce distinct taxa, either through (allo-) polyploidisation or the generation of clonally reproducing unisexual lineages, whereas introgression could lead to stable, independent lineages defined by unique genetic combinations (cf. Rieseberg 1997; Dowling and Secor 1997). A classical textbook example of a hybrid taxon is the common cord grass *Spartina anglica* (Gray et al. 1991), which originated from hybridization between *S. maritima* and *S. alterniflora*. In the early 19th century, the latter was introduced accidentally (probably as shipping ballast) from North America into Southampton, UK, where it hybridized with the local *S. maritima*, producing a sterile hybrid. A chromosome doubling event in the hybrid led to the generation of new fertile species, *S. anglica*. This very aggressive hybrid taxon now occupies large areas of the British Isles, and has successfully colonized a zone of mudflats not occupied by its parents, thereby endangering other species by its spread.

Recently, in a non-exhaustive review, Ellstrand and Schierenbeck (2000) compiled a list of 28 well-documented examples of the evolution of invasiveness in 12 plant families following a spontaneous hybridization event between a native and an introduced taxon. A striking common feature is that these examples all occur in human-disturbed areas. In all, 24 of the 28 examples are herbaceous species, the majority being outcrossing perennials. As the authors point out, these characteristics are also found to be frequent among cases of natural hybridization (Ellstrand et al. 1996), indicating that some groups seem to be more prone to generate new hybrid taxa.

Compared to plants, the formation of new hybrid taxa appears to be relatively rare in animals. However, as in the case of the rate of hybridization, this may be partly due to the negative attitudes toward hybridization in zoology (Dowling and Secor 1997). An interesting recent example in animals involves a freshwater sculpin of the genus *Cottus* (Nolte et al. 2005). Over the past two decades in the Rhine river system in Central Europe, a rapid upriver invasion of new habitats, previously free of sculpins and atypical for the known species *C. gobio*, has been observed. Genetic analysis revealed that the invasive fish are hybrids between two old *C. gobio* lineages, the one from the river Rhine drainage, the other from the river Scheldt drainage. It seems that artificial connections between the two river systems have provided the opportunity for hybridization between long-separated groups, leading to the emergence of a new, adaptationally distinct sculpin lineage.

16.5 Can we Predict Introgressive Hybridization and its Outcome?

16.5.1 Genetic Differentiation Between Taxa as an Indicator

In the case of biological invasions, it would be useful for managers to be able to predict whether an invader will hybridize with a local relative, and if so, to what extent. A recent review on experimental hybridization (Edmands 2002) clearly showed that, although pre- and postzygotic isolation are roughly correlated with divergence time, there is tremendous variation in divergence time, resulting in variable hybrid vigor, outbreeding depression, or only partial reproductive compatibility within and among different taxonomic groups. Thus, the extent of variation observed in this relationship does not allow one to predict the consequences of a specific encounter of invading and native taxa based on their genetic divergence.

16.5.2 Habitat Modifications

As can be easily deduced from the examples given in this text, anthropogenic disturbance of local habitats is a major factor promoting hybridization, and also affecting the outcome of hybridization. Rhymer and Simberloff (1996) discuss in detail three forms of habitat modifications.

1. The first form is local habitat modification, leading to a mixing of previously distinct gene pools. There are indeed many examples, in plants and in animals, reporting that introgressive hybridization occurs at higher frequency in disturbed or artificial habitats than in undisturbed natural sites (e.g., Bleeker and Hurka 2001; Riley et al. 2003).
2. As a second form, they defined regional habitat change promoting geographic range expansion of one taxon into the range of another, which would provide an opportunity for hybridization. One notable example is the genetic introgression between mallards and black ducks described above.
3. A third type of habitat modification promoting hybridization is the construction of permanent migration corridors between the ranges of allopatric taxa. For example, this has presumably led to the emergence of the new invasive hybrid *Cottus* taxon described above.

As a further important category of habitat modifications in this context, we should also mention habitat fragmentation. This modification does not promote hybridization by itself, but it makes small, local isolated populations more vulnerable to extinction through introgressive hybridization, as is the case for rare species (Levin et al. 1996).

16.5.3 Introduction Intensity

Even if we have some a priori knowledge on the fitness of hybrids or existing prezygotic barriers, e.g., from experimental crossings or natural hybrid zones, there is an important general difference between natural and most artificial cases of hybridization, making the prediction of hybridization nearly impossible. In human-mediated cases, the contacts between invading and native taxa usually do not occur along narrow contact zones, in contrast with many natural cases. Rather, exotic species are often introduced repeatedly across large parts of the native species range. This applies in particular to cultivated plant species, to many freshwater fish species, of interest to anglers and commercial fisheries, and also to game species including many birds and mammals. In such situations, the frequency of occasions for interbreeding is greatly enhanced. This artificially high rate of contact increases the probability that genetic incompatibilities are eventually broken down. Even if there is selection against hybrids or against the introduced species, this disadvantage is simply overcome by the fact that the invaders are reproduced artificially, e.g., in hatcheries. At the same time, declining numbers in local populations of the invaded species may not be compensated through immigration from neighboring 'pure' populations, due either to habitat fragmentation (e.g., migration obstacles, such as dams), or to the introductions being geographically widespread.

16.5.4 Differences Between Populations

Besides the two important factors of introduction intensity and habitat modifications affecting the intensity and outcome of introgressive hybridization between native and invading taxa, there is probably an additional modifying factor that has been greatly overlooked in the past. Local populations of a species may naturally greatly differ in their ecology and genetic composition, due to specific adaptations to their local environment or due to historical demographic processes, and this may also apply to the introduced species, which perhaps already represent a mixture of different source populations or are introduced from several different sources. Consequently, we may also expect these to differ in their susceptibility to introgression. Recently, experimental studies (Kodric-Brown and Rosenfield 2004) showed that males from different populations of Pecos pupfish (*Cyprinodon pecosensis*) differ in agonistic behavior, territoriality, and in mating success when competing with the introduced sheepshead minnow (*C. variegatus*). This suggests that in local populations of *C. pecosensis*, the rate of introgressive hybridization with *C. variegatus* may vary depending on differences in the competitive ability of males in these populations.

16.6 Conclusions

Anthropogenic introgressive hybridization is a widespread phenomenon, and it will certainly continue to increase in importance. Many habitat modifications are irreversible, and there are also many situations where the mixing of gene pools by introgression has occurred to a point where it is irreversible. Facing this reality, managers now need to answer important questions as to whether introgressed populations should also be conserved in some cases, and to what degree introgression is acceptable (Allendorf et al. 2001). Should we, for example, also conserve new taxa that have been created through hybridization between native and introduced species, or the hybrid swarm that has completely replaced the formerly native taxon? In this context, it is also worth mentioning that distinguishing between natural and man-induced hybridization is a difficult, but crucial task (Allendorf et al. 2001; Petit et al. 2004).

Predicting hybridization and its outcomes is probably an unsolvable task, due to the complexity of potential interactions between the factors involved. Nevertheless, for an improved management and conservation of native taxa that are threatened by hybridization with invading taxa, it is of primary importance to conduct more long-term studies, which systematically integrate information on environmental and biological (including genetic) characteristics at the level of local populations.

Finally, hybridization is an evolutionary process that plays an important part in the context of biological invasions, which in turn are a key driving force of current evolutionary change (Mooney and Cleland 2001). Thus, the recent progress toward integrating evolutionary biology into invasion biology (Lee 2002; Levin 2003) provides an important basis for further advance in research on anthropogenic hybridization.

References

Abbott RJ (1992) Plant invasions, interspecific hybridization and the evolution of new plant taxa. Trends Ecol Evol 7:401–405

Adams JR, Leonard JA, Waits LP (2003) Widespread occurrence of a domestic dog mitochondrial DNA haplotype in southeastern US coyotes. Mol Ecol 12:541–546

Allendorf FW, Leary RF, Spruell P, Wenburg JK (2001) The problems with hybrids: setting conservation guidelines. Trends Ecol Evol 16:613–622

Anderson EC, Thompson EA (2002) A model-based method for identifying species hybrids using multilocus genetic data. Genetics 160:1217–1229

Anttila CK, Daehler CC, Rank NE, Strong DR (1998) Greater male fitness of a rare invader (*Spartina alterniflora*, Poaceae) threatens a common native (*Spartina foliosa*) with hybridization. Am J Bot 85:1597–1601

Arnold ML (1992) Natural hybridization as an evolutionary process. Annu Rev Ecol Syst 23:237–261

Arnold ML, Buckner CM, Robinson JJ (1991) Pollen-mediated introgression and hybrid speciation in Louisiana irises. Proc Natl Acad Sci USA 88:1398–1402

Asmussen MA, Arnold J, Avise JC (1987) Statistics of natural populations. 4. Definition and properties of disequilibrium statistics for associations between nuclear and cytoplasmic genotypes. Genetics 115:755–768

Avise JC (1994) Molecular markers, natural history and evolution. Chapman & Hall, New York

Avise JC (2000) Cytonuclear genetic signatures of hybridization phenomena: rationale, utility, and empirical examples from fishes and other aquatic animals. Rev Fish Biol Fisheries 10:253–263

Avise JC, Ankney CD, Nelson WS (1990) Mitochondrial gene trees and the evolutionary relationship of mallard and black ducks. Evolution 44:1109–1119

Barilani M, Deregnaucourt S, Gallego S, Galli L, Mucci N, Piombo R, Puigcerver M, Rimondi S, Rodriguez-Teijeiro JD, Spano S, Randi E (2005) Detecting hybridization in wild (*Coturnix c. coturnix*) and domesticated (*Coturnix c. japonica*) quail populations. Biol Conserv 126:445–455

Beaumont M, Barratt EM, Gottelli D, Kitchener AC, Daniels MJ, Pritchard JK, Bruford MW (2001) Genetic diversity and introgression in the Scottish wildcat. Mol Ecol 10:319–336

Bleeker W, Hurka H (2001) Introgressive hybridization in *Rorippa* (Brassicaceae): gene flow and its consequences in natural and anthropogenic habitats. Mol Ecol 10:2013–2022

Butcher PA, Williams ER (2002) Variation in outcrossing rates and growth in *Eucalyptus camaldulensis* from the Petford Region, Queensland; evidence of outbreeding depression. Silvae Genetica 51:6–12

Davison A, Birks JDS, Griffiths HI, Kitchener AC, Biggins D, Butlin RK (1999) Hybridization and the phylogenetic relationship between polecats and domestic ferrets in Britain. Biol Conserv 87:155–161

Deregnaucourt S, Gyomarc'h JC, Belhamra M (2005) Comparison of migratory tendency in European quail *Coturnix c. coturnix*, domestic Japanese quail *Coturnix c. japonica* and their hybrids. Ibis 147:25–36

Dobzhansky T (1950) Genetics of natural populations. 19. Origin of heterosis through natural selection in populations of *Drosophila pseudoobscura*. Genetics 35:288–302

Dowling TE, Secor CL (1997) The role of hybridization and introgression in the diversification of animals. Annu Rev Ecol Syst 28:593–619

Edmands S (2002) Does parental divergence predict reproductive compatibility? Trends Ecol Evol 17:520–527
Ellstrand NC (1992) Gene flow by pollen – implications for plant conservation genetics. Oikos 63:77–86
Ellstrand NC (2003a) Current knowledge of gene flow in plants: implications for transgene flow. Philos Trans R Soc Lond B Biol Sci 358:1163–1170
Ellstrand NC (2003b) Dangerous liaisons: when crops mate with their wild relatives. John Hopkins University Press, Baltimore, MD
Ellstrand NC, Elam DR (1993) Population genetic consequences of small population-size – implications for plant conservation. Annu Rev Ecol Syst 24:217–242
Ellstrand NC, Schierenbeck KA (2000) Hybridization as a stimulus for the evolution of invasiveness in plants? Proc Natl Acad Sci USA 97:7043–7050
Ellstrand NC, Whitkus R, Rieseberg LH (1996) Distribution of spontaneous plant hybrids. Proc Natl Acad Sci USA 93:5090–5093
Ellstrand NC, Prentice H, Hancock JF (2002) Gene flow and introgression from domesticated plants into their wild relatives. In: Syvanen M, Kado CI (eds) Horizontal gene transfer. Academic Press, San Diego, pp 217–236
Estoup A, Angers B (1998) Microsatellites and minisatellites for molecular ecology: theoretical and empirical considerations. In: Carvalho GR (ed) Advances in molecular ecology. IOS Press, Amsterdam, pp 55–86
Garcia-Vazquez E, Perez J, Ayllon F, Martinez JL, Glise S, Beall E (2004) Asymmetry of post-F-1 interspecific reproductive barriers among brown trout (*Salmo trutta*) and Atlantic salmon (*Salmo salar*). Aquaculture 234:77–84
Goldberg TL, Grant EC, Inendino KR, Kassler TW, Claussen JE, Philipp DP (2005) Increased infectious disease susceptibility resulting from outbreeding depression. Conserv Biol 19:455–462
Grant PR, Grant BR (1996) Speciation and hybridization in island birds. Philos Trans R Soc Lond B Biol Sci 351:765–772
Gray AJ, Marshall DF, Raybould AF (1991) A century of evolution in *Spartina anglica*. Adv Ecol Res 21:1–62
Guyomarc'h JC, Combreau O, Puigcerver M, Fontoura P, Aebischer NJ. (1998) *Coturnix coturnix* quail. BWP Update 2:27–46
Halbert ND, Ward TJ, Schnabel RD, Taylor JF, Derr JN (2005) Conservation genomics: disequilibrium mapping of domestic cattle chromosomal segments in North American bison populations. Mol Ecol 14:2343–2362
Hubbs CL (1955) Hybridization between fish species in nature. Syst Zool 4:1–20
Huxel GR (1999) Rapid displacement of native species by invasive species: effects of hybridization. Biol Conserv 89:143–152
Kodric-Brown A, Rosenfield JA (2004) Populations of Pecos pupfish (*Cyprinodon pecosensis*) differ in their susceptibility to hybridization with sheepshead minnow (*C. variegatus*). Behav Ecol Sociobiol 56:116–123
Konishi M, Takata K (2004) Impact of asymmetrical hybridization followed by sterile F-1 hybrids on species replacement in *Pseudorasbora*. Conserv Gen 5:463–474
Kulikova IV, Zhuravlev YN, McCracken KG (2004) Asymmetric hybridization and sex-biased gene flow between Eastern spot-billed ducks (*Anas zonorhyncha*) and mallards (*A. platyrhynchos*) in the Russian Far East. Auk 121:930–949
Largiadèr CR, Scholl A (1995) Effects of stocking on the genetic diversity of brown trout populations of the Adriatic and Danubian drainages in Switzerland. J Fish Biol 47:209–225
Largiadèr CR, Scholl A (1996) Genetic introgression between native and introduced brown trout *Salmo trutta* L populations in the Rhone River Basin. Mol Ecol 5:417–426

Leary RF, Allendorf FW, Forbes SH (1993) Conservation genetics of bull trout in the Columbia and Klamath river drainages. Conserv Biol 7:856–865

Lecis R, Pierpaoli M, Biro ZS, Szemethy L, Ragni B, Vercillo F, Randi E (2006) Bayesian analyses of admixture in wild and domestic cats (*Felis silvestris*) using linked microsatellite loci. Mol Ecol 15:119–131

Lee CE (2002) Evolutionary genetics of invasive species. Trends Ecol Evol 17:386–391

Levin DA (2003) Ecological speciation: lessons from invasive species. Syst Bot 28:643–650

Levin DA, Francisco Ortega J, Jansen RK (1996) Hybridization and the extinction of rare plant species. Conserv Biol 10:10–16

Lodge DM, Taylor CA, Holdich DM, Skurdal J (2000) Nonindigenous crayfishes threaten North American freshwater biodiversity: lessons from Europe. Fisheries 25:7–20

Lynch M (1991) The genetic interpretation of inbreeding depression and outbreeding depression. Evolution 45:622–629

Lynch M (1996) A quantitative-genetic perspective on conservation issues. In: Avise JC, Hamrick JL (eds) Conservation genetics: case histories from nature. Chapman & Hall, New York, pp 471–501

Mallet J (2005) Hybridization as an invasion of the genome. Trends Ecol Evol 20:229–237

Mank JE, Carlson JE, Brittingham MC (2004) A century of hybridization: decreasing genetic distance between American black ducks and mallards. Conserv Gen 5:395–403

Maran T, Henttonen H (1995) Why Is the European mink (*Mustela lutreola*) disappearing – a review of the process and hypotheses. Ann Zool Fenn 32:47–54

Mooney HA, Cleland EE (2001) The evolutionary impact of invasive species. Proc Natl Acad Sci USA 98:5446–5451

Morin PA, Woodruff DS (1996) Noninvasive genotyping for vertebrate conservation. In: Smith TB, Wayne RK (eds) Molecular genetic approaches in conservation. Oxford University Press, Oxford, pp 298–313

Mullis K, Faloona F, Scharf S, Saiki R, Horn G, Erlich H (1986) Specific enzymatic amplification of DNA invitro - the polymerase chain reaction. Cold Spring Harbor Symp Quant Biol 51:263–273

Nijman IJ, Otsen M, Verkaar ELC, de Ruijter C, Hanekamp E, Ochieng JW, Shamshad S, Reges JEO, Hanotte O, Barwegen MW, Sulawati T, Lenstra JA (2003) Hybridization of banteng (*Bos javanicus*) and zebu (*Bos indicus*) revealed by mitochondrial DNA, satellite DNA, AFLP and microsatellites. Heredity 90:10–16

Nolte AW, Freyhof J, Stemshorn KC, Tautz D (2005) An invasive lineage of sculpins, *Cottus* sp (Pisces, Teleostei) in the Rhine with new habitat adaptations has originated from hybridization between old phylogeographic groups. Proc R Soc B Biol Sci 272:2379–2387

Olden JD, Poff NL, Douglas MR, Douglas ME, Fausch KD (2004) Ecological and evolutionary consequences of biotic homogenization. Trends Ecol Evol 19:18–24

Perry WL, Feder JL, Dwyer G, Lodge DM (2001) Hybrid zone dynamics and species replacement between *Orconectes* crayfishes in a northern Wisconsin lake. Evolution 55:1153–1166

Perry WL, Lodge DM, Feder JL (2002) Importance of hybridization between indigenous and nonindigenous freshwater species: An overlooked threat to North American biodiversity. Syst Biol 51:255–275

Petit RJ, Bialozyt R, Garnier-Gere P, Hampe A (2004) Ecology and genetics of tree invasions: from recent introductions to Quaternary migrations. Forest Ecol Manage 197:117–137

Pritchard JK, Stephens M, Donnelly P (2000) Inference of population structure using multilocus genotype data. Genetics 155:945–959

Randi E, Lucchini V (2002) Detecting rare introgression of domestic dog genes into wild wolf (*Canis lupus*) populations by Bayesian admixture analyses of microsatellite variation. Conserv Gen 3:31–45

Rhymer JM, Simberloff D (1996) Extinction by hybridization and introgression. Annu Rev Ecol Syst 27:83–109

Rhymer JM, Williams MJ, Braun MJ (1994) Mitochondrial analysis of gene flow between New-Zealand mallards (*Anas platyrhynchos*) and grey ducks (*A. superciliosa*). Auk 111:970–978

Rieseberg LH (1997) Hybrid origins of plant species. Annu Rev Ecol Syst 28:359–389

Riley SPD, Shaffer HB, Voss SR, Fitzpatrick BM (2003) Hybridization between a rare, native tiger salamander (*Ambystoma californiense*) and its introduced congener. Ecol Appl 13:1263–1275

Sagvik J, Uller T, Olsson M (2005) Outbreeding depression in the common frog, *Rana temporaria*. Conserv Gen 6:205–211

Scribner KT, Page KS, Bartron ML (2000) Hybridization in freshwater fishes: a review of case studies and cytonuclear methods of biological inference. Rev Fish Biol Fisheries 10:293–323

Tautz D (1989) Hypervariability of Simple Sequences as a general source for polymorphic DNA markers. Nucleic Acids Res 17:6463–6471

Utter F (2000) Patterns of subspecific anthropogenic introgression in two salmonid genera. Rev Fish Biol Fisheries 10:265–279

Vilà C, Walker C, Sundqvist AK, Flagstad O, Andersone Z, Casulli A, Kojola I, Valdmann H, Halverson J, Ellegren H (2003) Combined use of maternal, paternal and biparental genetic markers for the identification of wolf-dog hybrids. Heredity 90:17–24

Vilà M, Weber E, D'Antonio CM (2000) Conservation implications of invasion by plant hybridization. Biol Invasions 2:207–217

Williams CL, Brust RC, Fendley TT, Tiller GR, Rhodes OE (2005) A comparison of hybridization between mottled ducks (*Anas fulvigula*) and mallards (*A. platyrhynchos*) in Florida and South Carolina using microsatellite DNA analysis. Conserv Gen 6:445–453

Wolf DE, Takebayashi N, Rieseberg LH (2001) Predicting the risk of extinction through hybridization. Conserv Biol 15:1039–1053

17 Genetically Modified Organisms as Invasive Species?

ROSIE HAILS and TRACEY TIMMS-WILSON

17.1 Introduction

The release of genetically modified organisms (GMOs) is a controversial subject. Some perceive it to be the single most important development in biology since the discovery of natural selection. Others are concerned that the movement of genes with no reference to natural species boundaries could pose new ecological risks. One conjectural risk is that transgenes will either cause the host species to become invasive or they will escape from the original host species and cause other species to become invasive. Gene flow between species occurs naturally, although the frequency varies within and across kingdoms. Such gene flow is responsible for creating new combinations of genes, with the potential for introgression or speciation. Hybridisation has been proposed as a stimulus for the evolution of invasiveness in plants (Ellstrand and Schierenbeck 2000), suggesting that new combinations can create genotypes with different, and perhaps surprising ecological behaviours. Do transgenes pose particular risks in this respect? Is it possible to predict the probability that transgenes will cause invasiveness in recipient organisms?

17.2 Quantitative Measures of Invasion Risk

To answer this question, we present a quantitative framework for the estimation of the probability of invasiveness. The risk assessment of genetically modified organisms has adapted approaches developed for pesticides and other potentially toxic substances in the environment. In its simplest form, quantifying risk involves estimating the expression of toxicity and the likeli-

Ecological Studies, Vol. 193
W. Nentwig (Ed.)
Biological Invasions

hood of exposure. Risk (R) may then be defined as the product of the probability of toxicity (P_t) and the probability of exposure (P_e), i.e.

$$R = P_e \times P_t$$

This has been successfully applied to the case of the Monarch butterfly and its risk from *Bt* corn (containing insecticidal transgenes from the soil bacterium *Bacillus thuringiensis*) through ingestion of pollen expressing cry proteins, giving an estimate of 1 in 10,000 larval mortality if 20 % of corn grown was *Bt*, rising to 1 in 2,000 if the legal maximum uptake led to 80 % of corn being transgenic (Sears et al. 2001). It was concluded that *Bt* corn posed a negligible risk to the Monarch, and the real threats to this species lay elsewhere, for example, in habitat destruction. This quantitative approach to risk assessment has also been adapted for GM plants (Poppy 2004; Raybould 2004) and transgenic fish (Muir 2004). We use a similar framework here to facilitate our discussion of the invasive potential of transgenic organisms. We break down the risk of transgenes causing invasiveness through escape into new recipient species into three steps:

– P(Transgene escape) $\times$ P(Transgene spread/Escape) $\times$ P(Harm/Exposure)

This can be read as the probability of transgene escape, multiplied by the probability of transgene spread – given that escape has occurred, multiplied by the probability of ecosystem harm – given exposure to the transgene. Escape of transgenes would be through hybridisation (plants and animals) or transformation/conjugation/transduction (bacteria), moving organisms beyond the genetic context in which they were originally produced. The rate at which transgenes spread through a recipient population will then depend upon the fitness consequences of carrying those transgenes. Finally, organisms are usually classed as invasive only if their spread causes economic or ecological harm, and so we discuss possible measures of harm. These three steps will now be reviewed for three taxa of GMOs – bacteria, plants and animals – drawing particularly on soil- and plant-associated bacteria, oilseed rape and salmon as case studies.

The genes which have been introduced into bacteria are hugely varied, many of them acting as marker genes, but the majority of these have been created for laboratory purposes. One potential application of genetic modification (GM) is to use transgenes to alter existing metabolic pathways, for example, to degrade pollutants. Few naturally occurring microorganisms possess the pathways required to mineralise the more recalcitrant xenobiotic compounds such as pentachlorophenols (PCPs and PCBs; Johri et al. 1999). GM technology has the potential to improve existing catabolic pathways or to extend such pathways to include additional target compounds which may otherwise not be degraded (Timmis et al. 1994; Brazil et al. 1995), and may also be applied to overcome the toxic or inhibitory effect of a particular pollutant or a metabolite (Mason et al. 1997). These GM microorganisms, or GMMs, have yet to be released in the field. However, experimental field releases of GMMs of plant growth-promoting rhizobacteria containing marker genes (Bailey et al. 1994, 1997) and antifungal genes to explore the improved potential of biocontrol agents against fungal phytopathogens (Timms-Wilson et al. 2000; Glandorf et al. 2001) have taken place.

The majority of GM plants have been manipulated to contain transgenes for herbicide tolerance or insect resistance. As anticipated, it has been demonstrated that herbicide-tolerant transgenes do not enhance fitness or invasiveness in the absence of the herbicide (Crawley et al. 1993, 2001). Insect-resistance genes are usually derived from the soil bacterium *Bacillus thuringiensis*. *Bt* plants express partially activated toxins which provide resistance to a range of Lepidopteran and Coleopteran pests (gene-dependent). Other GM plants include those with resistance to viruses (using coat protein genes), fungi (using chitinases), and a range of other features. We will focus here on transgenes which confer resistance to natural enemies, as it is postulated that such transgenes may alter invasiveness.

Fish have been modified to express growth factor genes, which cause juvenile transgenic fish to grow as much as 4–6 times faster than their wild counterparts. Other genetic modifications have improved resistance to bacterial diseases or tolerance to cold temperatures (Muir 2004). In our case studies, we focus on growth hormone (GH) transgenic fish, as there are concerns that these will interbreed with, and outcompete wild fish.

17.3 Gene Flow: the First Step to Invasiveness of Transgenes

An early step in genetically modified organisms causing invasiveness is for the transgenes to move from the domesticated species in which they were first introduced into other species or habitats. Rates of gene flow have been increasingly measured in many systems only since there has been concern about the escape of transgenes, leading to a realisation that some species barriers are more 'leaky' than previously supposed.

17.3.1 Gene Escape in Bacterial Communities

In terms of movement of genes between genotypes, bacteria are the extreme case: genomic studies continue to reveal examples of natural gene exchange across huge phylogenetic distances. There are three ways in which they can transfer genes and acquire new DNA: (1) by transformation (direct uptake), (2) or by the transfer of genes by mobile genetic elements, which is termed conjugation when involving plasmids or transposons and (3) transduction when by bacteriophage. Laboratory and field experiments over the years and, more recently, phylogenetic data have revealed the extent and central role of natural gene transfer in bacterial ecology and evolution (Bailey et al. 1994; Lorenz and Wackernagel 1994), including horizontal gene transfer between evolutionary unrelated bacteria (for a review, see Gogarten and Townsend 2005).

17.3.1.1 Transformation

Most simply, bacteria can actively take up free DNA in the form of plasmids or fragments of chromosomal DNA from their external environment. The newly acquired DNA may then recombine into the recipients genome and be expressed to confer a novel or modified phenotype.

17.3.1.2 Conjugation

Plasmids are key in maintaining the fluidity of the horizontal gene pool in bacterial populations. They are autonomously replicating, extra-chromosomal genetic elements, ubiquitous in natural bacterial communities. Plasmids provide accessory, albeit generally non-essential functions such as antibiotic resistance to their host. Genetic traits which confer adaptations to local environments tend to form clusters on plasmids and, consequently, are often viewed as desirable elements (symbiotic), even though they can impose an energetic drain onto the host bacterium. The transfer of plasmids between bacteria is the process of conjugation, an active, regulated process in part stimulated by direct cell–cell contact. It is normally unidirectional (from donor to recipient) but occasionally reciprocal (Ankenbauer 1997).

17.3.1.3 Transduction

Transduction involves the movement of bacterial DNA by viral infection (bacteriophage; McKane and Milkman 1995; Chiura 1997). Transduction is a central, if not the most important gene transfer mechanism in the generation of genomic diversity and bacterial evolution between closely related taxa. Two types of transduction are recognised, these being (1) generalised (or unrestricted) and (2) specialised (or restricted). In generalised transduction, any genetic element (chromosomal and plasmid) within a host cell has equal probability of being transduced. Specialised transduction involves the transfer only of specific genetic elements. The exchange of genetic material by this mechanism has been shown in a variety of environmentally important bacteria.

Transgenes placed on the chromosome will have much lower rates of escape than those placed on plasmids within the bacteria, even though gene flow between chromosomes of unrelated bacteria does occur in evolutionary time, via conjugation or, more commonly, transduction.

17.3.1.4 Evidence for Gene Transfer from GMMs

Most, if not all studies of the field and microcosm release of GMMs have included a component to specifically study the possibility of gene transfer to or from the transgenic bacteria, allowing us to test the prediction posed above. To date, no investigation has revealed an outcome which had not been predicted and no evidence has been provided demonstrating the invasion of transgenes from genetically stable constructs. Therefore, from existing evidence, it is apparent that the likelihood of gene transfer directly equates to the method of genetic construction. When introduced genes were located on mobile genetic elements such as conjugative plasmids, transfer has been observed in laboratory investigations. However, as this outcome was entirely predictable, field releases have involved bacteria genetically modified to carry transgenes on their chromosomes. Insertions have been mediated by the use of transposons, disarmed transposons or by site-directed homologous recombination. The order in which these three approaches are listed represents their relative genetic stability and, therefore, the likelihood of being transferred (Bailey et al. 1995; Troxler et al. 1997). In all the reports of field testing of bacteria modified to carry transgenes on their chromosomes, none have found either the transfer or loss of the transgenes. Even in the most appropriately designed laboratory investigations, transfer frequencies were negligible or effectively zero (Bailey et al. 1995; Troxler et al. 1997).

17.3.2 Gene Escape in Plant Communities

Gene escape from a plant species involves two steps: hybridisation with a wild relative, and the survival and reproduction of resulting hybrids. Successful hybridisation between crop plants and wild relatives was earlier thought to be infrequent (Ellstrand et al. 1999). This is probably because gene flow was not perceived to be of significant concern – people had simply not looked. A comprehensive review of the world's most important food crops has found that most spontaneously hybridise somewhere in their range, although rates vary considerably (Ellstrand 2003a). Oilseed rape can hybridise with close relatives (reviewed in Hails and Morley 2005), principally wild turnip, *Brassica rapa*, with national estimates of 32,000±26,000 hybrids being formed annually in the UK (Wilkinson et al. 2003). Although this represents a hybridisation rate of only 0.019 (i.e. 19 hybrids for every 1,000 *B. rapa*), the sheer number of opportunities for hybridisation means that this translates into 1,000s of hybrid plants each year.

Hybridisation does not equate to escape of genes from a plant species, as both hybridisation rate and the relative fitness of subsequent generations of hybrids will determine gene flow. Those species crosses for which F1 hybrids are sterile represent only very limited opportunities for gene escape (wheat, for example, van Slageren 1994) but hybrids between *Brassica napus* and *B. rapa* can survive and reproduce. Conventional wisdom would suggest that the introduction of genes from a domesticated genotype would be deleterious to the wild relative, and that fitness would be context-dependent. Experimental data for *B. napus* × *B. rapa* hybrids support both of these expectations. If the hybrids are protected from plant competition, then the F1 generation is found to be intermediate in fitness vis-à-vis the parents (Hauser et al. 1998a). Under such conditions, crop genes could be positively advantageous. Even when subject to competition from either parent, however, one component of fitness – seed production – was enhanced in hybrids (Hauser et al. 2003). Without whole life-cycle estimates, it is not possible to conclude if this would translate into enhanced fitness, as there may be trade-offs between different parts of the life cycle. Later hybrid and backcross generations had reduced fitness relative to both parents (Hauser et al. 1998b) but, nevertheless, the evidence illustrates that the species barrier between *B. napus* and *B. rapa* is permeable. This permeability has been demonstrated for other crop–wild relative combinations, albeit to varying degrees (Ellstrand 2003b; Hails and Morley 2005).

17.3.3 Gene Escape in Animal Populations

Escape of transgenes from animals is more likely to occur through escape of the GM animal, followed by within-species gene flow, rather than through hybridisation with related species. Escape of GM fish from fish farms is a case in point, with transgenes conferring cold or salt tolerance allowing the transgenic fish to occupy new niches. Further gene flow could occur through mating with wild fish of the same species, so *P*(Transgene escape) in these cases equates to the probability that the fish will physically escape from the fish farm and subsequently breed. How ever fish pens are constructed, there will always be the possibility of destruction through extreme weather events or vandalism and, in fact, it is estimated that around 2,000,000 farmed salmon escape each year in the North Atlantic region (McGinnity et al. 2003). If transgenic fish are grown in conventional facilities, then GM fish would also escape, spread rapidly and interbreed. One solution to this is to ensure the transgenic fish are sterile. If eggs are subjected to heat or pressure shock, then they retain an extra set of chromosomes and the resulting triploid fish are sterile. Still, this is unlikely to be 100 % reliable. This term in the risk equation is possibly relatively high for within-species risk but low for between-species escape of the transgene.

17.4 Transgene Spread

Gene flow creates the opportunities for transgene spread but the rate at which this will occur depends upon the relative fitness of transgenic versus wild type genotypes.

17.4.1 Transgene Spread in Bacterial Populations

There is only one experiment which has directly compared the fitness of transgenic and non-transgenic bacteria in 'semi-field' conditions, when the transgene has been located on plasmids. The transgene in this case was a marker. This was conducted in an ecotron (http://www.cpb.bio.ic.ac.uk/ecotron/ecotron.html), a facility which houses replicated microcosms containing simplified ecological communities. The aim of the experiment was to compare the population dynamics of *Pseudomonas fluorescens* SBW25R (the control) with the same bacterium carrying a gene cassette which had been introduced into the genome in three different ways: directly inserted into the bacterial chromosome (SBW25R::KX), as a similar insertion but including a lysogenic phage (SBW25::KX-F101), and inserted into a conjugative plasmid (SBW25R pQBR11::KX). Following seed dressing, all bacterial strains successfully colonized the phytosphere of chickweed (*Stellaria media*) and became established. However, densities of the introduced strain in the lysogenic-phage treatment were consistently lower than in the other treatments, at times approaching the limits of detection. The lower density of this strain was attributed to cell deaths caused by lysis following phage induction, which was expected in the phytosphere (Ashelford et al. 1999). In the plasmid treatment, the densities of *P. fluorescens* SBW25R (pQBR11::KX) on day 53 were significantly lower on roots and leaves than was the case for the plasmid-free control, with plasmid carriage resulting in a marked reduction in colonizing fitness. However, from day 53 onwards, in each of the four chambers the population density of the plasmid-carrying strain increased significantly on roots (25-fold) and leaves (13-fold) until, by day 95, these densities either matched or slightly exceeded those of the control. These results can be explained if carriage of the plasmid were typically associated with a cost to the host (as has frequently been demonstrated; Lilley and Bailey 1997) but that, periodically, an unidentified plasmid gene (or genes) improves the fitness of the bacteria, perhaps by conferring the ability to utilise or tolerate certain substrates exuded by the plants as they mature. Variation in the fitness of bacteria carrying plasmids is correlated with plant growth stage in the field.

Because of the rates of horizontal gene flow expected, there has yet to be a field release of bacteria containing transgenic plasmids. However, there have

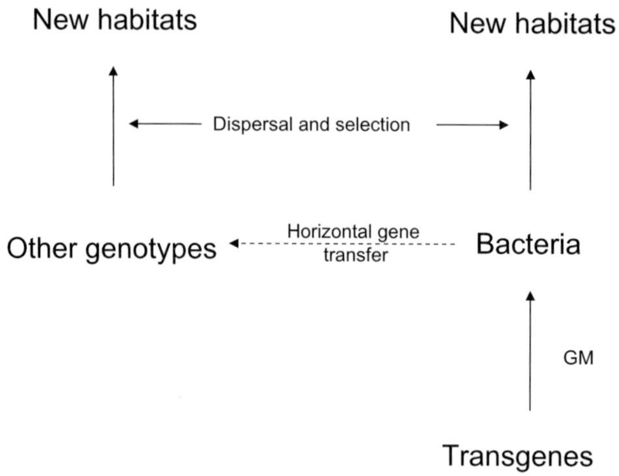

Fig. 17.1 Dispersal of transgenes through microbial ecosystems

been experimental field releases of bacteria with introduced genes on the chromosome, where there is still the potential for transgene spread through fitness advantages to the host bacterium, carrier proliferation and vertical transfer. A study of *P. fluorescens* SBW25 in wheat compared two transgenic bacteria, with one or two marker genes. As predicted, it was found that the genotype carrying two markers was less fit than the genotype carrying one marker because of the extra burden brought about by expressing the extra phenotype (DeLeij et al. 1995). Other field studies have involved transgenic lines with functional genes designed to promote growth or protect the plant. The use of transgenic *P. fluorescens* SBW25 in several field releases (Bailey et al. 1994) showed that the bacterium survived well in the plant phytosphere. During the growing season, the GMM flourished but, after harvest, it could no longer be detected, unable to persist once the host plant was removed. This means the use of these GMMs is very seasonal and would need to be reintroduced for a second growing season, if required. Thus, although there is a degree of transgene invasion and spread within the season, this is predictably transient in this case (Fig. 17.1).

17.4.2 Transgene Spread in Plant Populations

Empirical studies to determine the potential impact of transgenes on plant fitness fall into two categories: (1) those which measure fitness impacts under 'near-agricultural' conditions with manipulated densities of the plant's natural enemies (herbivores or pathogens) and (2) those which measure fitness impacts in semi-natural habitats. These two types of study are providing different classes of evidence.

In the first group, crop–wild relative hybrids containing the transgene were grown under experimental conditions often close to the conditions found in cultivation and, most particularly, with the experimental plants being released from plant competition. The densities of those natural enemies which were transgene targets were then manipulated, the results illustrating that, under appropriate conditions, those hybrids containing the transgenes have greater fitness. For example, F1 hybrids between *B. napus* and *B. rapa* containing *Bt* transgenes were found to have a fecundity advantage under high insect pressure (Vacher et al. 2004). These studies do little more than illustrate that the transgene will behave as expected in the hybrid as well as the crop plant, conferring a selective advantage when those plant populations are affected by the natural enemy (herbivore or pathogen) targeted by the transgene.

In the second group, far fewer studies have addressed the same questions under natural field conditions, without manipulating natural enemy pressure. One exception to this involved a *Bt* gene backcrossed into wild sunflower populations. The transgenic backcrossed line had significantly higher fecundity, compared to the backcrossed control line (Snow et al. 2003). All else being equal, this would lead to enhanced fitness of plants carrying this transgene.

The difference between these two classes of study is crucial in determining the frequency of those conditions under which the transgene would be expected to confer a selective advantage and, therefore, the rate at which it would spread. The difficulties associated with estimating the relative fitness of transgenic hybrids may in part explain why the first class of studies is so much more common than the second. Herbivores and pathogens occur sporadically in space and time in natural communities: over many years and at numerous sites, pathogen-resistance genes may provide little advantage until that one year when a new, virulent pathogen sweeps through an area. Thus, experiments conducted over a limited number of years and sites run the risk of being unable to detect any fitness differences. The temptation is then to manipulate natural enemy pressure to demonstrate the obvious – as long as some significant result is obtained. The key question is 'what role do natural enemies play in regulating existing plant populations?' So, the most informative studies will not necessarily involve transgenic plants at all but, rather, underpinning ecological processes in natural communities.

Another parameter in determining the relative selective advantage of transgenic plants is the cost of carrying the transgene in the absence of the target. Costs would contribute to the rate at which transgene frequency may decline when the population is not under that specific selection pressure. Again, studies of natural herbivore and pathogen resistance in non-transgenic plants can be highly informative. A recent review revealed that, if genetic background is controlled, then 82 % of studies demonstrated fitness costs associated with carrying herbivore-resistance genes (Strauss et al. 2002), either as direct costs, such as a trade-off in resources allocated to defence or

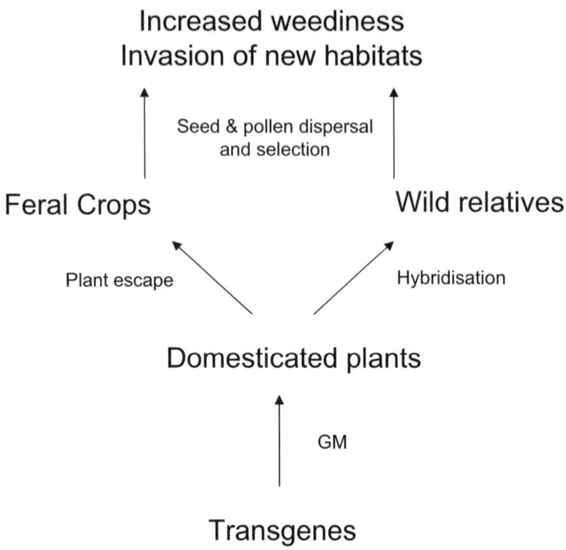

Fig. 17.2 Dispersal of transgenes through plant communities

reproduction, or indirect costs depending upon interactions with other species. However, there is yet to be a clear demonstration of costs of carrying transgenes in the absence of selection. As with studies of natural resistance, controlling the genetic background is important. Most studies are based on material from only one transformation event, so that the effect of the transgene is confounded with those of other factors, such as positional effects. Thus, the costs detected cannot be unequivocally attributed to the transgene (Hails et al. 1997; Snow et al. 1999). One study used replicate lines from multiple transformation events in *Arabidopsis thaliana*, and found that the presence of the transgene significantly reduced fecundity (Bergelson et al. 1996; Purrington and Bergelson 1997). However, this is most likely due to a disruption of genes by the insertion of the herbicide-resistance transgene, rather than to enhanced allocation of resources. It may also be that, compared to other species, such costs are most likely to be detected in *A. thaliana* because it has a very small genome, which may explain why these results have yet to be replicated in other genomes (Fig. 17.2).

17.4.3 Transgene Spread in Animal Populations

Empirical data on the relative fitness of transgenic and non-transgenic fish raise a number of issues of relevance to other sexually reproducing organisms. Growth hormone transgenic salmon have dramatically increased feeding rates, feed conversion efficiency and, ultimately, growth rates (Cook et al. 2000). If this growth potential were to be realised in the wild, then transgenic escapees could be superior competitors, directly threatening already vulnerable wild populations. Experiments in aquaria have illustrated that the higher feeding motivation of transgenic fish does make them superior competitors, transgenic coho salmon consuming 2.5 times more food than did the non-transgenic controls, and being significantly larger (Devlin et al. 1999). Other studies have suggested that the increase in feeding motivation comes with a cost. Transgenic fish are willing to incur an increased level of risk when foraging, so mortality rates from predators are likely to be higher (Abrahams and Sutterlin 1999). More recent studies have investigated the potential impact of food abundance and predation risk on growth and survival of transgenic salmon, attempting to simulate near-natural environments. Landscaped stream aquaria with live food and predators illustrated that the relative fitness of transgenic and non-transgenic salmon is dependent on the environment. The enhanced vulnerability of transgenic hatchlings to predation is amplified as food abundance decreases. Transgenic fish were also found to grow more slowly than non-transgenics at low food abundance. So, both food abundance and predation pressure will influence relative fitness, with the competitive superiority of growth hormone transgenic salmon most likely to be manifest at times of high food abundance (Fig. 17.3).

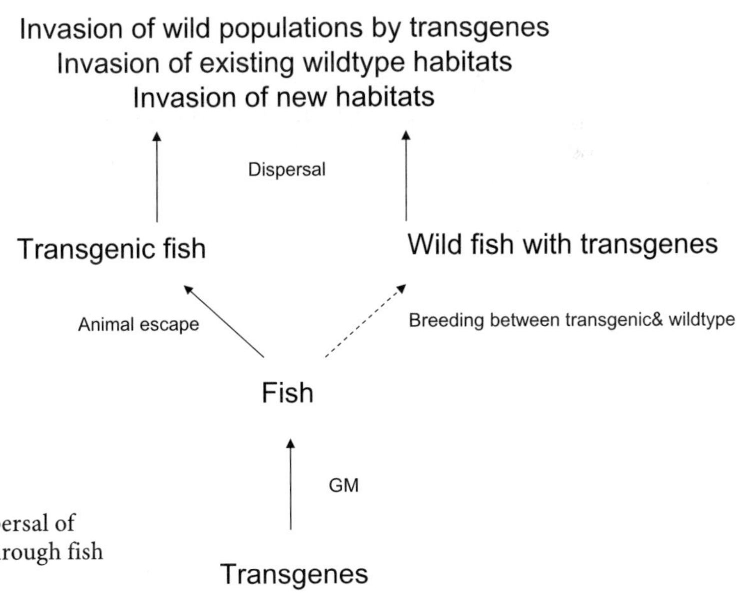

Fig. 17.3 Dispersal of transgenes through fish populations

17.5 Ecological Impact

A considerable amount of research effort has been expended in measuring the rates at which transgenes may spread into a population. A key question, however, is 'does transgene spread matter?' The answer will, of course, depend upon the transgene and the receiving environment but the circumstances under which transgene spread will be ecologically unacceptable (and, therefore, considered 'invasive') remain ill defined.

17.5.1 Detecting Impacts in Bacterial Populations

Bacteria are attributed with many key ecological roles. In soil, these include soil formation, the promotion of plant growth, plant protection, mineralisation of chemicals including pollutants, and the cycling of nitrogen, carbon, sulphur, iron and phosphorus. The question arises as to whether the altered behaviour of bacteria carrying or acquiring transgenes (especially those expressing traits of relevance in that habitat) may result in the sustained perturbation of the indigenous community to the detriment of its normal functions or displacement of specific species engaged in key activities (Tiedje et al. 1989).

A field release of a functionally active GMM showed minor changes in diversity. The GMM, a fluorescent pseudomonad containing a constitutively expressed antifungal compound, survived well throughout the growing season in the rhizosphere of spring wheat, and had a transient effect (in the first 2 months) on the microbial diversity measured by several culture-dependent and culture-independent methods (Timms-Wilson et al. 2002). Population function was measured by examining carbon utilisation profiles of the rhizosphere microbial community. Over the entire growing season, impacts were not significant and, once the plants were removed, the isolate was no longer detected. In fact, several field-based and laboratory-based studies have shown that the presence of phytopathogens has a greater effect on microbial diversity than do GMMs, and that individual plant species will also have a huge impact on the diversity of indigenous microbes (Timms-Wilson et al. 2002; Houlden 2005). So, although biodiversity impacts can be detected, these are small in magnitude and have not translated into functional differences.

Microbial communities are highly heterogeneous, consisting of large numbers of diverse populations (Dykhuizen 1998). It is often assumed in macrobiological scientific communities that populations occupy distinct niches, play distinct roles and may be displaced by more efficient competitors. However, it appears that microbial communities support coexisting populations occupying similar or heavily overlapping niches (Atlas 1984). This leads to functional redundancy where the loss of species may be compensated for by the activity of others (Kennedy and Smith 1995). Thus, if impact is measured in terms of a change in ecosystem function, then no significant impacts have been detected and, indeed, seem unlikely (Timms-Wilson et al. 2002; Griffiths et al. 2003).

17.5.2 Potential Impacts in Plant Populations

Transgenes which would alter the invasiveness of plants, ultimately causing the extinction of other species, would be unacceptable. Fitness and invasiveness are sometimes used interchangeably in the literature, yet this is not always appropriate. In fact, only when a species is invading a habitat for the first time is it appropriate to equate fitness and invasiveness – under those circumstances, population growth rate (fitness) can be used as a measure of how invasive that genotype is. When considering the invasion of transgenes into wild populations, enhanced fitness may result in a change in gene frequencies but not necessarily a change in invasiveness. This latter attribute depends upon the factors responsible for limiting and regulating the population. For example, the extent to which the enhanced fecundity of *Bt* wild sunflowers (Snow et al. 2003) will enhance invasiveness depends upon the extent to which wild sunflowers are seed-limited and the herbivores remain susceptible to the transgene products. Rates of co-evolution may be quite rapid in response to *Bt* genes (Shelton et al. 1993; Ferré and van Rie 2002), in which case any increases in fitness may be transient. If wild sunflower populations are seed-limited and herbivore populations remain susceptible, then enhanced fitness may translate into increased abundance. A review of seed sowing experiments aimed at unravelling the extent to which natural populations are seed-limited (Turnbull et al. 2000) demonstrated that approximately 50 % of seed augmentation experiments showed some evidence of seed limitation; this tends to occur in early successional habitats and with early successional species. In other words, if succession proceeds, any changes in abundance resulting from enhanced fecundity may also be transient. The most likely problems to occur are in disturbed (agricultural) habitats.

17.5.3 Potential Impacts in Animal Populations

Theoretical scenarios have been raised in which the release of transgenic fish could have very detrimental impacts on wild populations, and one of these is the Trojan gene hypothesis. This hypothesis illustrates how data on relative fitness for part of the life cycle could mislead as to the potential long-term impact of released transgenic salmon. If transgenic escapees have a mating advantage but viability of offspring is reduced, then models predict that the transgene will spread through the wild population but that reduced viability of offspring will cause local extinction of populations (Muir and Howard 1999). Thus, a transgene invasion could have the ultimately undesirable outcome of population extinction. The mating advantage of transgenics is based on the idea that mature transgenic fish would be larger than non-transgenics. However, faster growth does not necessarily lead to larger size at maturity, and the genotype-by-environment interactions discussed above should also be taken into account. If food abundance is low, then the lower survival of the risk-prone transgenic juveniles may prevent transgene spread in the population. Whether the Trojan gene is a real threat remains an open question, albeit a theoretical possibility (Reichhardt 2000).

17.6 Conclusions

Much of the current disquiet about the creation of GMOs is that their introduction into the natural environment could be irreversible, causing a perturbation to the ecological community which will propagate and possibly grow in magnitude as it ripples through the ecosystem. The pathways by which transgenes could invade ecosystems have been considered for microbes, plants and fish (Figs. 17.1 to 17.3). In bacteria, if genes are introduced into chromosomes, then horizontal gene flow will be low (hard to detect in ecological time) but dispersal of bacteria per se is very high. The traditional view is that the dispersal of bacteria is unhindered by geographic boundaries – the environment selects (Finlay 2002; Whitfield 2005). This view of ecological panmixis is peculiar to the microbiological world, and from this arise two possible consequences. Firstly, the phenotype of the transgenic bacteria is all important – it will determine the conditions under which selection will occur. Secondly, panmixis leads to the functional redundancy discussed above. So, perhaps counter-intuitively, greater mixing means less impact of altered phenotypes on ecosystem services.

Studies of natural systems can be used to illustrate the same principles of horizontal gene flow and selection which we have been discussing with GMMs, the introduction of *Lotus corniculatus* to New Zealand being one example. At first sowing, the seeds needed to be inoculated with a natural isolate, *Mesorhizobium loti*, a nitrogen-fixing symbiont which passes ammonium to the plant, in exchange for nutrients. Subsequent inoculation was not necessary. However, this did not correlate with the establishment of the original inoculated bacterium, as this strain could no longer be detected in the field. The symbiotic genes carried on a conjugative element had been acquired by indigenous *Mesorhizobium* species, better adapted to the prevailing soil conditions but now able to maximise use of a new niche – the roots of *L. corniculatus* (Sullivan and Ronson 1998). This dataset serves two purposes: it demonstrates the resilience of indigenous bacteria to invasion by an alien strain but it warns that, if selection is strong enough, then these indigenous strains will acquire beneficial traits.

A study of potential gene flow between alien North American gooseberry species (resistant to the co-evolved American gooseberry mildew) and mildew-susceptible native British gooseberries provides a very similar picture. Hybrid seedlings containing resistance genes were found to have a selective advantage (Warren and James 2006), illustrating the potential for crop gene escape to alter the ecological behaviour of native British gooseberries in much the same way as does transgenic disease resistance. Furthermore, plants containing the alien genes supported significantly more, albeit smaller invertebrates.

The key questions arising from these last two examples are the same as those we have discussed for transgenic organisms – what impact does the transgene target (for example, mildew) have on native populations, and what is the significance of the 'unpredicted' side effects of the introduced genes? We need to understand the mechanisms and impacts of naturally or deliberately introduced genes to predict the path of invasions. The principles and processes of gene flow, selection and invasion are essentially the same. Whether this results in significant changes in biodiversity or ecosystem function is yet to be documented, and the risk assessment of transgenic organisms remains to be resolved on a case-by-case basis.

References

Abrahams MV, Sutterlin A (1999) The foraging and antipredator behaviour of growth-enhanced transgenic Atlantic salmon. Anim Behav 58:933–942

Ankenbauer RG (1997) Reassessing forty years of genetic doctrine: retrotransfer and conjugation. Genetics 145:543–549

Ashelford KE, Day MJ, Fry JC (1999) Elevated abundance of bacteriophage infecting bacteria in soil. Appl Environ Microbiol 69:285–289

Atlas RM (1984) Diversity in microbial communities. Adv Microb Ecol 7:1–47

Bailey MJ, Kobayashi N, Lilley AK, Powell BJ, Thompson IP (1994) Potential for gene transfer in the phytosphere: isolation and characterisation of naturally occurring plasmids. In: Bazin MJ, Lynch JM (eds) Environmental gene release. Models, experiments and risk assessment. Chapman and Hall, London, pp 77–98

Bailey MJ, Lilley AK, Thompson IP, Rainey PB, Ellis RJ (1995) Site directed chromosomal marking of a fluorescent pseudomonad isolated from the phytosphere of sugar beet; stability and potential for marker gene transfer. Mol Ecol 4:755–763

Bailey MJ, Lilley AK, Ellis RJ, Bramwell PA, Thompson IP (1997) Microbial ecology, inoculant distribution, and gene flux within populations of bacteria colonizing the surface of plants: case study of a GMM field release in the United Kingdom. In: Van Elsas JD, Trevors JT, Wellington EMH (eds) Modern soil microbiology. Dekker, New York, pp 479–500

Bergelson J, Purrington CB, Palm CJ, Lopez-Gutierrez JC (1996) Costs of resistance: a test using transgenic *Arabidopsis thaliana*. Proc R Soc Lond B Biol Sci 263:1659–1663

Brazil GM, Kenefick L, Callanan M, Haro A, de Lorenzo V, Dowling DN, O'gara F (1995) Construction of a rhizosphere pseudomonad with potential to degrade polychlorinated-biphenyls and detection of bph gene-expression in the rhizosphere. Appl Environ Microbiol 61:1946–1952

Chiura HX (1997) Generalized gene transfer by virus-like particles from marine bacteria. Aquat Microb Ecol 13:75–83

Cook JT, McNiven MA, Richardson GF, Sutterlin AM (2000) Growth rate, body composition and feed digestibility/conversion of growth-enhanced transgenic Atlantic salmon (*Salmo salar*). Aquaculture 188:15–32

Crawley MJ, Hails RS, Rees M, Kohn D, Buxton J (1993) Ecology of transgenic oilseed rape in natural habitats. Nature 363:620–623

Crawley MJ, Brown SL, Hails RS, Kohn DD, Rees M (2001) Transgenic crops in natural habitats. Nature 409:682–683

DeLeij FAAM, Sutton EJ, Whipps JM, Fenlon JS, Lynch JM (1995) Impact of field release of genetically modified *Pseudomonas fluorescens* on indigenous microbial populations of wheat. Appl Environ Microbiol 61:3443–3453

Devlin RH, Johnsson JI, Smailus DE, Biagi CA, Jonsson E, Bjornsson BT (1999) Increased ability to compete for food by growth hormone-transgenic coho salmon *Oncorhynchus kisutch* (Walbaum). Aquacult Res 30:479–482

Dykhuizen DE (1998) Santa Rosalisa revisited: why are there so many species of bacteria? Antonie van Leeuwenhoek 73:25–33

Ellstrand NC (2003a) Current knowledge of gene flow in plants: implications for transgene flow. Philos Trans R Soc Lond B Biol Sci 358:1163–1170

Ellstrand NC (2003b) Dangerous liaisons? When cultivated plants mate with their wild relatives. John Hopkins University Press, Baltimore

Ellstrand NC, Schierenbeck KA (2000) Hybridization as a stimulus for the evolution of invasiveness in plants? Proc Natl Acad Sci USA 97:7043–7050

Ellstrand NC, Prentice HC, Hancock JF (1999) Gene flow and introgression from domesticated plants into their wild relatives. Annu Rev Ecol Syst 30:539–563

Ferré J, van Rie J (2002) Biochemistry and genetics of insect resistance to *Bacillus thuringiensis*. Annu Rev Entomol 47:501–533

Finlay BJ (2002) Global dispersal of free-living microbial eukaryote species. Science 296:1061–1063

Glandorf DCM, Verheggen P, Jansen T, Jorritsma JW, Smit E, Leeflang P, Wernars K, Thomashow LS, Laureijs E, Thomas-Oates JE, Bakker P, Van Loon LC (2001) Effect of genetically modified *Pseudomonas putida* WCS358r on the fungal rhizosphere microflora of field-grown wheat. Environ Microbiol 67:3371–3378

Gogarten PJ, Townsend JP (2005) Horizontal gene transfer, genome innovation and evolution. Nature Rev Microbiol 3:679–687

Griffiths RI, Whiteley AS, O'Donnell AG, Bailey MJ (2003) Physiological and community responses of established grassland bacterial populations to water stress. Appl Environ Microbiol 69:6961–6968

Hails RS, Morley K (2005) Genes invading new populations: a risk assessment perspective. Trends Ecol Evol 20:245–252

Hails RS, Rees M, Kohn DD, Crawley MJ (1997) Burial and seed survival in *Brassica napus* subsp. *oleifera* and *Sinapis arvensis* including a comparison of transgenic and non-transgenic lines of the crop. Proc R Soc Lond B Biol Sci 264:1–7

Hauser TP, Shaw RG, Ostergard H (1998a) Fitness of F-1 hybrids between weedy *Brassica rapa* and oilseed rape (*B. napus*). Heredity 81:429–435

Hauser TP, Jorgensen RB, Ostergard H (1998b) Fitness of backcross and F-2 hybrids between weedy *Brassica rapa* and oilseed rape (*B. napus*). Heredity 81:436–443

Hauser TP, Damgaard C, Jorgensen RB (2003) Frequency-dependent fitness of hybrids between oilseed rape (*Brassica napus*) and weedy *B. rapa* (Brassicaceae). Am J Bot 90:571–578

Houlden A (2005) Bacteria and fungal diversity effects and the activity of biocontrol agents in the rhizosphere of crop plants. PhD Thesis, Cardiff University, Cardiff

Johri AK, Dua M, Singh A, Sethunathan N, Legge RL (1999) Characterization and regulation of catabolic genes. Crit Rev Microbiol 24:245–273

Kennedy AC, Smith KL (1995) Soil microbial diversity and the sustainability of agricultural soils. Plant Soil 170:78–86

Lilley AK, Bailey MJ (1997) Impact of plasmid pQBR103 acquisition and carriage on the phytosphere fitness of *Pseudomonas fluorescens* SBW25: burden and benefit. Appl Environ Microbiol 63:1584–1587

Lorenz G, Wackernagel W (1994) Bacterial transfer by natural genetic transformation in the environment. Microbiol Rev 58:563–602

Mason J R, Briganti F, Wild JR (1997) Protein engineering for improved biodegradation of recalcitrant pollutants. In: Wild JR (ed) Perspectives in bioremediation. Kluwer, Dordrecht, pp 107–118

McGinnity P, Prodohl P, Ferguson A, Hynes R, Maoileidigh NO, Baker N, Cotter D, O'Hea B, Cooke D, Rogan G, Taggart J, Cross T (2003) Fitness reduction and potential extinction of wild populations of Atlantic salmon, *Salmo salar*, as a result of interactions with escaped farm salmon. Proc R Soc Lond B Biol Sci 270:2443–2450

McKane M, Milkman R (1995) Transduction, restriction and recombination patterns in *Escherichia coli.* Genetics 139:35–43

Muir WM (2004) The threats and benefits of GM fish. EMBO Rep 5:654–659

Muir WM, Howard RD (1999) Possible ecological risks of transgenic organism release when transgenes affect mating success: sexual selection and the Trojan gene hypothesis. Proc Natl Acad Sci USA 96:13853–13856

Poppy GM (2004) Geneflow from GM plants – towards a more quantitative risk assessment. Trends Biotechnol 22:436–438

Purrington CB, Bergelson J (1997) Fitness consequences of genetically engineered herbicide and antibiotic resistance in *Arabidopsis thaliana.* Genetics 145:807–814

Raybould AF (2004) A decade of gene flow research: improved risk assessments or missed opportunities? In: van Emden HF, Gray AJ (eds) GM crops – ecological dimensions. AAB Wellesbourne, Warwick, pp 27–33

Reichhardt T (2000) Will souped up salmon sink or swim? Nature 406:10–12

Sears MK, Hellmich RL, Stanley-Horn DE, Oberhauser KS, Pleasants JM, Mattila HR, Siegfried BD, Dively GP (2001) Impact of Bt corn pollen on monarch butterfly populations: a risk assessment. Proc Natl Acad Sci USA 98:11937–11942

Shelton AM, Robertson JL, Tang JD, Perez C, Eigenbrode SD, Preisler HK, Wilsey WT, Cooley RJ (1993) Resistance of diamondback moth (Lepidoptera: Plutellidae) to *Bacillus thuringiensis* in the field. J Econ Entomol 86:697–705

Snow AA, Andersen B, Jorgensen RB (1999) Costs of transgenic herbicide resistance introgressed from *Brassica napus* into weedy *B. rapa*. Mol Ecol 8:605–615

Snow AA, Pilson D, Rieseberg LH, Paulsen MJ, Pleskac N, Reagon MR, Wolf DE, Selbo SM (2003) A Bt transgene reduces herbivory and enhances fecundity in wild sunflowers. Ecol Appl 13:279–286

Strauss SY, Rudgers JA, Lau JA, Irwin RE (2002) Direct and ecological costs of resistance to herbivory. Trends Ecol Evol 17:278–285

Sullivan JT, Ronson CW (1998) Evolution of rhizobia by acquisition of a 500-kb symbiosis island that integrates phe-tRNA gene. Proc Natl Acad Sci USA 95:5145–5149

Tiedje JM, Colwell RK, Grossman YL, Hodson RE, Lenski RE, Mack RN, Regal PJ (1989) The planned introduction of genetically engineered organisms-ecological considerations and recommendations. Ecology 70:298–315

Timmis KN, Steffan RJ, Unterman R (1994) Designing microorganisms for the treatment of toxic wastes. Annu Rev Microbiol 48:525–557

Timms-Wilson TM, Ellis RJ, Renwick A, Rhodes DJ, Mavrodi DV, Weller DM, Thomashow LS, Bailey MJ (2000) Chromosomal insertion of phenazine-1-carboxylic acid biosynthetic pathway enhances efficacy of damping-off disease control by *Pseudomonas fluorescens*. Mol Plant-Microbe Interact 13:1293–1300

Timms-Wilson TM, Kilshaw K, Bailey MJ (2002) Risk assessment for engineered bacteria used in biocontrol of fungal disease in agricultural crops. Plant Soil 266:57–67

Troxler J, Azelvandre P, Zala M, Defago G, Haas D (1997) Conjugative transfer of chromosomal genes between fluorescent pseudomonads in the rhizosphere of wheat. Appl Environ Microbiol 63:213–219

Turnbull LA, Crawley MJ, Rees M (2000) Are plant populations seed-limited? A review of seed sowing experiments. Oikos 88:225–238

Vacher C, Weis AE, Hermann D, Kossler T, Young C, Hochberg ME (2004) Impact of ecological factors on the initial invasion of Bt transgenes into wild populations of birdseed rape (*Brassica rapa*). Theor Appl Genet 109:806–814

van Slageren MW (1994) Wild wheat: a monograph of *Aegilops* L. and *Amblyopyrum* (Jaub & Spach) Eig (Poacae). Veenman Druckers, Wageningen

Warren J, James P (2006) The ecological effects of exotic disease resistance genes introgressed into British gooseberries. Oecologia 147:69–75

Whitfield W (2005) Biogeography: is everything everywhere? Science 310:960–961

Wilkinson MJ, Elliott LJ, Allainguillaume J, Shaw MW, Norris C, Welters R, Alexander M, Sweet J, Mason DC (2003) Hybridization between *Brassica napus* and *B. rapa* on a national scale in the United Kingdom. Science 302:457–459

Section V

Economy and Socio-Economy of Biological Invasions

Short Introduction

Wolfgang Nentwig

With their limited view, natural scientists often do not realize that alien species not only interact with native species but that they also entail economic costs. The public, on the other hand, often recognizes some costs but does not attribute these to biological invasion by alien species. Additionally, with the exception of single instances, the general pattern of biological invasions has hardly been deciphered.

What do feral cats, feral domestic pigeons and imported red fire ants have in common? Each year, by killing millions of birds or fouling buildings or provoking other problems, each of these alien species causes damages or control costs in the range of billions of US$ worldwide. Feral domestic pigeons are vectors of many human and livestock diseases, analogous to alien aphids transferring pathogens to crop plants. Many human pathogens are also aliens, and the history of AIDS would arguably have developed differently without the spread of this virus from remote parts of Africa into the world (Chap. 18).

The perception of one and the same alien species may be extremely different in various sectors of our society, and the assessment of their effects as well. One well-known example is that of the Nile perch, intentionally introduced to African lakes and which resulted in the extirpation of several 100 native fish species. Although this also enabled a prospering, export-oriented fish industry, it promoted a degeneration of societal structures, and even increased malnutrition and starvation among local populations (Chap. 19).

Was the Nile perch really the cause of all these changes? Could these have been foreseen at the time, when a fisheries inspector intended only to "increase" the fish production of a lake? This example, including far-reaching effects in a general socio-economic sphere, cautions us to recognize that even a single event should not be considered as isolated. The framework of our society is highly complex, all happenings are widely interconnected and, ultimately, all effects of alien species can be expressed in terms of costs, notably on a monetary basis.

Short Introduction

[illegible]

18 Plant, Animal, and Microbe Invasive Species in the United States and World

David Pimentel, Marcia Pimentel, and Anne Wilson

18.1 Introduction

Approximately 50,000 plant, animal, and microbe invasive species are present in the United States, and an estimated 500,000 plant, animal, and microbe invasive species have invaded other nations of the world. Immediately, it should be pointed out that the US and world agriculture depend on introduced food crops and livestock. Approximately 99 % of all crops and livestock in all nations are intentionally introduced plants, animals, and microbes (Pimentel 2002). Worldwide, the value of agriculture (including beneficial non-indigenous species) is estimated to total $ 30 trillion per year. Other exotic species have been introduced for landscape restoration, biological pest control, sport, and food processing, also contributing significant benefits.

Unfortunately, some invasive species are causing major economic losses in the United States and worldwide in agriculture, forestry, fisheries, public health, and natural ecosystems. Documenting the full extent of the environmental and economic damages caused by exotic species, and the number of species extinctions is difficult because little is known about the estimated 750,000 species that exist in the US and the estimated 15 million that exist worldwide (McNeeley 1999). Only an estimated 2 million species have been described worldwide. In the US, an estimated 40 % of those species forced to extinction can be accounted for by the impacts of invasive species (Pimentel et al. 2005). In some regions of the world, as many as 80 % of endangered species have been threatened and forced to extinction due to the pressures of nonnative species (Armstrong 1995). In addition, many other species worldwide, even if they have not been forced to extinction or endangered status, are negatively affected by various alien species or ecosystem changes caused by alien species.

Calculating the negative economic impacts associated with the invasion of exotic species is difficult. For a few species, there are sufficient data to estimate

Ecological Studies, Vol. 193
W. Nentwig (Ed.)
Biological Invasions

some impacts on agriculture, forestry, fisheries, public health, and the natural ecosystem in the US and worldwide. In this article, we estimate the magnitude of the economic benefits, and environmental and economic costs associated with a variety of invasive species that exist in the United States and elsewhere in the world.

18.2 Agricultural and Forestry Benefits from Introduced Species

The value of the US food system is more than $ 800 billion per year (USCB 2004–2005), and the value of the world food system is estimated at more than $ 30 trillion per year. According to the World Health Organization (Pimentel 2004a), the world's food system is not providing adequate amounts of food for all people on earth, more than 3.7 billion of the current population of 6.5 billion being malnourished. In addition, food production per capita has been declining each year for the past 21 years (FAOSTAT 1960–2004). This assessment is based on cereal grains, since cereal grains provide about 80 % of the world's food. Clearly, more needs to be done to increase food production per capita, at the same time significantly reducing the rate of growth of the world population (Pimentel and Pimentel 2001).

18.3 Environmental Damages and Associated Control Costs

Most plant and vertebrate animal introductions in the US and world have been intentional, whereas most invertebrate animal and microbe introductions have been accidental. During the past 60 years, the total number of introductions of all species has nearly doubled in the world. The rate of introductions of exotic species has increased enormously because of high human population growth, rapid movement of people, and alteration of the environment everywhere in the world. In addition, significantly more goods and material are being exchanged among nations than ever before, creating greater opportunities for unintentional introductions (USCB 2004–2005). Some of the estimated 50,000 species of plants, animals, and microbes that have invaded the US, and 500,000 species of plants, animals, and microbes that have invaded the total world ecosystem provide significant benefits but also many types of damage to managed and natural ecosystems, as well as public health.

18.3.1 Plants

Most exotic plant species now established in the United States and elsewhere in the world were introduced for food, fiber, or ornamental purposes. An estimated 5,000 introduced plant species have escaped and now exist in US natural ecosystems (Morse et al. 1995), compared with a total of approximately 17,000 species of native plants (Morin 1995). In Florida, of the approximate 25,000 alien plant species (mostly introduced ornamental species), more that 900 have escaped and become established in neighboring natural ecosystems (Frank et al. 1997; Simberloff et al. 1997). More than 3,000 plant species have been introduced into California, and many of these have escaped into this natural ecosystem as well (Dowell and Krass 1992).

Worldwide, an estimated 30,000 species of exotic plants have been intentionally introduced as crops, and have escaped to become established in various natural ecosystems. Most of the non-indigenous plants that have escaped and become established have adapted well to the favorable living conditions characteristic of moist tropical regions in countries such as India, Brazil, and Australia.

Some of the invasive plants established in the US and world have displaced native plant species. In the United States, introduced plant species are spreading and invading approximately 700,000 ha of US natural ecosystems per year (Babbitt 1998). For instance, the European purple loosestrife (*Lythrum salicaria*), which was introduced in the early19th century as an ornamental plant (Malecki et al. 1993), has been spreading at a rate of 115,000 ha per year, strongly altering the basic structure of the wetlands that it has invaded (Thompson et al. 1987). Stands of purple loosestrife have reduced the abundance of 44 native plant species, and endangered many wildlife species, including turtles and ducks (Gaudet and Keddy 1988). Loosestrife is present in 48 states, and about $ 45 million are spent each year for control of the weed (ATTRA 1997).

Many of these exotic species have become established in national parks. In the Great Smokey Mountains National Park, for example, 400 of the 1,500 vascular plant species are exotic, and 10 of these are currently displacing and threatening native plant species (Hiebert and Stubbendieck 1993). The problem of introduced plants is particularly serious in Hawaii, where 946 of the total of 1,690 plant species on the island are non-indigenous (Elredge and Miller 1997).

In some cases, one exotic plant species may competitively overcome an entire ecosystem. In California, the yellow starthistle (*Centaurea solstitalis*), for example, dominates more than 4 million ha of northern grassland in the state, resulting in the total loss of this once productive forage system (Campbell 1994). In addition, the European cheat grass (*Bromus tectorum*) is dramatically altering the vegetation and fauna of many natural ecosystems in the western US. Cheat grass is an annual that has invaded and spread throughout the shrub-steppe habitat of the Great Basin in Idaho and Utah, predisposing the altered habitat to fires (Kurdila 1995). Before the invasion of cheat grass, fire burned once every 60 to 110 years, and shrubs in the region had a chance to become reestablished. Currently, fires occur once every 3 to 5 years, and this has led to a decrease in shrubs and other vegetation, and the occurrence of monocultures of cheat grass on more than 5 million ha in Idaho and Utah. The reason that the alteration of original vegetation is so significant is that all the animals and microbes that were dependent on the original vegetation have been reduced or totally eliminated.

Insufficient information exists concerning invasive plants in the United States and other countries. This is true even in countries that are dominated by invasive plants, such as the British Isles. For example, of the 27,515 total plant species on the British Isles, only 1,515 species are considered native (Crawley et al. 1996). More than 80 % of alien plant species in the British Isles are established in disturbed habitats (Clement and Foster 1994; Crawley et al. 1996).

One group of agriculturalists introduced 463 species of plants as potential forage species in Australia (Lonsdale 1994). Only 21 species of this group of 463 plant species turned out to be beneficial, many others had little impact, but several became serious pest weeds in Australia. In India, weeds are estimated to cause a 30 % loss in potential crop production each year (Singh 1996), amounting to about $ 90 billion in reduced crop yields. Assuming that 42 % of the weeds in crops are alien (Nandpuri et al. 1986), the total cost associated with the alien plants in India is about $ 37.8 billion per year.

18.3.2 Mammals

About 20 mammal species have been intentionally introduced into the United States, including dogs, cats, horses, cattle, sheep, pigs, and goats (Layne 1997). Several of these mammal species escaped into the wild, and have become pests by preying on native animals, grazing on native vegetation, or intensifying soil erosion. Goats (*Capra aegagrus hircus*), for instance, introduced on San Clemente Island, California, have caused the extinction of 8 endemic plant species and have endangered 8 others (Kurdila 1995).

Several small mammal species, especially rodents, have been introduced into the United States. These include the European rat (*Rattus rattus*), the Asiatic rat (*Rattus norvegicus*), the house mouse (*Mus musculus*), and the European rabbit (*Oryctolagus cuniculus*; Layne 1997). Some of the introduced rats and mice have become particularly abundant and destructive on farms. On poultry farms, there is about 1 rat per 5 chickens (Smith 1984; D. Pimentel, unpublished data). Using this ratio, it is estimated that the rat number is more than 1.8 billion on farms in the US. Another 250 million rats are estimated to be in homes and stores in cities and towns. If it is estimated that each rat causes $ 15 in damages each year, then the damage per year would be about $ 30 billion.

Although the cost of the impact of invasive mammals is relatively high, the percentage of alien mammals introduced into the United States is relatively low, or 6 %; in the United Kingdom, the percentage is relatively high, or 31 % (Pimentel et al. 2001). The UK introduced mammals include those species recorded in the US, plus many others.

Australia is another nation that has a large number of alien mammals. In Australia, pigs native to Eurasia and North Africa were introduced and now number from 4 to 20 million (Emmerson and McCulloch 1994). Feral pigs cause soil erosion, damage agricultural crops, fences, native plants and animals, and are a threat to livestock and humans; they also spread various animal diseases, including tuberculosis, brucellosis, rabies, and foot-and-mouth disease (Lever 1994). The estimate of pig damage in Australia is more than $ 80 million per year (Emmerson and McCulloch 1994).

Rodents, including the European and Asiatic rats and the house mouse, have invaded all countries in the world. In addition, domestic dogs, cats, and European rabbits have been introduced into all nations of the world. In Australia, feral cats are a serious problem, killing native bird, mammal, marsupial, and amphibian populations. The estimate is that there are 3 million pet cats, and 18 million feral cats in Australia (Anon 1996). The cats are considered responsible for having exterminated 23 native Australian species of animals (Low 1999). Assuming that each bird has a minimum value of $ 30 in the US (Pimentel et al. 2000), then the total impact from cats in Australia is $ 540 million per year. In the US, it is estimated that cats kill an estimated 570 million birds per year, with an estimated damage of $ 17 billion (Pimentel et al. 2000).

18.3.3 Birds

Of the 1,000 species of birds in the United States, nearly 100 are exotic (Temple 1992). Approximately 5 % of the introduced birds are beneficial, such as the chicken.

One of the bird pest species is the English sparrow (*Passer domesticus*), introduced in 1853 into the US for the control of canker worm and other pest caterpillars (Roots 1976). By 1900, English sparrows were reported to be a pest, consuming wheat, corn, and the buds of fruit trees (Laycock 1966). In addition, they harass native birds, including robins, Baltimore orioles, and the yellow-billed and black-billed cuckoos, and they displace bluebirds, wrens, purple martins, and cliff swallows (Long 1981). English sparrows are also associated with the spread of about 30 human and livestock diseases (Weber 1979).

One of the most serious bird pests is the common pigeon (*Columbia livia*), which has been introduced to all cities in the world (Robbins 1995). Pigeons present a nuisance because they foul buildings, statues, cars, and sometimes people, and they feed on grains (Smith 1992). It is estimated that pigeons cause an estimated $ 1.1 billion in damages per year in the United States. They also serve as reservoirs and vectors of more than 50 human and livestock diseases, including parrot fever, ornithosis, histoplasmosis, and encephalitis (Long 1981).

Another serious bird pest in the US is the European starling (*Sturnus vulgaris*), a species that in some cases occurs at densities of more than one per hectare in agricultural regions (Moore 1980). They are capable of destroying as much as $ 2,000 worth of cherries ha^{-1} in the spring (Feare 1980). They also destroy large quantities of grain crops (Feare 1980). The estimate is that they are responsible for damages amounting to $ 800 million per year (Pimentel et al. 2000).

Information on other bird species that have invaded other nations is not as abundant as one would expect. Of the other nations, the UK has some of the best data. Of the 542 species of birds in the UK, 47 are alien (Gooders 1982). Pigeons in the UK are as serious a problem as they are in the US. In the UK, pigeons are estimated to cause more than $ 270 million in damages each year (Alexander and Parsons 1986; Bevan and Bracewell 1986).

18.3.4 Amphibians and Reptiles

About 53 species of amphibian and reptile species have been introduced into the United States. These species invasions have all occurred in the warmer regions. For example, Florida is host to 30 species (Lafferty and Page 1997). The negative impacts of these invasive species have been enormous.

The brown tree snake (*Boiga irregularis*) is one of the worst. It was introduced into the US territory of Guam immediately after World War II, when military equipment was transferred to the island (Fritts and Rodda 1995). The snake population reached high densities of 100 snakes ha^{-1}, and dramatically reduced populations of native bird species, small mammals, and lizards. A total of 10 bird species and 9 lizard species were exterminated from Guam (Rodda et al. 1997). The brown tree snake also eats chickens, eggs, pet birds, and causes major problems to farmers. In some cases, the snake enters houses and bites small children in cribs and playpens (OTA 1993). Another costly impact is that the snake is causing power failures by damaging electric transformers. The estimate is that the brown tree snake causes more than $ 2 million in damages per year on Guam. A major worry is that the snake will invade Hawaii, and cause major extinctions of birds, mammals, and amphibians on the island.

An estimated 700 species of reptiles and amphibians exist in Australia (Fox 1995). However, only two of these are exotic. One of the introduced species is the cane toad (*Bufo marinus*), introduced from South America for insect control in cane fields. However, it was soon reported to be a serious pest (Fox 1995). The cane toad is poisonous to dogs, cats, and other mammals (Sabath et al. 1981). In South Africa, there have been 13 species of reptiles and 11 species of amphibians introduced (Siegfried 1989). One of the invasive species is the red-eared slider (*Chrysemys scripta elegans*) that was introduced from North America. This invasive turtle has become a major threat to the 12 native turtle species (Boycott and Bourquin 1988).

18.3.5 Fishes

A total of 138 invasive fish species have been introduced into the United States (Courtenay et al. 1991; Courtenay 1997). Most of the invaders are found in the warmer regions such as Florida, which has at least 50 of these species (Courtenay 1997). Introduced fish species frequently alter the ecology of aquatic ecosystems. In the Great Lakes, for instance, nearly 50 invasive species are found, and these invaders are causing an estimated $ 5 billion in damages to the fisheries per year (Pimentel 2005). In addition, most of the alien fish species in South Africa are regarded as pests (Bruton and Van As 1985). In total, alien fish species are responsible for the reduction or local extinction of at least 11 species of fish in South Africa (Bruton and Van As 1985).

18.3.6 Arthropods

An estimated 4,500 arthropod species (more than 2,500 species in Hawaii alone, and more than 2,000 in continental US) have been introduced into the United States (OTA 1993). Approximately 95 % of these introductions were accidental, the remainder being intentional for purposes of biological control and pollination. About 1,000 invasive species of insects and mites are crop pests in the US. Introduced insects account for 98 % of the crop insect pests in Hawaii (Beardsley 1991). Approximately 40 % of the insect and mite pests in crops in continental US are pests of agricultural crops. The major group of pests consists of native insects and mites that switched from feeding on native vegetation to feeding on crops (Pimentel et al. 2000). Pest insects are estimated to destroy $ 14 billion worth of crops per year. One ant species, the red imported fire ant, is alone causing $ 6 billion in damages and control costs (Linn 2005).

Of the 360 species of invasive species in US forests, about 30 % are now serious pests in these forests (Liebold et al. 1995), causing about $ 7 billion in losses each year (Hall and Moody 1994). A new introduction, the Asian longhorn beetle, is threatening maple and ash trees in New York and Illinois (Hajek 2005).

Of the 80,000 species of insects, and 6,000 species of spiders and numerous other arthropod species that exist in South Africa, several invasive species are causing problems (South Africa 1998). One of the most serious invaders is the Argentine ant (*Linepithema humile*), which is destroying native vegetation, including endangered plants (Macdonald et al. 1986). This ant species is also negatively affecting native ants and other beneficial arthropod species. In addition, the Argentine ant is a serious pest in agriculture.

18.3.7 Mollusks

A total of about 88 species of mollusks have been introduced and established in United States aquatic ecosystems (OTA 1993). The two most serious pest species introduced are the zebra mussel, *Dreissena polymorpha*, and the Asian clam, *Corbicula fluminea* (see also Chaps. 5 and 15).

The zebra mussel was introduced from Europe, and probably gained entrance via ballast water released into the Great Lakes by ships traveling from Europe (Benson and Boydstun 1995). The mussel was first noted in Lake St. Clair, has spread into most of the Great Lakes and most aquatic ecosystems in the eastern United States, and is expected to invade most freshwater habitats throughout the nation. Large mussel populations (up to 700,000 m^2; Griffiths et al. 1991) reduce food and oxygen for the native fauna. Zebra mussels have been observed covering native mussel, clams, and snails, and threatening the survival of these and other species (Benson and Boydstun 1995; Keniry and Marsden 1995).

In addition to ecological effects on other aquatic organisms, the zebra mussel also invades and clogs water intake pipes in water infiltration and electric power plants. It is estimated that the mussels will cause $ 5 billion in damages and associated control costs in the US. In the Great Lakes alone, they are reported to cause $ 1 billion in damages and control costs (Pimentel 2005). Although the Asian clam grows and disperses less quickly than the zebra mussel, it also causes significant damage to native organisms and damage to water filtration plants and electric power plants. Costs associated with this animal are estimated to be more than $ 1 billion per year (OTA 1993). In various US coastal bay regions, the introduced shipworm (*Teredo navalis*) is estimated to cause from $ 205 million to $ 750 million in damages per year (Cohen and Carlton 1995; D. and M. Pimentel, unpublished data).

Unfortunately, there are not data available on mollusk invaders in other nations. This is due to the general lack of knowledge concerning the ecology and systematics of mollusks in the world; they appear to be causing a relatively small amount of damage to aquatic ecosystems in other regions worldwide, and/or few biologists have investigated these organisms.

18.4 Livestock Pests

For a start, it should be pointed out that the majority of livestock worldwide are introduced species. For example, in the United States more than 99 % of the livestock species are introduced (Pimentel 2004b). Microbial and other parasitic organisms have generally been introduced when the livestock species have been introduced. In addition, to the more than 100 species of pest microbes and other parasitic species that have already invaded the United States (Pimentel 2005), there are more than 60 additional microbes and other parasitic species that could easily invade the United States and become serious pests of US livestock (Pimentel 2005). A conservative estimate of the losses to US livestock from exotic microbes and other parasitic species is more than $ 9 billion per year.

Australia already has several species of alien diseases infecting and causing losses to livestock. In addition, there are an estimated 44 exotic diseases in other regions of the world that could infect Australian livestock, if they were introduced (Meischke and Geering 1985). At present, 3 alien insect and mite species already cause $ 228 million per year damage to the wool and sheep industry (Slater et al. 1996).

In India, there are more than 50 exotic species of disease and parasitic organisms that are causing major problems for the introduced livestock and native wildlife. Already present in India is the serious foot-and-mouth disease. Recently, it was reported that there were more than 50,000 cases of foot-and-mouth disease (Foot-and-Mouth Disease Leak 2004), treatment costs being about $ 20,000 per year.

South Africa also reports problems with introduced livestock pests. The exotic diseases include tuberculosis, brucellosis, East Coast fever, anthrax, and rinderpest. Estimates are that Brucellosis alone is causing livestock losses of more than $ 100 million per year (Coetzer et al. 1994). In Brazil and other Latin American countries, imported bovine tuberculosis has become a serious threat to the beef and dairy industry. These losses are estimated to be about $ 100 million per year (Cosivi et al. 1998).

18.5 Human Diseases

Various influenza virus types, originating mostly in the Far and Near East, have quickly spread to the United States and other nations in the past. Recent disease epidemics have been associated with SARS, and now there is the major threat of bird flu that is infecting some people in the Far and Near East. The current influenza strains are responsible for nearly 10 % of all human deaths in the US (USCB 2004–2005). The costs of hospitalization for a single outbreak of influenza, such as type A, can exceed \$ 500 million per year.

One of the most notorious of all alien human disease is HIV/AIDS. The pathogen is reported to have originated in East Africa, probably from some species of monkey. The disease now occurs in all parts of the world. The costs of treatment of HIV/AIDS in the world today are estimated to be \$ 100 billion per year. In addition to influenza and HIV/AIDS, there are numerous other diseases infecting humans in various parts of the world. These include syphilis, Lyme disease, and tuberculosis. These diseases are causing an estimated \$ 20 billion in losses and damages per year.

New influenza strains in the UK are reported to cause from 3,000 to 4,000 deaths per year (Kim 2002). In total, both influenza and HIV/AIDS claim the lives of more than 4,000 people per year. The treatment costs are in excess of \$ 1 billion per year. Influenza and tuberculosis in India are reported to cause more than 3 million deaths per year (Kim 2002). Several non-indigenous human diseases threaten people in South America. These diseases include HIV/AIDS, influenza, malaria, cholera, yellow fever, and dengue. More than 2 million people are infected per year, associated with more than \$ 100 billion in damages and treatment costs per year.

18.6 The Situation Today and Projections for the Future

The number of invading species worldwide has been increasing rapidly, an estimated tenfold increase having been recorded in the past 100 years. Some countries with a rapidly increasing population, growing population movement, and increasing global trade, such as the United States, are suffering a greater problem from invaders than is the case for other nations. Approximately 500,000 species of plants, animals, and microbes have invaded the nations of the world, with about 50,000 in the US alone. It must be pointed out that, for all nations combined, about 5 % of all these species were intentionally introduced as crops and livestock. Unfortunately, an estimated 10–20 % of the introduced species are, or have become, pests and are causing major environmental problems. Although relatively few of these species become really serious pests, some species do inflict significant damage to natural and managed ecosystems, and cause serious public health problems. Various ecological factors help exotic species become abundant and emerge as serious ecological threats in their new habitat. These factors include exotic plant and animal species being introduced without their natural enemies (e.g., purple loosestrife); the existence of favorable predator–prey conditions in the new habitat (e.g., for house cats); the development of new associations between alien parasites and hosts (e.g., HIV/AIDS and humans); the occurrence of disturbed habitats that promote invasion by some species (e.g., crop weeds); the occurrence of favorable, newly created artificial habitats for invasives (e.g., cheat grass); and the occurrence of species-specific traits promoting invasion by highly adaptable alien species (e.g., the water hyacinth and zebra mussel).

This investigation reports on various economic damages associated with invasive species in various nations of the world that total more than \$ 1.4 trillion per year (Pimentel 2002). This amounts to about 5 % of the world GNP (USCB 2004–2005). Unfortunately, precise economic costs associated with some of the most ecologically damaging species of invasives are not available. For example, cats and pigs have been responsible for the extinction of various animals, and perhaps some plants. For these invasive animals, however, only minimal cost impact data are available. In addition, it is impossible to assess the value attached to various species that have been forced to extinction. If economic values could be assigned to species forced to extinction, then in terms of losses in biodiversity, ecosystem services, and esthetics, the costs of destructive invasive species would be extremely high. The value of \$ 1.4 trillion cited above already suggests that exotic species are extracting major environmental and economic tolls worldwide.

As mentioned above, 95–99 % of all crop and livestock are introduced species. These alien crops (e.g., corn and rice) and livestock (e.g., cattle and poultry) are vital to maintaining world agriculture and the food system. The food system has an estimated value of \$ 30 trillion worldwide. However, these benefits do not compensate for the enormous negative impacts of exotic pest species.

A real challenge lies in preventing further damage from invading exotic species to natural and managed ecosystems of the world. This is especially true in view of rapid population growth and increasing global trade. The United States has taken a few steps to protect and prevent the invasion of exotic species into the nation. Many governments of other nations have taken, and are taking, additional steps to combat non-indigenous species. Evidently, it is being increasingly recognized that investing a few million dollars to prevent future introduced species from invading a country, where they might cause billions of dollars worth of damage and control costs, is worthwhile.

Specific laws are needed in all nations to diminish or prevent invasive species introductions. All introductions of exotic species of plants, animals, and microbes – for whatever purpose – should be strictly regulated. In addition, governments should make efforts to inform the public concerning the serious environmental and economic threats that are associated with the invasion of exotic species.

18.7 Biological Control of Invasives

Introducing a new species into a nation for the control of a plant, animal, or microbe pest invasive species is sometimes criticized as being a hazardous technology. In the past, where vertebrate species such as mammals, amphibians, birds, and fishes were introduced for biological control, several became pests themselves (Chaps. 2 and 23). For instance, the Indian mongoose, introduced for rat control in the West Indian Islands and Hawaiian Islands, and the English sparrow, introduced into the US for caterpillar control, have both turned out to be disasters. However, introductions of insect species, such as the vedalia beetle *Rodolia cardinalis* into the US, and of a virus species for the control of the European rabbit in Australia, have been notable successes. Controls of cacti in Australia, knapweed in the US, and the cassava mealy bug in Africa, all employing biocontrol insects, have also been successful.

The first response after detecting an invasive pest in a country should be to immediately travel to the country of origin of the pest, and attempt to introduce natural enemies of the pest. This is sometimes successful, but not always. There have been almost as many successful biological controls employing new associated biocontrol agents. In new associated biocontrol, the biological control agents are sought from a related species of the pest invasive in another country. The new association biocontrol agent offers an ecological advantage because the biocontrol agent has never interacted with the invasive pest species, and often this advantage makes the new biocontrol agent highly pathogenic to the invasive pest species. The advantage of biological controls is that they reduce the invasive pest species without the need for using pesticides in the new ecosystem, and with minimal or no damage to the new ecosystem (Hokkanen and Pimentel 1989). Details on the pros and cons of biological control are given in Chap. 23.

References

Alexander DJ, Parsons G (1986) Pathogenicity for chickens of avian paramyxovirus type I isolates obtained from pigeons in Great Britain 1983-1985. Avian Pathol 15:487–494

Anon (1996) Menace of moggies. The Christchurch Press, Christchurch

Armstrong S (1995) Rare plants protect Cape's water supplies. New Scientist 11:8

ATTRA (1997) Purple loosestrife: public enemy #1 on federal lands. Interior Helper, Washington, DC

Babbitt B (1998) Statement by Secretary of the Interior on invasive alien species. In: Proc National Weed Symp. Bureau of Land Management, Denver, CO

Beardsley JW (1991) Introductions of arthropod pests into the Hawaiian Islands. Micronesica suppl 3:1–4

Benson AJ, Boydstun CP (1995) Invasion of the zebra mussel into the United States. In: LaRoe ET, Farris GS, Puckett CE, Doran PD, Mac MJ (eds) Our living resources: a report to the nation on the distribution, abundance, and health of U.S. plants, animals, and ecosystems. US Dept Interior, National Biological Service, Washington, DC, pp 445–446

Bevan BJ, Bracewell CD (1986) Chlamydiosis in birds in Great Britain. 2. Isolations of *Chlamydia psittaci* from birds sampled between 1976 and 1984. J Hygiene 96:453–458

Boycott RC, Bourquin O (1988) The South African tortoise book: a guide to South African tortoises, terrapins and turtles. South African Book Publishers, Johannesburg

Bruton MN, Van As J (1985) Faunal invasions of aquatic ecosystems in Southern Africa, with suggestions for their management. In: Macdonald AIW, Kruger FJ, Ferrar AA (eds) Proc National Synthesis Symp Ecology and Management of Biological Invasions in Southern Africa. Oxford University Press, Cape Town, pp 47–61

Campbell FT (1994) Killer pigs, vines, and fungi: alien species threaten native ecosystems. Endangered Species Tech Bull 19:3–5

Clement EJ, Foster MC (1994) Alien plants of the British Isles. Botanical Society of the British Isles, London

Coetzer JAW, Thomson GR, Tustin RC (1994) Infectious disease of livestock with special reference to Southern Africa. Oxford University Press, Cape Town

Cohen AN, Carlton JT (1995) Nonindigenous aquatic species in a United States estuary: a case study of the biological invasions of the San Francisco Bay and Delta. US Fish and Wildlife Service, Washington, DC

Cosivi O, Grange JM, Daborn CJ, Raviglione MC, Fujikura T, Cousins D, Robinson RA, Huchzermeyer HFAK, Kantor I, Meslin FX (1998) Zoonotic tuberculosis due to *Mycobacterium bovis* in developing countries. Emerg Infect Diseases 4:1–16

Courtenay WR (1997) Nonindigenous fishes. In: Simberloff D, Schmitz DC, Brown TC (eds) Strangers in paradise. Island Press, Washington, DC, pp 109–122

Courtenay WR, Jennings DP, Jennings DP, Williams JD (1991) Appendix 2. Exotic fishes of the United States and Canada. In: Robins CR (ed) A list of common and scientific names of fishes from the United State and Canada. American Fisheries Society, Bethesda, MD, Spec Publ 20

Crawley MJ, Harvey PH, Purvis A (1996) Comparative ecology of the native and alien floras of the British Isles. Philos Trans R Soc Lond Biol Sci 351:1251–1259

Dowell RV, Krass CJ (1992) Exotic pests pose growing problem for California. California Agriculture 46:6–10

Eldredge LG, Miller SE (1997) Numbers of Hawaiian species: supplement 2, including a review of freshwater invertebrates. Bishop Mus Occas Pap 48:3–32

Emmerson G, McCulloch J (1994) Feral peril. Queensland's introduced plants and animals. Queensland Parliamentary Library, Brisbane

FAOSTAT (1960–2004) Quarterly bulletin of statistics. FAO, Rome

Feare CJ (1980) The economics of starling damage. In: Wright EN, Inglis IR, Feare CJ (eds) Bird problems in agriculture. The British Crop Protection Council, Croydon

Foot-and-Mouth Disease Leak (2004) Foot-and-mouth disease leak at Plum Island BSL-3 Lab. GeneWatch 17(5/6):22 (www.gene-watch.org)

Fox MD (1995) Conserving biodiversity: impact and management of exotic organisms. In: Bradstock RA, Auld TD, Keith DA, Lunney RUT (eds) Conserving biodiversity: threats and solutions. Surrey Beatty, Chipping Norton, NSW, pp 177–183

Frank JH, McCoy ED, Hall HG, O'Meara F, Tschinkel WR (1997) Immigration and introduction of insects. In: Simberloff D, Schmitz DC, Brown TC (eds) Strangers in paradise. Island Press, Washington, DC, pp 75–100

Fritts TH, Rodda GH (1995) Invasions of the brown tree snake. In: LaRoe ET, Farris GS, Puckett CE, Doran PD, Mac MJ (eds) Our living resources. A report to the nation on the distribution, abundance, and health of US plants animals and ecosystems. US Dept Interior, National Biological Service, Washington, DC, pp 454–456

Gaudet CL, Keddy PA (1988) Predicting competitive ability from plant traits: a comparative approach. Nature 334:242–243

Gooders J (1982) Collins British birds. Harper Collins, London

Griffiths DW, Schloesser DW, Leach JH, Koalak WP (1991) Distribution and dispersal of the zebra mussel (*Dreissena polymorpha*) in the Great Lakes Region. Can J Fish Aquat Sci 48:1381–1388

Hall JP, Moody B (1994) Forest depletions caused by insects and diseases in Canada 1982-1987. Forest insect and disease survey information. Forest Insect and Disease Survey, Canadian Forest Service, Natural Resources, Ottawa, Rep ST-X-8

Hajek AE (2005) Asian longhorned beetle: ecology and control. In: Pimentel D (ed) Encyclopedia of pest management. Dekker, New York

Hiebert RD, Stubbendieck J (1993) Handbook for ranking exotic plants for management and control. US Dept Interior, National Park Service, Denver, CO

Hokkanen HMT, Pimentel D (1989) New associations in biological control: theory and practice. Can Entomol 121:828–840

Keniry T, Marsden JE (1995) Zebra mussels in southwestern Lake Michigan. In: LaRoe ET, Farris GS, Puckett CE, Doran PD, Mac MJ (eds) Our living resources: a report to the nation on the distribution, abundance, and health of US plants, animals, and ecosystems. US Dept Interior, National Biological Service, Washington, DC, pp 445–448

Kim Y (2002) World exotic diseases. In: Pimentel D (ed) Biological invasions: economic and environmental costs of alien plant, animal and microbe species. CRC Press, Boca Rotan, FL, pp 331–354

Kurdila J (1995) The introduction of exotic species into the United States: there goes the neighborhood. Environ Affairs 16:95–118

Lafferty KD, Page CJ (1997) Predation of the endangered tidewater goby, *Eucyclogobius newberryi*, by the introduced African clawed frog, *Xenopus laevis*, with notes on the frog's parasites. Copeia 3:589–592

Laycock G (1966) The alien animals. Natural History Press, New York

Layne JN (1997) Nonindigenous species. In: Simberloff D, Smitz DC, Brown TC (eds) Strangers in paradise. Island Press, Washington, DC, pp 157–186

Lever C (1994) Naturalized animals: the ecology of successfully introduced species. Poyser Natural History, London

Liebold AM, MacDonald WL, Bergdahl D, Mastro VC (1995) Invasion by exotic forest pests: a threat to forest ecosystems. Forest Sci 41(2):1–49

Linn M (2005) Ant warfare: science fights fire with flies: natural enemy central to effort to limit spread of the insects. USA Today (Nation), 18 November 2005, p 3A

Long JL (1981) Introduced birds of the world: the worldwide history, distribution, and influence of birds introduced to new environments. Universe Books, New York

Lonsdale WM (1994) Inviting trouble: introduced pasture species in northern Australia. Austral J Ecol 19:345–354

Low T (1999) Feral future: the untold story of Australia's exotic invaders. Viking Penguin Books, Melbourne

Macdonald IAW, Kruger FS, Ferrar AA (1986) Ecology and management of biological invasions in Southern Africa. Oxford University Press, Cape Town

Malecki RA, Blossey B, Hight SD, Schroeder D, Kok LT, Coulson JR (1993) Biological control of purple loosestrife. BioScience 43:680–686

McNeely JA (1999) Biodiversity: man is one among 15 million species. IUCN, Gland

Meischke HRC, Geering WA (1985) Exotic animal diseases. In: Gibbs A, Meischke R (eds) Pests and parasites as migrants. Cambridge University Press, Cambridge, pp 28–39

Moore NW (1980) How many wild birds should farmland support? In: Wright EN, Inglis IR, Feare CJ (eds) Bird problems in agriculture. The British Crop Protection Council, Croydon, pp 2–6

Morin N (1995) Vascular plants of the United States. In: LaRoe ET, Farris GS, Puckett CE, Doran PD, Mac MJ (eds) Our living resources: a report to the nation on the distribution, abundance, and health of US plants, animals, and ecosystems. US Dept Interior, National Biological Service, Washington, DC, pp 200–205

Morse LE, Cartes JT, Center L (1995) Native vascular plants. In: LaRoe ET, Farris GS, Puckett CE, Doran PD, Mac MJ (eds) Our living resources: a report to the nation on the distribution, abundance, and health of US plants, animals, and ecosystems. US Dept Interior, National Biological Service, Washington, DC, pp 205–209

Nandpuri KS, Singh B, Kolar JS, Kang MS, Chahal BS (1986) Field problems of important crops. Punjab Agricultural University, Ludhiana, India

OTA (1993) Harmful non-indigenous species in the United States. Office of Technology Assessment, United States Congress, Washington, DC

Pimentel D (2002) Biological invasions: economic and environmental costs of alien plant, animal, and microbe species. CRC Press, Boca Raton, FL

Pimentel D (2004a) Agriculture: changing genes to feed the world. Science 306:815

Pimentel D (2004b) Livestock production and energy use. In: Cleveland C (ed) Encyclopedia of energy. Elsevier, Amsterdam, pp 671–676

Pimentel D (2005) Aquatic nuisance species in the New York State Canal and Hudson River system and the Great Lakes Basin: an economic and environmental assessment. Environ Manage 35:692–701

Pimentel D, Pimentel M (2001) Agriculture, technology, and natural resources. In: Dorf R (ed) Technology, humans and society: toward a sustainable world. Academic Press, San Diego, pp 356–363

Pimentel D, Lach L, Zuniga R, Morrison D (2000) Environmental and economic costs of non-indigenous species in the United States. BioScience 50:53–65

Pimentel D, McNair S, Janecka J, Wightman J, Simmons C, O'Connell C, Wong E, Russel L, Zern J, Aquino T, Tsomondo T (2001) Ecological and economic threat of alien plant, animal, and microbe invasions in the world. Agric Ecosyst Environ 84:1–20

Pimentel D, Zuniga R, Morrison D (2005) Update on the environmental and economic costs associated with alien-invasive species in the United States. Ecol Econ 52:273–288

Robbins CC (1995) Non-native birds. In: LaRoe ET, Farris GS, Puckett CE, Doran PD, Mac MJ (eds) Our living resources: a report to the nation on the distribution, abundance, and health of US plants, animals, and ecosystems. US Dept Interior, National Biological Service, Washington, DC, pp 437–440

Rodda GH, Fritts TH, Chiszar D (1997) The disappearance of Guam's wildlife. BioScience 47:565–574

Roots C (1976) Animal invaders. Universe Books, New York

Sabath MD, Boughton WC, Easteal S (1981) Expansion of the range of the introduced toad *Bufo marinus* in Australia 1935-1974. Copeia 1981:676–680

Siegfried WR (1989) Preservation of species in Southern Africa nature reserves. In: Huntley BJ (ed) Biotic diversity in Southern Africa: concepts and conservation. Oxford University Press, Oxford, pp 186–201

Simberloff D, Smitz DC, Brown TC (1997) Strangers in paradise. Island Press, Washington, DC

Singh SP (1996) Biological control. In: Paroda RS, Chadha KL (eds) 50 years of crop science research in India. Indian Council of Agricultural Research, New Delhi, pp 88–116

Slater PD, Lewington D, Prathey JE (1996) An investigation into the effects of redlegged earth mite and Lucerne flea on the performance of subterranean clover in annual pasture in S. New South Wales. Plant Protect Quarant 11:6–8
Smith R (1984) Producers need not pay startling "rodent tax" losses. Feedstuffs 56(22):13–14
Smith RH (1992) Rodents and birds as invaders of stored-grain ecosystems. In: Jayas DS, White NDG, Muir WE (eds) Books in soils, plants, and the environment: stored-grain ecosystems. Dekker, New York, pp 289–323
South Africa (1998) South Africa government online (http://www.gov.za)
Temple SA (1992) Exotic birds, a growing problem with no easy solution. The Auk 109:395–397
Thompson DG, Stuckey RL, Thompson EB (1987) Spread, impact, and control of purple loosestrife (*Lythrum salicaria*) in North American wetland. US Fish and Wildlife Service, and Fish and Wildlife Research 2, Washington, DC
USCB (2004–2005) United States statistical abstracts. US Census Bureau, Washington, DC
Weber WJ (1979) Health hazards from pigeons, starlings and English sparrow: disease and parasites associated with pigeons, starlings, and English sparrows. Thomson Publications, Fresno, CA

19 Socio-Economic Impact and Assessment of Biological Invasions

Rosa Binimelis, Wanda Born, Iliana Monterroso,
and Beatriz Rodríguez-Labajos

19.1 Introduction

Biological invasions have been object of ecological research for years. As one objective, natural scientists investigate the effects of invasive species on ecosystems and their functioning (Levine et al. 2003). However, impacts on ecosystems are also of relevance for society. Changes in ecosystems affect humans insofar as ecosystems provide goods and services, such as fresh water, food and fibres or recreation, which might be altered due to invasive species. Therefore, impacts of biological invasions should be an object of socio-economic interest, which is also demanded by the Convention on Biological Diversity (2002).

This chapter aims at providing elements for the analysis of impacts of invasive species from the socio-economic point of view. Such an analysis is politically relevant, since impacts are the focal point of every decision to establish an appropriate management regime. For an all-encompassing analysis, an integrative framework is needed to structure the information on impacts. For that purpose, the concept of ecosystem services (Chap. 13) is introduced (Sect. 19.2). Alternative decisions on the appropriate management of invasive species face trade-offs between outcomes and impacts. For handling such trade-offs, evaluation is needed. As discussed in Sect. 19.3, perception presents the prerequisite of an explicit evaluation. Finally, different evaluation methods are introduced so as to value the information about impacts during the decision-making process (Sect. 19.4).

Ecological Studies, Vol. 193
W. Nentwig (Ed.)
Biological Invasions

19.2 Impacts on Ecosystems from the Perspective of Human Wellbeing

Identifying the impacts of invasive species is required in order to evaluate the consequences of invasion processes and to implement management measures. The purpose of this section is to present an integrated framework for structuring the information on impacts in order to describe what happens if an invasion occurs. First, this is done by defining what type of impacts can be associated with bioinvasions. Second, the concept of ecosystem services is used for classifying these impacts. As humans depend on ecosystems and ecosystem processes, effects caused by biological invasions can be of high socio-economic relevance. Perceptions and assessment of these effects will determine policy-making.

From a socio-economic point of view, impacts caused by biological invasions are changes of recipient ecosystems which are perceived by humans. In addition to impacts on ecosystem services, biological invasions can have impacts on human-made goods and services, such as road systems or artificial waterways and reservoirs. Although damages to human-made infrastructure can be considerable, in the following the focus is on impacted services supplied by natural or semi-natural ecosystems (Kühn et al. 2004).

Two types of impacts can be identified. The first type includes direct impacts of invasions on ecosystem functions and on human wellbeing. The second type refers to indirect impacts which stem from the implementation of response actions, such as control costs or side effects of the introduction of biological control agents (Tisdell 1990). A comprehensive decision-making process demands reviewing both types of impacts. However, impact assessment studies do not always distinguish between the two.

By affecting the ecological processes at the level of genes, species and ecosystems, biological invasions modify the provision of ecosystem services. Defined as "the conditions and the processes through which natural ecosystems, and the species that make them up, sustain and fulfil human life" (Daily 1997), ecosystem services are foundations of human wellbeing. Thus, ecosystem services encompass both ecological and socio-economic aspects of ecosystems, illustrating the human dependence on ecosystem functioning. Impacts of biological invasions on ecosystems are of socio-economic concern because they alter the benefits provided by ecosystems for human life.

The Millennium Ecosystem Assessment (2003) is based on a taxonomy of ecosystem services encompassing four main categories (Fig. 19.1):

1. Supporting services are those necessary for the production of all other ecosystem services;
2. Provisioning services refer to the products obtained from ecosystems;

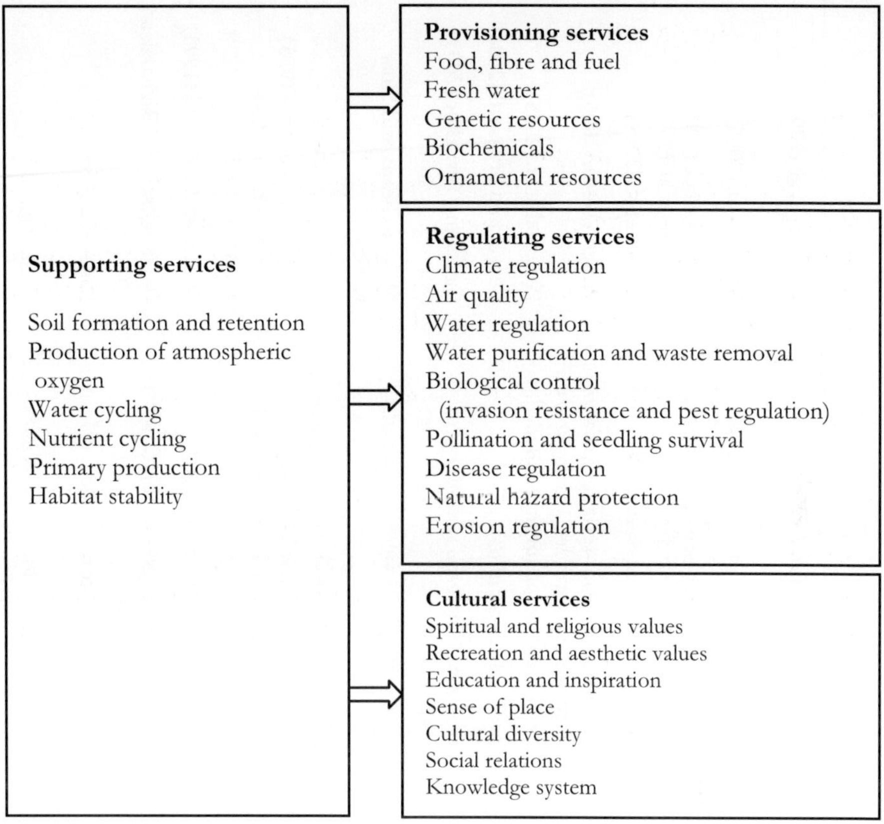

Fig. 19.1 Classification of ecosystem services according to the Millennium Assessment categories

3. Regulating services are benefits supplied by self-maintenance properties of ecosystems;
4. Cultural services generate non-material benefits derived from ecosystems.

Table 19.1 compiles examples illustrating the impacts of various well-known invasive species. It reveals the impacts of invaders on certain ecosystem services by describing their alteration.

As can be noted, there are many mechanisms by which biological invasions can impact different types of ecosystem services. The most evident examples are effects on the provisioning of food. For instance, agricultural and forestry yields are affected by pests such as the Russian wheat aphid (*Diuraphis noxia*; Brewer et al. 2005), the sirex wasp (*Sirex noctillo*) and the skeleton weed (*Chondrilla juncea*; Cullen and Whitten 1995). Other impacts, such as those caused by the zebra mussel (*Dreissena polymorpha*), affect human-made

Table 19.1 Impacts of biological invasions on ecosystem services

	Ecosystem service	Impact description/effect	Associated species (examples)	Reference
Supporting services	Soil formation	Changes in biochemical characteristics of soils	Grand fir (*Abies grandis*)	Griffiths et al. (2005)
		Increase in soil aggregation	Barb goatgrass (*Aegilops triuncialis*)	Batten et al. (2005)
	Nutrient cycling	Reduction of food and oxygen availability	Zebra mussel (*Dreissena polymorpha*)	Minchin et al. (2002)
		Alteration of soil nitrogen levels	Grand fir (*Abies grandis*)	Griffiths et al. (2005)
			Black wattle (*Acacia mearnsii*)	De Wit et al. (2001)
	Primary production	Alteration of biomass production of native plants	European purple loosestrife (*Lythrum salicaria*); black wattle (*Acacia mearnsii*)	Pimentel et al. (2005)
		Reduction in aquatic vegetation	Grass carp (*Ctenopharyngodon idella*)	Pimentel et al. (2005)
		Competition for grazing primary production	Horse (*Equus caballus*)	Beever and Brussard (2004)
	Habitat stability	Changes in vegetation cover affecting community assemblages	Green alga (*Caulerpa taxifolia* and *C. racemosa*)	Cavas and Yurdakoc (2005)
			Common reed (*Phragmites australis*)	Maheu-Giroux and Blois (2005)
Provisioning services	Food	Loss in commercial production and harvest (agriculture, forestry, fisheries, aquaculture)	Russian wheat aphid (*Diuraphis noxia*)	Brewer et al. (2005)
			Skeleton weed (*Chondrilla juncea*)	Cullen and Whitten (1995)
			Rice field rat (*Rattus argentiventer*)	Stenseth et al. (2003)
			Comb jelly (*Mnemiopsis leidyi*)	Knowler (2005)
	Fuel, wood	Loss of forest products	Gypsy moth (*Lymantria dispar*)	Sharov and Liebhold (1998)
	Fresh water	Losses in water catchments	Acacia (*Acacia longifolia*); black wattle (*Acacia mearnsii*)	Galatowitsch and Richardson (2005)
	Genetic resources	Threat to the viability of endangered species	Indo-Pacific soft coral (*Stereonephthya* aff. *curvata*)	Lages et al. (2006)
		Genetic hybridization	Baculoviruses (*Autographa californica nucleopolyhedrovirus*, AcNPV)	Hails et al. (2002)

Regulating services	Water regulation	Choking waterways	Hydrilla (*Hydrilla verticillata*)	Pimentel et al. (2005)
	Water purification	Reduction of water quality	Acacia (*Acacia longifolia*); black wattle (*Acacia mearnsii*)	Galatowitsch and Richardson (2005)
		Increase in water filtration	Zebra mussel (*Dreissena polymorpha*)	Minchin et al. (2002)
	Waste regulation	Colonization of industrial waste dumps	Bacterivorous nematodes (*Acrobeloides nanus*; *Panagrolaimus rigidus*)	Hánel (2004)
	Biological control	Displacement of native and endemic species	Brown trout (*Salmo trutta*)	Quist and Hubert (2004)
	Pollination	Reduction in the reproductive success of flora	Argentine ant (*Linepithema humile*)	Blancafort and Gomez (2005)
	Seedling survival	Depression of the diversity and abundance of seedlings	Shrub (*Lonicera maackii*)	Webster et al. (2005)
	Disease regulation	Infection of native fauna	Chytrid fungus (*Batrachochytrium dendrobatidis*)	Beard and O'Neill (2005)
		Production of toxic substances	Green alga (*Caulerpa racemosa*)	Cavas and Yurdakoc (2005)
		Vectors of human and livestock diseases (e.g. dengue)	Mosquito (*Aedes aegypti*)	Takahashi et al. (2005)
	Natural hazard protection	Disruption in flood control mechanisms	Salt cedar (*Tamarix* sp.)	Lesica and Miles (2004)
		Increase predisposition to fires	Cheat grass (*Bromus tectorum*)	Vitousek et al. (1996)
	Erosion regulation	Intensification of soil erosion	Goat (*Capra aegagrus hircus*)	Pimentel et al. (2005)
Cultural services	Recreational	Reduction of recreational use of rivers and lakes	Black wattle (*Acacia mearnsii*)	De Wit et al. (2001)
		Emerging sport fisheries	Brown trout (*Salmo trutta*)	Quist and Hubert (2004)
	Aesthetics	Changes in the character of rural and urban landscapes	Rhododendron (*Rhododendron ponticum*)	Dehnen-Schmutz et al. (2004)
			Horse chestnut leaf-miner (*Cameraria ohridella*)	Gilbert et al. (2003)
		Use as ornamental flora	Salt cedar (*Tamarix ramosissima*)	Knowler and Barbier (2005)
		Residential weeds	Dandelion (*Taraxacum officinale*)	Pimentel et al. (2005)
	Education	Threat to the value of protected areas	Salt cedar (*Tamarix ramosissima*)	Lesica and Miles (2004)
	Cultural diversity	Loss of subsistence fisheries which shaped local cultures	Brown trout (*Salmo trutta*)	Quist and Hubert (2004)

goods and services, damaging many different hydraulic infrastructures worldwide (Minchin et al. 2002). Further examples and discussion on these issues are provided by Chaps. 13 and 18.

Table 19.1 also illustrates that one single species can have a variety of effects. For instance, the black wattle (*Acacia mearnsii*) affects the regional water table, local vegetation cover, i.e. species composition, and also alters the recreational function of the Cape region in South Africa, since people gain less access to rivers and lakes (Galatowitsch and Richardson 2005).

By structuring the information about impacts using the ecosystem services categories, two general characteristics can be outlined: (1) the variety of impacts caused by invasive species, and (2) the complexity of impacts on ecosystem services. Ecosystem services and impacts on these are not only manifold but also complex, as can be illustrated with the example of the Nile perch (*Lates niloticus*). Its intentional introduction to Lake Victoria in Africa for aquaculture and sport fishing resulted in the extirpation of 200 native fish species (Kasulo 2000). This led to a shift of the whole ecosystem, as the availability of phytoplankton changed, altering the local fish species composition (Chu et al. 2003). This introduction favoured a prospering fish industry in the vicinity of the lake, due to increased profits from perch exports. However, relatively cheap native fish was no longer available, and local inhabitants could not afford the more expensive perch and, therefore, could not complement their diet. Additionally, the availability of fuel wood decreased because this was used to dry the perch, necessary to preserve it. By contrast, the smaller native fish could be sun-dried, rather than being smoked. In this example, the intentional modification of an ecosystem to improve the services of recreation (sport fishing) and the provisioning of food for exports (aquaculture) had important side effects, such as the decrease of habitat stability. Furthermore, cultural practices and social relations changed, and the basic diet of the local inhabitants deteriorated, rather than being improved (www.darwinsnightmare.com).

The Nile perch example serves to highlight the complexity of affected ecosystem services. It also shows the interlinked ecological and socio-economic dimensions of impacts – in this case, some impacts show a direct influence on human wellbeing, such as the alteration of the provisioning service of food and fuel.

19.3 Perception as a Prerequisite for Valuation

Invasive species cause manifold effects. How these are valued depends on human perception at a given point in time. Interests embedded within cultural contexts and production patterns configure the personal attribution of either a positive or negative character to a given effect. Thus, when including these individual or collective appraisals into the decision-making process, their context dependency should be taken into account (Sect. 19.4).

Certain impacts of invasive species are of public concern, such as health problems, e.g. asthma and allergies caused by the rag weed (*Ambrosia artemisiifolia*; Zwander 2001). Others, such as alterations in ecosystem integrity, are not a subject of public discussion. For instance, ecosystem integrity in Canada is strongly affected by the common reed (*Phragmites australis*; Maheu-Giroux and Blois 2005). Although this changes habitat conditions, these impacts generally lie outside the set of social concerns. As the linkage between these impacts on the ecosystem and human wellbeing is not obvious, people who are not involved in conservation issues care little. Indeed, invasions in waters take place mostly in a hidden manner (Nehring 2005). Lack of social concern about the ecologically damaging green alga *Caulerpa racemosa* is a good example (Cavas and Yurdakoc 2005; Piazzi et al. 2005; Ruitton et al. 2005). In fact, plant invaders (not only aquatic) which affect ecosystem integrity are often not of public concern.

Another aspect of perception is that, from a utilitarian point of view, not all the effects are damages. For instance, soil aggregation is enhanced by barb goatgrass (*Aegilops triuncialis*; Batten et al. 2005), and black wattle (*Acacia mearnsii*) increases nitrogen levels in soils (De Wit et al. 2001; Le Maitre et al. 2002). Whereas ecologically concerned people may regard these changes as undesirable, farmers might take advantage of them. In fact, many introduced species are valued both positively and negatively by different stakeholders. An example is brown trout (*Salmo trutta*), which displaces native species and affects cultural practices dependent on these but also promotes economic activities related to recreational angling (Quist and Hubert 2004). Indeed, invasive fish species favouring emergent sport fisheries are often associated with a positive public rating, and this despite their adverse ecological impacts. This example illustrates that personal or social interest can give importance to some effects of an invasive species but neglect others.

As explained above, valuation is dependent on perception. The perception of impacts is heterogeneous, context-dependent and dynamic. The alien invasive acacia (*Acacia* sp.) was introduced for pulp production and tanning-compound extraction in plantations in South Africa (De Wit et al. 2001). Its spread out of control has been associated with changes in water regulation. Different positions taken by the stakeholders reflect the heterogeneous character of this species' impacts – on the one hand, communities suffer from water scarcity and, on the other, they benefit from increased access to fuel wood and timber for building materials. The example also shows the dynamic and context-dependent character of valuation. The effects of acacia growth on water regulation is a main concern of the affected communities. Information on the problem allowed the creation of social partnerships for the control of the acacia. In South Africa, the fight against plant invaders has been boosted by means of the Working for Water Program (www.dwaf.gov.za/wfw) – in this case, information evidently led to higher awareness.

The reasoning presented above demonstrates the need of identifying the stakeholders and their roles as prime perceivers and promoters of impacts. Due to the reflexive nature of the invasion processes (new relevant attributes are continuously added to the relationship between people and invasive species), the participation of stakeholders in both the identification of outcomes and the analysis of priorities is needed in the evaluation processes. The advantage of the concept of ecosystem services lies in the structuring of information about impacts. Further analysis can be done to discuss stakeholder perception of the impacts. Such impacts can be taken into account in the valuation concerning the appropriate management of the species.

By revealing the direct and indirect influence of invasive species on human wellbeing, the ecosystem service concept also supports a reflection on uncertainty and ignorance. Uncertainty exists if outcomes are known but the distribution of probabilities cannot be identified. Ignorance can be defined as the situation where the probability neither of the potential outcome nor of the outcome itself is known. In other words, "we don't know what we don't know" (Wynne 1992). One key feature of invasive species processes is often the lack of knowledge. Due to the complexity of interlinked ecological processes, the predictive power of information available about dispersal rates, traits and ecological behaviour is small (Williamson 1996). Furthermore, often there is no such information available, especially not on the social impacts of invasive species. However, for decision making it is necessary to structure the available information on impacts. The use of the ecosystem services concept can serve this aim because this reveals whether the information about impacts is available or not. Under conditions of uncertain outcomes and irreversible effects, a precautionary approach should be employed concerning management decisions on invasive species.

19.4 Alternatives for the Evaluation of Impacts: from Valuation to Deliberation

Decision making requires evaluation because trade-offs between different management options occur, e.g. if a certain management option promotes one impact and concurrently diminishes another. For instance, eradicating the black wattle (*Acacia mearnsii*) in the Cape region on the one hand implies diminished access to fuel wood for the local population and, on the other, it increases fresh water availability. Furthermore, decisions about invasive species management should take the perceptions of affected people into account. The acceptance and outcome of these decisions will be highly dependent on the individual or social perception of the impacts caused by invasive species.

Table 19.2 Overview of evaluation approaches for the management of invasive species

	Risk assessment	Cost-benefit analysis	Cost-effectiveness	Multi-criteria analysis	Scenario development
Management purpose	Introduction	Introduction and/or control	Control	Introduction and/or control	Introduction and/or control
Purpose of the evaluation	Risk level	Ranking (optimisation)	Ranking (optimisation)	Deliberation and ranking	Deliberation and prospective storylines
Type of impacts	Associated with invasion species (hazards)	Caused directly by invasive species and those derived from management responses (cost of damage, cost of control and benefits)	Associated with management responses (cost of control)	Associated with invasive species and/ or those derived from management (criteria)	Associated with invasive species and/or those derived from management (reference indicators)
Type of information used	Quantitative and qualitative	Quantitative (monetary)	Quantitative (monetary and physical units)	Quantitative and qualitative	Quantitative and qualitative
Participation potential	Low	Low/medium	Medium	High	High
Consideration of uncertainty	Uncertainty reduced to probability or precautionary approach	Sensitivity analysis	Sensitivity analysis	Robustness analysis, accounting for fuzzy data	Integrated set of assumptions
Operative constraints	Low cost and time requirement	Low–medium cost and time requirement	Low–medium cost and time requirement	Medium–high cost and time requirement	Medium–high cost and time requirement
Methodological constraints	Intrinsic uncertainties, risk thresholds	Trade-offs between natural capital and human-made capital, use of discount rate	Definition of thresholds	Definition of thresholds	Lack of precise results, non-replicable results
References	OTA (1993), Landis (2003), Andersen et al. (2004), Simberloff (2005)	Bertram (1999), De Wit et al. (2001), Le Maitre et al. (2002), McConnachie et al. (2003), Pimentel et al. (2005)	De Groote et al. (2003), Buhle et al. (2005), Dehnen-Schmutz et al. (2004)	Maguire (2004), Monterroso (2005)	Chapman et al. (2001), Rodriguez-Labajos (2006)

Management is essentially concerned with how to deal with impacts of biological invasions. This takes place at different stages of the invasion process, either preventing an introduction (accidental or intentional) or managing an invasive species once it is established. Uncertainties linked to the process will vary depending on the invasion stage. A sound decision-making process should also reflect on this (Born et al. 2005).

The purpose of this section is to introduce five approaches to the evaluation of management alternatives concerning invasive species. In this context, operational implications of assessing impacts of biological invasions by means of these approaches are discussed. Table 19.2 presents the main characteristics of each approach. However, it is important to note that every approach features a variety of specific methodologies and techniques. Therefore, specific processes and operational constraints can differ depending on the specificities of the implementation process. Alternatively, a combination of methods is sometimes advisable.

19.4.1 Risk Assessment

One of the approaches most used as a predictive tool concerning biological invasions is risk assessment. This aims at measuring risk by determining the likelihood of an introduction and the potential adverse effects, given available knowledge about alien invasive species and the recipient ecosystem. Risk assessment for invasive species is generally adopted in order to assess decisions regarding the introduction of potentially invasive species, their pathways and vectors before establishment. However, it might also be used for allocating resources to management measures once the species is already established. For instance, the US Environmental Protection Agency developed a framework for using three main steps: (1) problem formulation; (2) analysis of exposure and effects, and (3) risk characterisation (EPA 1998). For invasive species exposure, the analysis involves estimating the likelihood of introduction, establishment and/or spread, taking into account the quantity, timing, frequency, duration and pathways of exposure as well as number of species, their characteristics and the characteristics of the recipient ecosystem (Andersen et al. 2004). As this approach is based on expert judgement, participation of other interested groups is not foreseen. Results from the assessment can be both quantitative and qualitative, although the former is usually the goal (Simberloff 2005). Expenditure and time requirements usually remain low, since mainly standard procedures are involved (e.g. guidelines established by the European and Mediterranean Plant Protection Organization, EPPO, www.eppo.org).

19.4.2 Cost-Benefit Analysis

Cost-benefit analysis is the traditional evaluation instrument within the framework of welfare economics analysis. It assesses current and future costs and benefits in monetary units, associated with a range of alternatives, projects or policy instruments. It intends to consider all impacts of invasive species which can be valued in monetary terms, including the direct costs and benefits of invasives. This implies that the valuation of environmental damages as well as of environmental services has to be conducted in monetary units, guaranteeing the substitutability between ecosystem services and human-made goods and services, even if no markets exists for the service at hand. This method provides an "optimal solution" by ranking the alternatives. Participation of social groups is not necessary but might be considered, for instance, in the assessment of their willingness to pay. Time and cost requirements will depend on the specific techniques employed in the assessment. For instance, carrying out a contingent valuation (assessing the willingness to pay or willingness to accept) will be associated with increased costs, compared to the use of secondary source data. A representative example of this method is the extensive work on the fynbos biome of the Cape Floristic Region in South Africa, where cost-benefit analysis was used to investigate the consequences of plant invasions (e.g. *Acacia* sp., *Eucalyptus* sp.) on water supply (Enright 2000; De Wit et al. 2001; McConnachie et al. 2003). Another contribution consistent with this approach is the highly referenced work developed by Pimentel et al. (2005). To consider all impacts, again uncertainty must be ruled out. Essentially, cost-benefit analysis is a monetisation of risk assessment to generate substitutability. Thus, it allows one to obtain optimal solutions.

19.4.3 Cost-Effectiveness Analysis

When benefits of control actions of invasive species are difficult to assess, economics can use cost-effectiveness analysis to find the policy instrument or alternative best suited to avoid surpassing a given threshold of invasion. To reach the defined goal, several alternatives are compared so as to obtain an optimal solution by evaluating the direct and indirect costs associated with the implementation of these management options. The costs of keeping the invasion below the threshold are expressed in monetary units but the threshold itself is in physical terms (Baumol and Oates 1988). Assuming the objective is to diminish the presence of an invasive species by 50 %, this method reveals the cheapest control option – the most "cost-effective instrument" – to decrease current infestation level to this socially desired threshold. Reduction thresholds are established from outside strict economic reasoning, so this approach can require a higher level of participation. Expenditure and time associated with the implementation of this method may vary according to the techniques employed. This approach has been used by Dehnen-Schmutz et al. (2004) to analyse private and public expenditure allocated to different control options to manage *Rhododendrum ponticum* in the British Isles. All ignorance/uncertainty around the definition of the threshold lies outside the methodology. For the impacts of the management options, again uncertainty is assumed not to exist (otherwise, no well-defined optimum exists).

19.4.4 Multi-Criteria Analysis

Limitations in achieving monetary accountings of impacts, existence of conflicting values and uncertainties inherent to the invasion and the decision-making process are challenging conditions to assess invasive species. A methodological response is multi-criteria analysis, a family of methods rooted in operational research. This compares different alternatives by contrasting the performance of a set of alternatives according to different criteria (Munda 2004). In the context of invasive species, alternatives exist concerning the choice of management options to encounter impacts. The multi-criteria approach allows us to incorporate multiple dimensions of effects, and to include both qualitative and quantitative information associated with impacts of invasive species and those related to the implementation of management responses. Results from most multi-criteria methods provide a ranking of feasible alternatives. These can be achieved either by a vertical approach where no compensability exists (i.e. no trade-offs; e.g. lexicographic methods) or by a horizontal approach which encompasses varying degrees of compensability (e.g. multi-attribute theory, outranking methods). This approach has been used by Maguire (2004) to analyse trade-offs among conflicting objectives for controlling feral pigs (*Sus scrofa*) in Hawaii. In multi-criteria evaluation, the selection of alternatives and criteria may be decided during a participative deliberation exercise; therefore, attention is placed on the learning process and achieving a compromise solution, rather than an optimal solution. Application will usually require longer time periods and higher costs.

19.4.5 Scenario Development

Another analytical technique which has been used to face uncertainty and to integrate different values is scenario development. As opposed to predictions implying no uncertainties, this method is designed to deliver results in situations characterized by uncertainty. A variety of methods employ the term "scenario" referring to possible outcomes of different management alternatives. However, scenario development is also a method in itself. In this approach, scenarios are descriptions of alternative images of the future, created from mental models which reflect different perspectives on past, present and future events (Rotmans et al. 2000). These provide representations of plausible futures and typically include a narrative element called storyline, sometimes supported by quantitative indicators (Berkhout et al. 2002). Impacts of alien invasive species and effects associated with the implementation of response measures can be included when conducting deliberation on causal processes and outcomes of biological invasions. Social participation is desired to increase internal coherence of scenario development and to incorporate different perspectives. Its main purpose is to decrease uncertainty by discourse-based decisions. Cost and time requirements can vary depending on the specific process – as in other methods which pursue participation, these can be high. For instance, Chapman et al. (2001) used this approach to analyse different management scenarios of invasive species in South Africa to improve decision support.

19.5 Concluding Remarks

This chapter illustrates impacts of invasive species from the socio-economic point of view, within the integrative framework of ecosystem services. This framework facilitates a comprehensive review of the variety of impacts caused by invasive species. It links ecological effects of invasive species with the foundations of human wellbeing, as humans are dependent on ecosystems and their functioning in supplying special services to society. Invasive species can disrupt such ecosystem services.

Throughout the variety of examples displayed in the chapter, it can be seen that both the effects and the response impacts are perceived differently by various social groups. Individual or social perception is considered to be a prerequisite for the valuation of impacts in the context of decision making for appropriate management. Using ecosystem service categories helps to organize impacts when presenting information to interest groups, and it can help to include many perspectives during the valuation processes. In this way, the multidimensional character of impacts is highlighted.

Additionally, assessment approaches deal with impacts differently. Every method has different potentials and constraints which shape its use for supporting decision making. Choosing the most suitable approach may rest on different reasons, such as the type of information employed, the participation potential, the consideration of uncertainty and, especially, the type of impacts which are taken into account. In fact, the further away the impact is from holding a market price, the more relevant is social participation in the deliberation process.

Acknowledgements. We would like to thank Ingo Bräuer, Ingolf Kühn, Joan Martinez Alier, Johannes Schiller and Joaquim Spangenberg for useful comments on an earlier draft. The ICTA-UAB group on the socio-economics of biological invasions is funded by the EC within the FP 6 integrated project ALARM (COCE-CT-2003-506675), and Wanda Born by the project "Invasions: The Invasion Potential of Alien Species – Identification, Assessment, and Risk Management", within the BioTEAM-Programme (Grant 01 LM 0206), financed by the German Federal Ministry of Education and Research.

References

Andersen MC, Adams H, Hope B, Powell M (2004) Risk analysis for invasive species: general framework and research needs. Risk Analysis 24:893–900

Batten KM, Six J, Scow KM, Rillig MC (2005) Plant invasion of native grassland on serpentine soils has no major effects upon selected physical and biological properties. Soil Biol Biochem 37:1177–1183

Baumol W, Oates W (1988) The theory of environmental policy. Cambridge University Press, Cambridge

Beard KH, O'Neill EM (2005) Infection of an invasive frog *Eleutherodactylus coqui* by the chytrid fungus *Batrachochytrium dendrobatidis* in Hawaii. Biol Conserv 126:591–595

Beever EA, Brussard PF (2004) Community- and landscape-level responses of reptiles and small mammals to feral-horse grazing in the Great Basin. J Arid Environ 59:271–297

Berkhout F, Hertin J, Jordan A (2002) Socio-economic futures in climate change impact assessment: using scenarios as 'learning machines'. Global Environ Change 12:83–95

Bertram G (1999) The impact of introduced pests on the New Zealand economy. In: Hackwell K, Bertram G (eds) Pests and weeds, a blueprint for action. New Zealand Conservation Authority, Wellington, pp 45–71

Blancafort X, Gomez C (2005) Consequences of the Argentine ant, *Linepithema humile* (Mayr), invasion on pollination of *Euphorbia characias* (L.) (Euphorbiaceae). Acta Oecol 28:49–55

Born W, Rauschmayer F, Bräuer I (2005) Economic evaluation of biological invasions – a survey. Ecol Econ 55:321–336

Brewer MJ, Noma T, Elliott NC (2005) Hymenopteran parasitoids and dipteran predators of the invasive aphid *Diuraphis noxia* after enemy introductions: temporal variation and implication for future aphid invasions. Biol Control 33:315–323

Buhle ER, Margolis M, Ruesink JL (2005) Bang for buck: cost-effective control of invasive species with different life histories. Ecol Econ 52:355–366

Cavas L, Yurdakoc K (2005) A comparative study: Assessment of the antioxidant system in the invasive green alga *Caulerpa racemosa* and some macrophytes from the Mediterranean. J Exp Mar Biol Ecol 321:35–41

Chapman RA, Le Maitre DC, Richardson DM (2001) Scenario planning: understanding and managing biological invasions in South Africa. In: McNeely JA (ed) The great reshuffling. Human dimensions of invasive alien species. IUCN, Gland, pp 195–208

Chu D, Strand R, Fjelland R (2003) Theories of complexity. Common denominators of complex systems. Complexity 8:19–30

Convention on Biological Diversity (2002) Decision VI/23. Alien species that threaten ecosystem, habitats or species. Proc 6th Ordinary Meet Conf Parties to the Convention on Biological Diversity, http://www.biodiv.org

Cullen JM, Whitten MJ (1995) Economics of classical biological control: a research perspective. In: Hokkanen H, Lynch MT, James M (eds) Biological control: benefits and risks. Cambridge University Press, Cambridge, pp 270–276

Daily GC (ed) (1997) Nature's services. Societal dependence on natural ecosystems. Island Press, Washington, DC

De Groote H, Ajuonu O, Attignon S, Djessou R, Neuenschwander P (2003) Economic impact of biological control of water hyacinth in Southern Benin. Ecol Econ 45:105–117

Dehnen-Schmutz K, Perrings C, Williamson M (2005) Controlling *Rhododendron ponticum* in the British Isles: an economic analysis. J Environ Management 70:323–332

De Wit MP, Crookes DJ, Van Wilgen BW (2001) Conflicts of interest in environmental management: estimating the costs and benefits of a tree invasion. Biol Invasions 3:167–178

Enright WD (2000) The effect of terrestrial invasive alien plants on water scarcity in South Africa. Phys Chem Earth 25:237–242

EPA (1998) Guidelines for ecological risk assessment. Risk Assessment Forum, US Environmental Protection Agency Press, Washington, DC, Rep no EPA/630/R-95/002F

Galatowitsch S, Richardson DM (2005) Riparian scrub recovery after clearing of invasive alien trees in headwater streams of the Western Cape, South Africa. Biol Conserv 122:509–521

Gilbert M, Svatos A, Lehmann M, Bacher S (2003) Spatial patterns and infestation processes in the horse chestnut leafminer *Cameraria ohridella*: a tale of two cities. Entomol Exp Appl 107:25–37

Griffiths R, Madritch M, Swanson A (2005) Conifer invasion of forest meadows transforms soil characteristics in the Pacific Northwest. Forest Ecol Manage 208:347–358

Hails RS, Hernandez-Crespo P, Sait SM, Donnelly CA, Green BM, Cory JS (2002) Transmission patterns of natural and recombinant baculoviruses. Ecology 83:906–916

Hánel L (2004) Colonization of chemical factory wastes by soil nematodes. Pedobiologia 48:373–381

Kasulo V (2000) The impact of invasive species in African lakes. In: Perrings C, Williamson M, Dalmazzone S (eds) The economics of biological invasions. Edward Elgar, Cheltenham, pp 183–207

Knowler D (2005) Reassessing the costs of biological invasion: *Mnemiopsis leidyi* in the Black Sea. Ecol Econ 52:187–199

Knowler D, Barbier E (2005) Importing exotic plants and the risk of invasion: are market-based instruments adequate? Ecol Econ 52:341–354

Kühn I, Brandenburg M, Klotz S (2004) Why do alien plant species that reproduce in natural habitats occur more frequently? Diversity Distrib 10:417–425

Lages BG, Fleury BG, Ferreira CEL, Pereira RC (2006) Chemical defense of an exotic coral as invasion strategy. J Exp Mar Biol Ecol 328:127–135

Landis WG (2003) Ecological risk assessment conceptual model formulation for nonindigenous species. Risk Analysis 24:847–858

Le Maitre DC, Van Wilgen BW, Gelderblom CM, Bailey C, Chapman RA, Nel JA (2002) Invasive alien trees and water resources in South Africa: case studies of the costs and benefits of management. Forest Ecol Manage 160:143–159

Lesica P, Miles S (2004) Ecological strategies for managing tamarisk on the C.M. Russell National Wildlife Refuge, Montana, USA. Biol Conserv 119:535–543

Levine JM, Vilà M, D'Antonio CM, Dukes JS, Grigulis K, Lavorel S (2003) Mechanisms underlying the impacts of exotic plant invasions. R Soc Rev 270:775–781

Maguire LA (2004) What can decision analysis do for invasive species management? Risk Analysis 24:859–868

Maheu-Giroux M, Blois SD (2005) Mapping the invasive species *Phragmites australis* in linear wetland corridors. Aquat Bot 83:310–320

McConnachie AJ, De Wit MP, Hill MP, Byrne MJ (2003) Economic evaluation of the successful biological control of *Azolla filiculoides* in South Africa. Biol Control 28:25–32

Millennium Ecosystem Assessment (2003) Ecosystems and human well-being. A framework for assessment. Island Press, Washington, DC

Minchin D, Lucy F, Sullivan M (2002) Zebra mussel: impacts and spread. In: Olenin S (ed) Invasive aquatic species of Europe – distribution, impact and management. Kluwer, Dordrecht, pp 135–146

Monterroso I (2005) Comparison of two socio-economic assessment methods for the analysis of the invasion process of *Hydrilla verticillata* in Lake Izabal, Guatemala. Master Thesis, Universitat Autònoma de Barcelona, Bellaterra (http://selene.uab.es/brodriguezl)

Munda G (2004) Social multi-criteria evaluation: methodological foundations and operational consequences. Eur J Operat Res 158:662–677

Nehring S (2005) International shipping - A risk for aquatic biodiversity in Germany. In: Nentwig W, Bacher S, Cock MJW, Dietz H, Gigon A, Wittenberg R (eds) Biological invasions – from ecology to control. Neobiota 6:125–143

OTA (1993) Harmful non-indigenous species in the United States. US Congress, Office of Technology Assessment, Washington, DC

Piazzi L, Balata D, Ceccherelli G, Cinelli F (2005) Interactive effect of sedimentation and *Caulerpa racemosa* var. *cylindracea* invasion on macroalgal assemblages in the Mediterranean Sea. Estuarine Coastal Shelf Sci 64:467–474

Pimentel D, Zuniga R, Morrison D (2005) Update on the environmental and economic costs associated with alien-invasive species in the United States. Ecol Econ 52:273–288

Quist MC, Hubert WA (2004) Bioinvasive species and the preservation of cutthroat trout in the western United States: Ecological, social, and economic issues. Environ Sci Policy 7:303–313

Rodríguez-Labajos B (2006) Interlinked biological invasions in the Ebro River. A multiscale scenario approach. Master Thesis, Universitat Autònoma de Barcelona, Bellaterra (http://selene.uab.es/brodriguezl)

Rotmans J, van Asselt M, Anastasi C, Greeuw S, Mellors J, Peters S, Rothman D, Rijkens N (2000) Visions for a sustainable Europe. Futures 32:809–831

Ruitton S, Javel F, Culioli JM, Meinesz A, Pergent G, Verlaque M (2005) First assessment of the *Caulerpa racemosa* (Caulerpales, Chlorophyta) invasion along the French Mediterranean coast. Mar Pollut Bull 50:1061–1068

Sharov AA, Liebhold AM (1998) Bioeconomics of managing the spread of exotic species with barrier zones. Ecol Appl 8:833–845

Simberloff D (2005) The politics of assessing risk for biological invasions: the USA as a case study. Trends Ecol Evol 20:216–222

Stenseth NC, Leirs H, Skonhoft A, Davis SA, Pech RP, Andreassen HP, Singleton GR, Lima M, Machangu RM, Makundi RH, Zhang Z, Brown PB, Shi D, Wan X (2003) Mice and rats: the dynamics and bioeconomics of agricultural rodents pests. Frontiers Ecol Environ 1:1–12

Takahashi LT, Maidana NA, Ferreira J, Wilson Castro P, Yang HM (2005) Mathematical models for the *Aedes aegypti* dispersal dynamics: travelling waves by wing and wind. Bull Math Biol 67:509–528

Tisdell CA (1990) Economic impact of biological control of weeds and insects. In: Mackauer M, Ehler LE, Roland J (eds) Critical issues in biological control. Interdept, Andover, pp 301–316

Vitousek PM, D'Antonio CM, Loope LL, Westbrooks R (1996) Biological invasions as global environmental change. Am Sci 84:468–478

Webster CR, Nelson K, Wangen SR (2005) Stand dynamics of an insular population of an invasive tree, *Acer platanoides*. Forest Ecol Manage 208:85–99

Williamson M (1996) Biological Invasions. Chapman and Hall, London

Wynne B (1992) Uncertainty and environmental learning. Reconceiving science and policy in the preventive paradigm. Global Environ Change 2:111–127

Zwander H (2001) Der Pollenflug im Klagenfurter Becken (Kärnten) 1980 bis 2000. Carinthia II 191:117–134

Section VI

Prevention and Management of Biological Invasions

Short Introduction

Wolfgang Nentwig

We have plenty of possibilities to manage and control, or prevent and avoid, or sometimes even to undo or reverse a biological invasion. The first approach in such a project is usually a discussion on the nature of the means which should be used. Economists conclude that biological invasions are simply the result of basic economic principles, and they predict that redefining the rules of the game will prevent further alien species from spreading or, at least, identify someone who would pay for any damages (Chap. 20).

Scientists involved in the often frustrating, everyday business of food quarantine and transportation stowaways have their own ideas about more intensive inspections (Chap. 21). The new ghost of globalization with unlimited trade around the globe could stigmatise protective measures as trade impediment. Is this the end of control possibilities?

In principle, conservation biologists are able to eradicate an invasive alien species once it has been established. There are, however, narrow limits to such techniques, making a successful eradication dependent on very specific circumstances (Chap. 22). In addition, biological control, once considered as a unique tool to eradicate alien species, initially dispersed even more aliens and caused more problems than it solved. Today, the situation has completely changed and modern biocontrol is a valuable tool (Chap. 23). Still, is it realistic to assume that at least one biological control agent would be available against each invasive alien species?

Would the precautionary principle not be a more appropriate reaction to the global threat of biological invasions? This would mean, of course, that stricter steps are urgently needed to manage the increasing hazard stemming from biological invasions. Public information and awareness are critical in this respect, and we certainly need specific education programs for the general public. The precautionary principle also includes a full spectrum of measures, from economically based tools to all control and eradication means suitable. It is easy to set up such long to-do lists but more difficult, and probably also frustrating, to work for their implementation. This, however, is exactly what is needed now (Chap. 24).

20 Economic Analysis of Invasive Species Policies

JULIA TOUZA, KATHARINA DEHNEN-SCHMUTZ, and GLYN JONES

20.1 Introduction

The economic aspects of invasive alien species (IAS) are increasingly being recognised as highly significant (Perrings et al. 2000; McNeely 2001). Even though the economics of invasive species is often associated solely with economic consequences of species, economics is equally important for the analysis of reasons for invasions. That IAS impose costs upon society is unchallenged. Pimentel et al. (2000, 2005 and Chap. 18) assess the costs of IAS to the US, and the latter paper estimates these at $ 120 billion per annum. These costs are borne by the whole of society and not only those responsible for the initial introductions.

IAS introductions are typically unintended or intended consequences of economic activities. These are not only responsible for first bringing species into an area where they are non-native but they also influence the direction and frequency of repeated introductions, and the pattern of spread of established species. Whether a species is introduced deliberately or unintentionally, research has shown that trade plays a significant role (Chaps. 2 and 3). For deliberately introduced species, it has been shown that trade variables seem to be a good explanatory variable for the successful invasion of species (Cassey et al. 2004; Semmens et al. 2004; Dehnen-Schmutz et al. 2006; Duggan et al. 2006). For unintentional introductions, Levine and D'Antonio (2003) found a positive relationship between the rate of establishment of unintentionally introduced alien species and import volumes in the USA. Economic pathways for unintentional introductions are, for example, ship ballast water (Chap. 4), wood packaging, and ornamental plants. In addition, that the invasibility of host ecosystems is affected by human impacts is widely accepted, although there are only a few studies exploring the link directly (Dalmazzone 2000; Vilà and Pujadas 2001).

Perrings et al. (2002) argue that, since the causes of the problem are primarily economic, they also require economic solutions. The role of economics

Ecological Studies, Vol. 193
W. Nentwig (Ed.)
Biological Invasions

in the control of IAS is multi-facetted. It includes the analysis of the economic drivers of biological invasions, their impact (by single-species impact analyses and cost-benefit appraisals), as well as constructing and analysing policy options for prevention and control.

There are several strategies to counter invasive species: (1) prevention of the introduction of potential IAS, (2) eradication in the early invasion stages and (3) control of fully established IAS (Mack et al. 2000; see also Chaps. 21–23). Among these, prevention – by identifying future invaders before introduction – and eradication – by detecting harmful invasions soon after initial establishment – are often seen as the most successful approaches. In economic terminology, policies to manage IAS are termed public goods (Perrings et al. 2000). This means that the provision of IAS policy is "non-rival" in that additional beneficiaries from its provision do not increase the cost. That is, one person's "consumption" of control of IAS is not at the expense of another's. Furthermore, IAS policies are "non-excludable" in that no one can be prevented from consuming the benefits, i.e. both payers and non-payers receive the benefits. This means that there is potentially a strong incentive for individuals (people, businesses or countries) to free-ride on the management efforts of others. These two characteristics ("non-rival" and "non-excludable") of public goods will result in an under-provision of policies for managing biological invasions if policies are left to market forces, and the two partly explain why the responsibility for environmental protection lies with national governments. The level of provision of some of these policies is also influenced by the manner in which they are provided. IAS management provision is only as good as the least effective provider, i.e. the weakest link determines the total provision of the IAS protection (Perrings et al. 2002). For example, the effectiveness of international measures to control the spread (or eradication) of a plant pathogen depends on the country with the weakest biosecurity policy, as the failure of its biosecurity measures would prevent any other country from controlling (or eradicating) the pathogen because of persistent reintroductions.

This chapter examines economic studies in terms of their contribution to increasing the understanding of policy decisions on biological invasions. It considers three areas where economics can contribute most to the management of IAS. First, we introduce different economic instruments to protect against IAS which have been suggested in the literature. The second area explored is the cost-effectiveness of policy. Although the prevention of the introduction of IAS intuitively seems to be the best way to avoid any IAS problems, it may not necessarily be the most cost-effective strategy, and any decisions as to how to allocate limited resources between prevention and reactive policies are subject to an analysis of the trade-offs between the two. The third area is the inclusion of uncertainty in economic analysis of IAS policy options. All policy decisions have to take into account the high level of uncertainty surrounding biological invasions, which makes it all but impossible to

Table 20.1 Summary of economic studies of policies to manage biological invasions

Study	Content
Costello and McAusland (2003)	Evaluation of the relationship between traded goods, import tariffs, and the damages arising from accidental introductions
Eiswerch and van Kooten (2002)	Identification of optimal strategies using expert opinion to incorporate uncertainty into the management decisions
Heikkila and Peltola (2004)	Evaluation of cost-effectiveness of the Finnish protection system against Colorado potato beetle (*Leptinotarsa decemlineata*)
Horan et al. (2002)	Analysis of biosecurity measures taken by firms under uncertainty
Horan and Lupi (2005)	Investigation into the possible effectiveness of tradable permits as an alternative IAS control mechanism
Finnoff et al. (2005)	Dynamic ecological-economic model of interactions between the environment and society's responses to IAS (private agents and social manager)
Jensen (2002)	Economic evaluation of invasion risk and strategies for prevention versus control efforts
Knowler and Barbier (2005)	Modelling of the risk of deliberate introduction of plants by the horticulture industry and analysis of the possibility of using taxes to regulate this industry
Leung et al. (2002)	Bioeconomic model to evaluate optimal investment in prevention and control strategies
Leung et al. (2006)	Evaluation of "rules of thumb" to guide prevention and control expenditures
Margolis et al. (2005)	Model of tariff formation taking into account invasion externality and the influence of lobby groups
McAusland and Costello (2004)	Analysis of the optimal combination of import tariffs and inspections to reduce the risk of accidental introductions
Mumford et al. (2000)	Examine the rationale of the UK Government's Plant Health Programme and consider its effectiveness in the deployment of resources
Perrings (2005)	Analysis of efficiency and effectiveness of prevention and control based on the stochastic process of IAS
Olson and Roy (2002)	Evaluation of eradication and control strategies taking into account that the spread of the invader is subject to environmental disturbances
Waage et al. (2004)	Estimation of the nature and magnitude of future non-native species risks in order to evaluate an efficient allocation of resources between these

predict, for example, which species will establish in new environments, start to spread, and what sort of impacts they will have (Williamson 2001; Andersen et al. 2004). Table 20.1 offers an overview of the studies reviewed in this chapter.

20.2 Economic Instruments as Measures for Preventing Invasions

In economics, invasions are externalities because they occur from the failure of markets or regulatory institutions to account for all damages the invasive species may cause to society (Perrings et al. 2000). An externality occurs when the decisions of one agent have an impact on the welfare or profit of another agent(s) in an unintended way, and when neither compensation nor payment is made by the generator of the impact to the affected party (Perman et al. 2003). This means that the market prices of potential IAS, or of species acting as host for pest/pathogens, do not reflect societal preferences about avoiding the costs of invasion. In the absence of instruments to correct the externality, responsibility for protection against IAS lies with national governments and their prevention programmes, which can include a set of regulatory measures (black and white lists, inspections, quarantine, etc.) which are conventionally applied to prevent/lower the risk of invasions. The regulator also has the option of applying economic instruments (taxes, tradable permits, etc.) as management tools which are coherent with the "Polluters Pay Principle", to try to guarantee the optimal level of prevention. These instruments directly address the effects of invasion-externalities because they confront those causing the problem with the social costs of their activities. So far, few economic studies have considered the use of such economic instruments to reduce the risk of invasions. The economic instruments explored here are risk-related taxes (at a national level), risk-related import tariffs (at an international level), and tradable permits.

20.2.1 Risk-Related Taxes

At a national level, Knowler and Barbier (2005) consider the possible use of taxes in the horticultural market. These authors model the horticulture industry as a source and pathway for the deliberate introduction of potential invasive plants. They recognise that the commercial sale of non-native plants implies a risk that invasions may occur but that they also provide benefits for the nursery industry and for consumers. The risk of invasion is assumed to depend on the characteristics of the plants and on the number of nurseries selling the plants. Therefore, the calculation of the socially optimal number of nurseries takes into account both the contribution to the probability that an ornamental plant becomes invasive (from allowing one more nursery to sell the plant) and the losses to the industry if the invasion occurs (additional control costs and restrictions on stock movements). Knowler and Barbier (2005) conclude that this socially optimal number of nurseries is lower than that of the existing nursery market, and they evaluate the use of taxes to restrict the number of nurseries to the social optimum. The optimal level of taxes was shown to be highly dependent on how the probability of invasion changes with even a marginal increase in the number of nurseries.

At the moment, the use of risk-related taxes to regulate the national trade of potential IAS and species which may act as a source of IAS is only a theoretical proposal. For deliberate introductions, voluntary codes of conduct are used as an alternative policy option. In the gardening sector, they have been applied in Australia, New Zealand, the USA and Britain (Baskin 2002; DEFRA 2005; Moss and Walmsley 2005). They aim to encourage risk-aware behaviour among the different stakeholder groups (e.g. the gardening public, the nursery industry and landscape gardeners), leading to the voluntary removal of invasive plants from the trade. Their voluntary nature makes them largely dependent on the effort with which they are promoted to the public and the industry. However, the pure public good nature of the environmental awareness created by these codes implies that it is in the best interest of the contributors to sit back and free-ride on the efforts of others in applying the codes of conduct. In addition, these policy schemes are applied without specific targets or time frames, and without the option for decision makers to introduce a regulatory approach if compliance with the codes is low. Therefore, voluntary schemes have so far failed to have a significant impact on the scale and range of invasive plants sold (Moss and Walmsley 2005).

20.2.2 Risk-Related Import Tariffs

At an international level, the use of market-based instruments such as import tariffs is limited by international trade agreements such as the World Trade Organization's General Agreement on Tariffs and Trade, GATT (Werksman 2004; Perrings et al. 2005). Nevertheless, the use of import tariffs has formally been examined in the theoretical economic literature. Costello and McAusland (2003) analyse the relationships among volume of traded goods, import tariffs, and impacts of accidental introductions. They argue that, although tariffs (i.e. higher protectionism) may reduce imported volumes of risky products, they may also change the composition of traded products, which may have an effect on the level of the disturbance in ecosystems in the importing country, and on their susceptibility to IAS. For example, a higher tariff may imply an increase in the volume of domestic agricultural output which, in turn, implies a potentially higher quantity of crops susceptible to invasions and of disturbed land. A more recent paper by the same authors examines import tariffs in combination with regulatory measures, such as import inspections (McAusland and Costello 2004). They show that the interaction between inspections and tariffs is determined by the level of infection/infestation of the traded goods. Both policies (inspections and tariffs) increase with the proportion of traded goods which may become invasive but this proportion may reach a point after which inspections are decreasing. This means that, when the level of infection/infestation of the imported material reaches a given level, it is optimal to inspect less and to have tariffs as the dominant prevention policy. McAusland and Costello (2004) also conclude that importers should apply a tariff which covers the inspection costs, and the expected damage from infested goods received which were undetected during inspections. Furthermore, the optimal level of these tariffs should depend on the characteristics of the trade partners in terms of the risk and potential damage of accidentally introduced IAS. This view, requiring an analysis of the risk characteristics of the trading partners, surfaces again in research on IAS and tradable permits discussed below.

Such tariffs are not possible under the non-discriminatory policies (i.e. discriminating against foreign goods are prohibited) which characterise international trade agreements. Margolis et al. (2005) argue that, if national governments could select tariffs freely, import tariffs should include the potential invasive species damages. However, tariffs may be used as instruments of disguised protectionism if pressure from lobby groups lead to tariffs that exceed the optimum level. Such pressure for protectionism could be avoided if there were an international agreement on how to measure damage from invasive species. However, the lack of such agreement makes it difficult to recognise when disguised protectionism may be occurring. It should also be noted that regulatory measures (inspections, quarantine, black and white lists, etc.) are not free from the potential influence of interest groups. For example, the appropriateness of quarantine regulations depends on the complex economic interests of stakeholders, different attitudes to risks, the uncertainties associated with these risks, and the considerable costs for the trading partners of over-controlled national borders (Mumford 2002). Even the efficiency of risk assessments is compromised because of political influence and regulators' competing requirements to, for example, facilitate exports and control invasion risk (Simberloff et al. 2005).

20.2.3 Tradable Permits

This section deals with an increasingly popular economic instrument to control environmental "bads": systems of tradable permits. Such systems are based on the principle that any increase in "emissions" must be offset by an equivalent decrease elsewhere (Perman et al. 2003). There is usually a limit set on the amount of emissions allowed. In the case of IAS, no "emissions" are allowed but there is always a risk of "emissions". In their analysis of introductions of IAS to the Great Lakes of North America via the ballast water of marine vessels, Horan and Lupi (2005) consider control of IAS by the use of tradable risk permits. Since IAS emissions cannot generally be directly measured (particularly for ballast water), they cannot be directly traded. Horan and Lupi consider a system where permits are denominated in terms of the probability of an IAS invasion. They list a series of problems particular to IAS which compromise the most efficient outcome from the use of tradable permits thus defined. The system would be too complex, due to the information requirements that all potential invaders and the likelihoods of invasion (the ex-ante nature of the permit denomination) be known. In addition, the potential expected damages from invasions would need to be known. The optimal solution would also require an excessive number of different permit types (one per IAS) which must be traded at vessel-specific rates. As it would be impossible to know exactly the species transported and their potential damage, a permit system at the species level could not be considered. Therefore, Horan and Lupi (2005) relax the denomination of a permit to one which restricts the probability of invasion from any species, as opposed to different permits for different species. This reduces efficiency when different IAS have different damage impacts. The degree of inefficiency is dependent upon (amongst other things) the heterogeneity of the marginal damage impacts of IAS. Nevertheless, this form of permit denomination does significantly reduce the information requirements. Horan and Lupi's application of a multiple-species permit approach to shipping and the risk of IAS via ballast in the Great Lakes suggests that risk reductions can be achieved at lower costs than those associated with uniform technology standards applied to all participants. The gains depend upon the agreed level of aggregate invasion risk, and arise from the heterogeneity in invasion risks and biosecurity cost structures associated with different market agents. As the agreed levels of invasion risk in the model are reduced, the potential savings using the permit system are decreased.

20.3 Trade-offs Between Prevention and Control Strategies

Prevention is often defined as the first and, usually, the most cost-effective line of protection against invasive species (Mack et al. 2000; Meyerson and Reaser 2002). However, a completely effective prevention strategy which reduces the invasion risk to zero is unrealistic and, therefore, policies for tackling invasions include control or post-invasion actions. Whereas prevention measures are expected to influence the expected probability of successful invasion, control tools focus on reducing the impact of the invasion on the environment.

From an economic perspective, it would be efficient to select these strategies such that the expected marginal benefits of the policy equal its marginal social costs (Perrings 2005). Therefore, the optimal combination of prevention and control efforts depends on the conditions of invasion (Jensen 2002; Finnoff et al. 2005; Leung et al. 2006).

Leung et al. (2006) derive some "rules of the thumb" relating to optimal expenditure on prevention and control. Their study shows that optimal prevention expenditure depends on the probability of invasions and it should decrease as invasions become more unpreventable; optimal control resources should increase with the value of the invaded habitat and decrease with uncontrollable damages (i.e. impacts which can not be reduced, regardless of control efforts). Jensen (2002) include the effect of time to explore the interactions between current prevention to protect against invasions and future control to reduce the damages. Similarly to Leung et al. (2006), he shows that prevention expenditures should be smaller for those invasions more likely to occur (i.e. higher natural hazard rate). In addition, they are also influenced by the relative weight of future benefits in present decisions. The lower the rate is at which future benefits are discounted, the higher is the investment in prevention because future wellbeing, when the invasion may occur, has a higher weight in policy decisions. Jensen (2002) also argues that it is optimal to undertake prevention expenditures if – and only if – the damage costs are high enough.

Finnoff et al. (2005) include the fact that policies often involve private and public actions, and explore their possible interactions. They argue that if private agents' beliefs over the environment are incomplete (i.e. they behave as if there is no change), the social planner either free-rides on private investments or is sole responsible for control actions (limiting the resources on prevention) and, therefore, the risk and abundance of IAS increase. This means that, if there is no invasion, private agents will never apply control measures and, if invasions occur in the future, then the social planner would be forced to employ greater collective resources in control at the expense of an investment into prevention policies. This, in turn, would cause an increase in invasions. By contrast, if the ecosystem is highly invaded, then the private agents' control will always be high, the social planner would therefore free-ride on the private control efforts, public prevention and control would be neglected, and invasions would also increase. However, if the social planner ignores or is not aware of the actions of the private agents, then he will either over-prevent and over-control (if he believes that private agents behave as if there were no invasion) – and invasions decrease – or he will neglect prevention and control (if he believes that private agents behave as if there were an invasion) – and invasions increase. Finnoff et al. (2005) conclude that neglecting these potential feedbacks can have a strong impact on the policy outcomes.

The cost-effectiveness of prevention and control efforts has also been studied, and results so far show that pre-invasion strategies are to be preferred.

Leung et al. (2002) focus on the invasion of zebra mussel (*Dreissena polymorpha*) in North America, and show that the costs of optimal *control* in an invaded lake reduce the social welfare (measured as the benefits from economic activities minus the costs of managing IAS) by one-half relative to welfare in a lake in which optimal *prevention* measures were adopted before the invasion. Therefore, they conclude that prevention is the best investment. More recently, Heikkila and Peltola (2004) analysed strategies to manage the Colorado potato beetle (*Leptinotarsa decemlineata*) in Finland. These authors compare the current public policy based on eradication or pre-emptive control, preventing the pest from establishing permanent populations, with a potential alternative in which reactive control, to limit the damage costs, is left to private agriculture producers. Their results show that, for most of their simulations, the total costs of pre-emptive control are smaller than those for the reactive control, so protection against pest establishment is recommended. Reactive policy is preferred only when the magnitude of the invasion is low and there are moderate–low damages. In an evaluation of the economic effectiveness of the UK's Plant Health Programme, Mumford et al. (2000) consider the costs and benefits of five single-organism case studies and one case study of protecting potatoes as a commodity. The lowest ratio of benefits to costs of the plant health programme (the publicly funded UK plant protection system) was 3.1:1, rising to almost 30:1. For all but one of the organism-specific studies, no change in the exclusion/eradication policy (pre-invasion policy) was recommended. For the other, a review of this policy was recommended due to expected falls in the benefits and expected rises in the costs.

20.4 Uncertainty Surrounding Invasion Risk

Uncertainty is a key feature of biological invasions and, therefore, determines the appropriateness and feasibility of the responses to invasions (Williamson 2001; Perrings 2005; Caley et al. 2006). Uncertainty surrounds the risk of introduction, establishment and spread of IAS, the potential severity of their impacts in the environment, and even the effectiveness of management instruments. For example, for the tradable permit system described above to function, one needs to know a significant number of these uncertain parameters, and the system provides gains only when stakeholders are prepared to accept a higher level of invasion risk.

Horan et al. (2002) compare prevention strategies when there is full information with those when there is ignorance or uncertainty. They focus on biosecurity measures by firms which may release invasive species into the environment. They show that, when there is information about the probability of invasion and its potential damages, the firms minimize damages plus control costs by using prevention measures up to the point where the cost of the last unit of prevention equals the additional benefits. This condition would include the uncertain effects which the measures taken have on the invasion. Their analysis implies that, when there are a large number of firms, it is not optimal for any of the firms to undertake biosecurity actions because chances are that the species will invade anyway. When there is uncertainty, however, it is optimal to employ more resources in preventing high-damage events which are considered possible. The key factor is thus the level of surprise that the decision makers will expect if the invasion takes place, and prevention expenditures focus on those potential invasions with low levels of surprise.

From a post-invasion perspective, Olson and Roy (2002) include the uncertainty of the effects of environmental disturbances on the spread of an invader when examining the conditions under which it is optimal to eradicate. They recognise that there may be disturbances which make the invasion small enough for eradication to be inevitable. They conclude that eradication is optimal if the marginal costs of controlling a small invasion are less than the marginal damage (i.e. marginal benefits of controlling), including future damages due to the expected spread of the invader. Note that because future damages are included in this condition, eradicating may be optimal even if the marginal costs of control are larger than the current damages.

Using expert opinions may reduce uncertainty in management decisions. Eiswerch and van Kooten (2002) identified optimal control management by consulting experts about the state of the invasion, its spread and its impacts. Their study of the agricultural weed yellow starthistle (*Centaurea solstitialis*) demonstrates that, as the productivity of the land increases, the optimal management strategy should have higher levels of control activities but, in this case, it would not be optimal to eradicate the species. Waage et al. (2004) use Monte Carlo analysis simulation to include uncertainty in a range of biological and economic parameters (which may reflect a range of expert opinion or historic evidence). The uncertain parameters include likelihood of entry, the intrinsic rate of spread, potential yield losses, and export losses. This paper seeks to develop tools to aid governments in their need to know the identity and magnitude of future non-native species risks, and uses this information to anticipate and allocate resources efficiently between IAS.

20.5 Discussion

Given the importance of human interactions in determining the scale and speed of invasions, economists have become increasingly interested in the analysis of policies to manage IAS. In this chapter, we surveyed this economic literature mainly from the perspective of prevention measures. This analysis shows that, so far, few economic studies have concentrated on the assessment of market-based instruments (e.g. national taxes, import tariffs and tradable permits) to directly address invasion-externalities. These instruments create incentives for those trading in risky material to avoid the risk of invasions and, therefore, the costs to the society which these may cause. They create signals in the market so that private and society interests coincide and, therefore, induce changes in the individual's behaviour to reduce the likelihood of invasion. Economic instruments should be combined with existing regulatory measures (e.g. standards, inspections, quarantine, black/white lists, etc.) to tackle the causes of invasions (Perrings et al. 2005). Therefore, economic studies which further explore these policies in the context of invasive species are required. Furthermore, other economic instruments to be investigated in the future are graduated license fees, cost-sharing instruments, and environmental bonds. The first capture the risk of invasion by applying more expensive licences to more risky products. The second split up the responsibility between the government and industry. Examples of this type of mechanism include government compensations, lobbies, cost funding approaches, and insurance. Currently, the possibility of insuring commercially against future environmental effects is limited by the uncertainty surrounding the risk of invasion and the expected high value of invasion damages.

Environmental bonds have been proposed for those industries where there is a high level of uncertainty about the nature or severity of the damages which they may cause (Costanza and Perrings 1990). The bond will be equal to the best estimate of the potential future damages, and it would be returned (plus some interests) when the firm could prove that the damages have not occurred or would not occur. This system stimulates research on the consequence of firms' activities and technologies which reduce their environmental impacts. At international level, Perrings et al. (2002, 2005) propose institutional changes to support counter-invader measures in developing countries, given that invasion policies in these countries (i.e. the weakest members of the society) determine the level of protection against invasive species at a global scale. The budget of developing countries in prevention efforts is limited or non-existent (Mumford 2002). Therefore, international agreements are necessary to guarantee the provision of rich countries' resources for poor countries' policies against IAS.

Responses to prevent the threat of invasions range from government agencies' policies to multiple individual actions. This review shows that the potential interactions between these responses need to be taken into account in the development of counter-invader policies. In the 1980s, the interaction between private agents and the regulator in the case of the invasion of the UK by the western flower thrips (*Frankliniella occidentalis*), about which neither group was well informed, illustrates the sometimes strained relationship between regulators and regulated. To eradicate the new arrival, very high costs were imposed (in the form of prescribed control and marketing restrictions) upon a small number of private agents. It is therefore necessary to establish mechanisms for the exchange of information, cooperation and coordination among government agencies and the private sector, and even among governments of neighbouring countries and trading partners (Meyerson and Reaser 2002). This chapter also shows that most papers on the economics of IAS examine the risk of accidental introductions, i.e. when traded goods are a source of IAS. However, studies in which the traded good itself may be a potential invader are lacking (with the exception of Knowler and Barbier 2005), despite the increasing evidence that the availability of alien species in national and international trade is a significant factor for the success of the invasions.

Economic instruments have proven to be useful in other areas of environmental problems. For example, tradable permits have been applied to sulphur dioxide (Stavins 1998; Burtraw 1999) emissions in the US, resulting in pollution reduction at lower costs than those for command and control policies. Environmental taxes have been used more recently, for example, in the area of waste management (landfill taxes) and vehicle emissions (differential tax rates). However, the uncertainty surrounding IAS and limitations of international rules make the construction of economic instruments for biological invasion problems more difficult and partly explain why they have not been employed in this field to date. Nevertheless, the increasing collaboration between economists and invasion biologists, and the growing economic literature on the subject are steps in the right direction.

Acknowledgements. We are grateful to Charles Perrings and Mark Williamson for their comments and suggestions. JT and KDS acknowledge funding by the Leverhulme Trust, and GJ thanks ESRC/NERC for financial support.

References

Andersen MC, Adams H, Hope B, Powell M (2004) Risk assessments for invasive species. Risk Analysis 24:787–793

Baskin Y (2002) The greening of horticulture: new codes of conduct aim to curb plant invasions. BioScience 52:464–471

Burtraw D (1999) Cost savings, market performance and economic benefits of the US Acid Rain Program. In: Sorrell S, Skea J (eds) Pollution for sale: emissions trading and joint implementation. Edward Elgar, Cheltenham, pp 43–62

Caley P, Lonsdale WM, Pheloung PC (2006) Quantifying uncertainty in predictions of invasiveness, with emphasis on weed risk assessment. Biol Invasions 8:277–286

Cassey P, Blackburn TM, Russell GJ, Jones KE, Lockwood JL (2004) Influences on the transport and establishment of exotic bird species: an analysis of the parrots (Psittaciformes) of the world. Global Change Biol 10:417–426

Costanza R, Perrings C (1990) A flexible assurance bonding system for improved environmental management. Ecol Econ 2:57–75

Costello C, McAusland C (2003) Protectionism, trade, and measures of damage from exotic species introductions. Am J Agric Econ 85:964–975

Dalmazzone S (2000) Economic factors affecting vulnerability to biological invasions. In: Perrings C, Williamson M, Dalmazzone S (eds) The economics of biological invasions. Edward Elgar, Cheltenham, pp 17–30

DEFRA (2005) Helping to prevent the spread of invasive non-native species. Horticultural code of practice. Dept Environment, Food and Rural Affairs, UK, pp 1–20

Dehnen-Schmutz K, Touza J, Perrings C, Williamson M (2006) The horticultural trade and ornamental plant invasions in Britain. Conserv Biol (in press)

Duggan IC, Rixon CAM, MacIsaac HJ (2006) Popularity and propagule pressure: determinants of introduction and establishment of aquarium fish. Biol Invasions 8:377–382

Eiswerch ME, van Kooten GC (2002) Uncertainty, economics, and the spread of an invasive plant species. Am J Agric Econ 84:1317–1322

Finnoff D, Shogren JF, Leung B, Lodge D (2005) The importance of bioeconomic feedback in invasive species management. Ecol Econ 52:367–381

Heikkila J, Peltola J (2004) Analysis of the Colorado potato beetle protection system in Finland. Agric Econ 31:343–352

Horan RD, Lupi F (2005) Tradable risk permits to prevent future introductions of invasive alien species into the Great Lakes. Ecol Econ 52:289–304

Horan RD, Perrings C, Lupi F, Bulte EH (2002) Biological pollution prevention strategies under ignorance: the case of invasive species. Am J Agric Econ 84:1303–1310

Jensen R (2002) Economic policy for invasive species. Dept Economics, University of Notre Dame, http://www.nd.edu/%7Erjensen1/workingpapers/InvasiveSpecies.pdf

Knowler D, Barbier E (2005) Importing exotic plants and the risk of invasion: are market-based instruments adequate? Ecol Econ 52:341–354

Leung B, Lodge DM, Finnoff D, Shogren JF, Lewis MA, Lamberti G (2002) An ounce of prevention or a pound of cure: bioeconomic risk analysis of invasive species. Proc R Soc Biol Sci 269:2407–2413

Leung B, Finnoff D, Shogren JF, Lodge DM (2006) Managing invasive species: rules of thumb for rapid assessment. Ecol Econ 55:24–36

Levine JM, D'Antonio CM (2003) Forecasting biological invasions with increasing international trade. Conserv Biol 17:322–326

Mack RN, Simberloff D, Lonsdale WM, Evans H, Clout M, Bazzaz FA (2000) Biotic invasions: causes, epidemiology, global consequences, and control. Ecol Appl 10:689–710

Margolis M, Shogren JF, Fischer C (2005) How trade politics affect invasive species control. Ecol Econ 52:305–313

McAusland C, Costello C (2004) Avoiding invasives: trade related policies for controlling unintentional exotic species introductions. J Environ Econ Manage 48:954–977

McNeely JA (ed) (2001) The great reshuffling. Human dimensions of invasive alien species. IUCN, Gland

Meyerson LA, Reaser JK (2002) Biosecurity moving toward a comprehensive approach. BioScience 52:593–600

Moss W, Walmsley R (2005) Controlling the sale of invasive garden plants: why voluntary measures alone fail. WWF Australia discussion paper. World Wide Fund for Nature Australia, Sydney, http://wwf.org.au/publications/InvasivesVoluntaryMeasures

Mumford JD (2002) Economic issues related to quarantine in international trade. Eur Rev Agric Econ 29:329–348

Mumford JD, Temple ML, Quinlan MM, Gladders P, Blood-Smyth JA, Mourato SM, Makuch Z, Crabb RJ (2000) Economic policy evaluation of MAFF's plant health programme. Report to Ministry of Agriculture, Fisheries, and Food, London

Olson LJ, Roy S (2002) The economics of controlling a stochastic biological invasion. Am J Agric Econ 84:1311–1316

Perman RJ, Ma Y, McGilvray J, Common M (2003) Natural resource and environmental economics. Pearson Education, Harlow

Perrings C (2005) Mitigation and adaptation strategies for the control of biological invasions. Ecol Econ 52:315–325

Perrings C, Williamson M, Dalmazzone S (eds) (2000) The economics of biological invasions. Edward Elgar, Cheltenham

Perrings C, Williamson M, Barbier EB, Delfino D, Dalmazzone S, Shogren J, Simmons P, Watkinson A (2002) Biological invasion risks and the public good: an economic perspective. Conserv Ecol 6:1, http://www.consecol.org/vol6/iss1/art1/

Perrings C, Dehnen-Schmutz K, Touza J, Williamson M (2005) How to manage biological invasions under globalization. Trends Ecol Evol 20:212–215

Pimentel D, Lach L, Zuniga R, Morrison D (2000) Environmental and economic costs of nonindigenous species in the United States. BioScience 50:53–56

Pimentel D, Zuniga R, Morrison D (2005) Update on the environmental and economic costs associated with alien-invasive species in the United States. Ecol Econ 52:273–288

Semmens BX, Buhle ER, Salomon AK, Pattengill-Semmens CV (2004) A hotspot of non-native marine fishes: evidence for the aquarium trade as an invasion pathway. Mar Ecol Prog Ser 266:239–244

Simberloff D, Parker IM, Windle PN (2005) Introduced species policy, management, and future research needs. Frontiers Ecol Environ 3:12–20

Stavins RN (1998) What can we learn from the grand policy experiment? Lessons from SO_2 allowance trading. J Econ Perspect 12:69–88

Vilà M, Pujadas J (2001) Land-use and socio-economic correlates of plant invasions in European and North African countries. Biol Conserv 100:397–401

Waage JK, Fraser RW, Mumford JD, Cook DC, Wilby A (2004) A new agenda for biosecurity. Faculty of Life Sciences, Imperial College London

Werksman J (2004) Invasive alien species and the multilateral trading system. In: Miller ML, Fabian RN (eds) Harmful invasive species: legal responses. Environmental Law Institute, Washington, DC, pp 203–214

Williamson M (2001) Can the impacts of invasive species be predicted? In: Groves RH, Panetta FD, Virtue JG (eds) Weed risk assessment. CSIRO Publishing, Collingwood, pp 20–33

21 Phytosanitary Measures to Prevent the Introduction of Invasive Species

Guy J. Hallman

21.1 Introduction

Regulatory animal and plant protection strives to safeguard agricultural species from pests, diseases, and competition from foreign non-beneficial species. The traditional concern for protecting only economic species (crops, livestock, grazing lands) has recently been broadened to include endangered native species and ecosystems in general, upon the acceptance that ecosystems other than agricultural also provide tangible economic benefits to humanity (Perrings et al. 2000), not to mention what are considered the more intangible benefits arising from our innate affinity with nature (Wilson 1984).

The problem with invasive species is as critical now as it has ever been. It could be argued that regulatory animal and plant protection has not done a good job at protecting economic species, let alone performing its more recent task of protecting ecosystems. Still, that apparent failure in the prevention of invasive species must be measured in light of the magnitude of the ever increasing problem, the modest amount of resources directed toward it, and educated guesses about what shape the world would be in without regulatory protections.

Changes in climate, habitats, soil nitrogen levels, atmospheric carbon dioxide levels, trade, and travel make the job of regulating potential invasive species more difficult (Schwalbe and Hallman 2002). Changes in global climates may exacerbate problems with invasive species by more than simply making temperatures more amenable to the survival of organisms from warmer climates (Chap. 12). Parker et al. (2006) argue that invasive "meltdown" occurs when exotic herbivores replace native ones that were better at controlling invasive plants. In essence, exotic invasive plants thrive not by escaping their natural herbivores but by following them. Atmospheric turbulence from hurricanes in Florida in 2005, the increased intensity of which is

Ecological Studies, Vol. 193
W. Nentwig (Ed.)
Biological Invasions

thought to be due to climate change, resulted in the spread of citrus canker, *Xanthamonas axonopodis* pv. *citri*, and the abandonment of that eradication program.

The tropics hold more potentially invasive species than do temperate regions. Conversion of subtropical areas to more tropical-like environments will result in greater varieties of invasive species moving into new territories than if the reverse were happening; i.e., temperate areas were growing at the expense of tropical regions. Predictions of the limits of establishment of tropical and subtropical invasive species may be outdated as changes in climate and host availability facilitate a greater poleward spread of these organisms (Coley 1999; Parmesan et al. 1999).

Trade places burdens on the struggle against biological invaders. Massive amounts of commodities are shipped across oceans and skies every day, and regulations and ongoing programs must be continually in place to prevent pests from being transported along with these commodities. For example, each year the European Union imports over € 20 billion worth of horticultural products, including fresh fruits and vegetables, nuts, dried fruits, and flowers, from all over the world. Getting these products into the European Union involves strict controls in the growing areas to reduce quarantine pest levels to virtually undetectable levels, safeguarding of the commodity after harvest until packing to prevent infestation of the commodities from a variety of potentially invasive species that may or may not actually feed on the commodity, and possible phytosanitary measures, such as trapping, inspection, treatment, and certification. The number of potentially invasive species accompanying these products could be staggering, although the abundance of control and regulatory methods applied keeps levels of invasive species remarkably low in commercial trade.

Compounding the problem, the integration of the European economy widens the risk factor, as commodities entering one country that may not be of risk for a certain invasive species may be transported more freely than before to another country that is at risk. For example, tropical fruit flies (Tephritidae) were historically not considered a risk for Scandinavian countries, whereas they are definite risks for Mediterranean countries.

Smuggling of agricultural commodities carries a high risk of introduction of invasive species. The availability of legal methods to transport these commodities across quarantine barriers may not alleviate smuggling significantly because it usually requires a significant expense, and is often done in quantities considered too small for "importers" to avail themselves of these methods.

Invasive species arrive in non-biological commodities. Ceramic and marble tiles from Italy are a major source of finds of quarantined land snails at ports of entry to the United States, and brass from India often contains the quarantined grain-infesting Khapra beetle, *Trogoderma granarium*, found in accompanying packing material.

Pallets, packing crates, and packing materials are significant sources of invasive species. Wooden crates and pallets have been the suspected routes of entry for tree-infesting pests, such as the Asian long-horned beetle, *Anoplophora glabripennis.* International guidelines for regulating wood packaging materials in international trade have been developed (FAO 2002).

Packing materials may comprise a variety of organic and inorganic materials, such as paper, plant fiber, and a variety of plastics. Besides directly infesting organic materials, invasive species may be casually collected with the dunnage, or enter the containers during or after packaging. Packing under bright artificial lighting at night in semi-open facilities often results in flying insects being included in the packages.

21.2 International Regulatory Organizations

The International Plant Protection Convention (IPPC) is an international treaty in force since 1952 to prevent the spread and introduction of pests of plants and plant products, and promote measures for their control. Four-fifths of the nations of the world are members. It adopts International Standards for Phytosanitary Measures and sets standards for settling disputes among member countries. Member countries have national plant protection organizations established according to the IPPC with authority in areas of quarantine control, risk analysis, and measures required to prevent the establishment and spread of invasive alien species that are pests of plants. Parties agree to cooperate on information exchange and on the development of International Standards for Phytosanitary Measures.

There are nine regional organizations that concentrate on phytosanitary and sometimes animal health issues within their own regions, and coordinate with the IPPC to gather information and implement phytosanitary measures. Some countries are members of overlapping regional organizations. For example, Mexico is member of the Caribbean Plant Protection Commission, the Organismo Internacional Regional de Sanidad Agropecuario (comprises Central America and Mexico), and the North American Plant Protection Organization.

The Convention on Biological Diversity, in force since 1993, expands the management of invasive species to include not strictly economic products. It calls on member countries to "prevent the introduction of, control or eradicate those alien species which threaten ecosystems, habitats or species." This convention has been very popular among United Nations members, being ratified by all except the small countries of Andorra, Brunei Darussalam, and East Timor, as well as Somalia, Iraq, and the United States of America.

The Agreement on the Application of Sanitary and Phytosanitary Measures (SPS Agreement) is a supplementary to the World Trade Organization

(WTO), and, therefore, adhered to by WTO members, which now number 149 since the WTO entered into force in 1995. The SPS Agreement provides a uniform interpretation of measures governing food safety and plant and animal health regulations, and is applicable to all like measures affecting international trade. These measures include any applied to protect animal or plant health within a member's territory from the entry, establishment, or spread of pests (broadly interpreted to include diseases and weeds). Regulations must be based on transparent science, and should be applied only to the extent necessary to protect human, animal, or plant health. They should not discriminate between countries where similar conditions prevail. Member countries are encouraged to use international standards and recommendations, and may use measures that result in greater levels of control, given scientifically defensible justification.

In the WTO, a specific food safety, or animal or plant health requirement established by one country and that leads to a trade restriction, can be challenged by another country if the latter believes that there is not sufficient scientific evidence supporting the need for the restriction. Challenges to phytosanitary trade barriers have been made a number of times since the WTO came into effect a little over a decade ago. Failure of a country to accept a WTO-brokered decision may result in retaliatory trade practices by other countries.

21.3 Phytosanitary Measures

The IPPC definition of a phytosanitary measure includes legislation, regulations, or official procedures, including treatments, to prevent the introduction and/or spread of quarantine pests (FAO 2004). A variety of methods are available to reduce the risk of invasive species accompanying commodities shipped across natural barriers to invasion. These can roughly be subdivided into two groups – methods that avoid the need to treat the shipped commodity, and those that require treatment.

21.3.1 Phytosanitary Measures that Do not Involve Commodity Treatment

Non-treatment methods include demonstration that a previously quarantined commodity is in fact of negligible risk for certain invasive species, possibly under specific circumstances, or a risk management system that reduces the overall risk to acceptable levels before the commodity is shipped.

21.3.1.1 Non-Host Status

Devising a list of hosts of any pest is one of the fundamental steps of pest management, and has repercussions for regulatory agriculture and trade. Host lists form one of the primary sources for deciding which commodities are quarantined because of certain pests. If host lists contain errors, then commodities may be needlessly quarantined. Armstrong (1994) gives several examples of quarantines arising from dubious host lists.

The IPPC defines the host range of a plant pest as the "Species of plants capable, under natural conditions, of sustaining a specific pest" (FAO 2004). Key words in this definition are "under natural conditions", as some pests can infest some commodities under forced conditions, but may never have been found infesting these in the field.

Armstrong (1986) presents a broad definition of host: "A quarantine host is any commodity which, at one or more of its growth stages, can be naturally infested by a quarantine pest in the field and on, or in which the quarantine pest either can complete its life cycle or otherwise use the commodity for transportation to any area where [the pest] does not already exist and become established as an economic pest." This definition is appropriate to phytosanitary issues because it includes the possibility that a quarantine pest may not use the host for sustenance, but be transported on it casually. This type of organism is referred to as a "contaminating" or "hitch-hiker" pest (FAO 2004), and they comprise a significant and diverse group of quarantined organisms.

The IPPC definition of a pest risk analysis is "The process of evaluating biological or other scientific or economic evidence to determine whether a pest should be regulated and the strength of any phytosanitary measures to be taken against it" (FAO 2004). A pest risk analysis is usually done at some point when a commodity is being considered for export (FAO 1996, 2003). Via the pest risk analysis, it may be discovered that a commodity traditionally thought to be at risk of carrying a quarantine pest is not, in fact, a significant risk. False identification as a pest may have resulted from infestation under unnatural conditions, dubious literature citations, or misidentification of the organism and/or commodity. In that case, no regulatory action is needed against that pest on that commodity, and the commodity can be considered a non-host for that pest.

If a commodity is reported in the literature as a host but that status seems questionable, additional research may be needed to support non-host status. Hennessey et al. (1992) determined that 'Tahiti' lime fruits were not at appreciable risk for infestation by the Caribbean fruit fly, *Anastrepha suspensa*, although several publications listed the fruit as a host of the fly. As a precaution, researchers should be careful and relatively sure that an organism really belongs on a pest list before placing it there, to avoid creating unnecessary trade barriers. Cowley et al. (1992) provide guidelines for determination of host status of fruits to tephritid fruit flies. They argue for precise terms in the literature to describe host status; terms such as "rarely infested" are not helpful in defining host range.

21.3.1.2 Systems Approach

A systems approach to achieving quarantine security is "[t]he integration of different pest risk management measures, at least two of which act independently, and which cumulatively achieve the appropriate level of phytosanitary protection" (FAO 2004). Pest risk management measures are available options to reduce the risk of introduction of a pest, and may be applied at appropriate times when they will have an effect in reducing pest risk at any point from before a commodity is planted to before it arrives at market (Table 21.1). These measures usually form part of an official protocol that must be followed, and achieve a specific goal, such as maintain pest trapping numbers below a predetermined level. If that level is exceeded, then export may be halted until corrective measures restore risk to acceptable levels. Jang and Moffitt (1994) present a thorough discussion of the systems approach.

Pre-plant or pre-season measures include those that define the pest prevalence in the area through trapping and sampling, host suitability, and the pest population level when susceptible stages of the host part to be exported are present. Off-season pest mitigation measures often must be carried out even though the commodity in its exported stage is not present. The use of attractants to detect and suppress pests is discussed by Robacker and Landolt (2002).

During the growing stages when the exportable part of the commodity is present in the fields, attention is focused directly on preventing infestation of the commodity. Survey trapping may be increased, and toxic baits may be employed to keep population levels from exceeding the limit of tolerance. Field applications of pesticides may be employed.

After harvest, the only methods for reducing pest infestation levels are culling of infested commodities, or a disinfestation treatment. If a disinfestation treatment is used as part of a systems approach to phytosanitary security, theoretically the treatment would not need to achieve the same level of control as a stand-alone phytosanitary treatment. Cowley et al. (1991) developed a methyl bromide fumigation treatment for watermelons that depended on the poor host status of watermelons, fruit fly control in the field, and culling of damaged, softened or misshapen fruit to reduce infestation levels of the fruit fly *Bactrocera xanthodes* to levels that would be controlled by a lower than usual dose of methyl bromide. For this scheme to be a true systems approach, each of the steps in the system would be precisely defined in the protocol.

Although systems approaches avoid the expense of treatment, they require a continual expense of the pest risk management measures required to keep

Table 21.1 Systems approach for exporting pink tomatoes from Morocco and the Western Sahara to the US. Risk of infestation by the Mediterranean fruit fly, *Ceratitis capitata* (Medfly), is reduced to acceptable levels

Step	Role in reducing risk
Limit production to provinces El Jadida, Safi (Morocco) and Dahkla (W. Sahara)	Sparse vegetation is poor habitat for Medfly
Only from "insect-proof" greenhouses inspected by and registered with Moroccan regulatory agency	Restricts Medfly from entering production area
Export between 1 Dec. and 30 April	Period of low activity for Medfly
Maintain Medfly traps from 1 Oct. to 30 April	Detects populations of Medfly
Capture of one Medfly in trap in greenhouse shuts it down until re-registration	Avoids shipping infested tomatoes
Capture of flies outside of greenhouse leads to increased number of traps and bait sprays	Reduce risk that outside fly population will enter greenhouse
Must be packed within 24 h of harvest	Reduce time exposed to Medfly infestation in packing house
Safeguarded in insect-proof covering in transit	Prevent access by Medfly
Tomatoes must be pink when packed	Pink tomatoes at less risk than red ones of having Medfly
Packed in insect-proof boxes	Prevent Medfly infestation after packing

pest risk below a predetermined level for an entire region, including areas that are not cultivated for the commodities being regulated. If the risk should exceed the predetermined level, then a phytosanitary treatment may be the only solution to uninterrupted export. Therefore, it may be wise to have a phytosanitary treatment available to back up a systems approach to overcoming a phytosanitary barrier to trade. A systems approach can result in more cost per unit of exported commodity than does a phytosanitary treatment. The entire cost might not be directly to the exporter; some of it may be born by publicly funded regulatory agencies who, for example, conduct sampling programs and sterile insect releases.

21.3.2 Phytosanitary Treatments

Phytosanitary treatments are done directly to the commodity at some point before it is released to the market in the importing country or region to reduce the risk of infestation of that commodity by invasive species to acceptable levels. Virtually any physical, chemical, or biological technique that can be used to kill an organism or prevent its reproduction could theoretically be used as a phytosanitary treatment (Table 21.2). Commercially used treatments are limited by several concerns:

1. The controlling factor must reach the organism, which may be inside the commodity with no easy access from the outside. For example, tephritid fruit fly and weevil (Curculionidae) larvae mine deep inside fruit and other plant parts with no opening to the outside of the plant, making their control more difficult than surface-infesting organisms.
2. The level of control must be near 100 %. The risk of an imported commodity resulting in the establishment of an invasive species depends on, among other things, the infestation rate of the species in the commodity, shipment size, and level of treatment efficacy. Because it takes only one mating pair or one parthenogenic organism to start an infestation, the level of control should ensure that this does not occur within an appropriate margin of error. Landolt et al. (1984) were among the first to consider using pest risk to determine the necessary level of treatment efficacy, rather than setting the level arbitrarily high. Follett and McQuate (2001) review the intervening literature on this topic, and give citations for calculating pest risk associated with treatments, along with real-world examples.
3. The treatment cannot harm the commodity significantly, or pose a health or environmental hazard. Virtually any physical treatment designed to kill organisms on fresh commodities, which are also alive, could harm or kill that commodity or accelerate decomposition and shorten shelf life. Biocidal chemical treatments may harm a live vegetative commodity, and are limited by health and environmental concerns. Many approved phytosanitary treatments cause some recognizable, albeit tolerable, damage to the commodity, and commercial interests should check tolerance of the items they wish to treat before investing in the treatment technology. More treatment options are available for durable commodities, such as bulk grain, lumber, hay, and tobacco, which are harder to damage than fresh produce.
4. The treatment must be commercially viable from the standpoint of price and logistics. This criterion may not keep a phytosanitary treatment out of the regulations, but it will keep it from being used commercially for very long.

Quarantine treatments are presented below in approximate chronological order of their development and commercial application.

Table 21.2 Comparison of phytosanitary treatments for various factors

Treatment	Commodity tolerance	Cost	Speed	Logistics	Accepted by organic growers
Cold	Moderate	Low	Very slow	Easy	Yes
Heated air	Moderate	Moderate	Moderate	Moderate	Yes
Methyl bromide fumigation	Moderate	Low	Fast	Easy	No
Sulfuryl fluoride fumigation	Low	Low	Fast	Easy	No
Hot water immersion	Moderate	Low	Fast	Moderate	Yes
Pesticide dips or sprays	High	Low	Fast	Easy	No
Ionizing irradiation	High	Moderate	Fast	Moderate	No
Low oxygen/ high CO_2	Moderate	Moderate	Slow	Moderate	Yes
Radiofrequency heating	Moderate	Moderate	Fast	Moderate	Yes

21.3.2.1 Cold Treatment

Sustained temperatures in the range of –0.6 to 3.3 °C for 7 to 90 days are one of the oldest and most widely used treatments. Temperate insects, which may undergo diapause during winter, require longer treatment times than do tropical insects that do not diapause. For example, at 2.2 °C or below, the Mediterranean fruit fly, *Ceratitis capitata*, requires 15 days, while the apple maggot, *Rhagoletis pomonella*, requires 42 days (FAO 1984).

Advantages to cold treatment are its tolerance by a wide variety of fruits, including many tropical ones. Some fruits, such as apples, are stored for months at temperatures that are lethal to apple pests. Cold treatment can be applied to fruits after packing and during lengthy transport in ships. The chief disadvantage is the long treatment times; no other treatment requires such long time periods to conduct. The lengthy treatment period exposes the treatment to greater risk of interruption caused by equipment or power failures. If a cold treatment is interrupted and the temperature rises by as little as 1 °C for even a relatively short period of time, then the treatment may have to be initiated again.

Although cold is one of the most widely used phytosanitary treatments, it has not been researched to any significant degree for use on cut flowers and foliage, even though many of these commodities tolerate the temperatures and time periods required to kill insects (Hardenburg et al. 1986). Temperatures that cause freezing of commodities that will be further processed, such as fruit pulp for juices, are used as phytosanitary treatments in limited cases. Freezing for about 1 day kills most insects that are not in diapause. Quick freezing at =–15 °C will usually result in quick kill of insects, including those in diapause.

21.3.2.2 Heated Air

The first commercial heated air treatment was done in 1929 during the first Mediterranean fruit fly outbreak in Florida (Hallman and Armstrong 1994). Early treatments were at a relatively low temperature (43.3 °C), humidity near saturation, and for long periods of time (14–16 h). The heated air was circulated in a room stacked with citrus fruits in field boxes. Subsequent research showed that many fruits could tolerate higher temperatures, and treatments today use air temperatures as high as 52 °C. Higher temperature results in shorter treatment time periods; some current heated air treatments can be done in as little as 3 h.

Reducing the humidity was shown to decrease damage to fruits in many instances, by keeping the fruit from getting wet due to condensation from the hot, moist air contacting the cooler fruit (Jones et al. 1939). Wetting the fruit was thought to restrict fruit respiration and promote decay. In many of today's heated air treatments, the dew point is kept below the surface temperature of the fruit to prevent condensation on the fruit. In the last 20 years, heated air treatments have been modified by forcing air through a fruit load to achieve faster and more uniform temperatures, rather than having circulating air in a treatment chamber gradually and slowly penetrating to the most protected fruits.

Heated air treatments are one of the most challenging groups of phytosanitary treatments to manage because many variables may affect efficacy and damage to treated commodities. Speed of treatment is dependent on the temperature, moisture content of the air, size of treatment chamber, air speed and flow-through load, size density, arrangement of individual commodities, and packaging. The speed of treatment may affect efficacy and damage to the commodity. A slower speed might allow for pests to accommodate to the raised temperature through heat-shock proteins (Denlinger and Yocum 1998), as well as produce less damage to treated commodities (McGuire 1991).

In general, heated air treatments are not tolerated well by temperate fruits, such as apples, pears, peaches, and plums. They are currently used to treat some tropical fruits, such as papaya and mango, imported by Japan, and for some papaya shipped from Hawaii to the continental US. A heated air treatment has been used to ship citrus fruits from Mexico to the US, although quality problems have been reported. Dry air treatments up to 100 °C are used to treat meal, grain, straw, and dried plants. Steam treatments are used to disinfest rice straw and packaging materials of fungal spores.

21.3.2.3 Hydrogen Cyanide Fumigation

Hydrogen cyanide (HCN) is one of the oldest fumigants; as early as the 1870s, it was used to fumigate museum specimens for control of dermestid beetles and other insects. It was used as a phytosanitary treatment for cut flowers, dormant nursery stock, and dried plant products until methyl bromide replaced it. Compared with methyl bromide and other fumigants, HCN has several disadvantages: it reacts with many substances, such as paper, paint, and oils, it is extremely soluble in water, making it readily absorbed and retained by moist commodities, its low vapor pressure makes it difficult to use in large areas, and it has a short shelf life, about 6 months in the cylinder. Uncertainty about the future of methyl bromide has resulted in some research being refocused on HCN for insects on cut flowers and foliage (Hansen et al. 1991a, 1991b; Weller and Graver 1998).

21.3.2.4 Methyl Bromide Fumigation

Fumigation with methyl bromide has been one of the major phytosanitary treatments since its development in the 1950s. Its importance increased in the 1980s when it was used to replace some of the uses of ethylene dibromide fumigation, which was banned as a probable carcinogen and mutagen. It was generally not as favorable as ethylene dibromide, requiring often a doubling of doses to achieve the same effect (up to 50 g m^{-3} for some applications), and sometimes resulting in damage to fresh commodities. For example, methyl bromide cannot be used on mangoes or papayas without the fruits suffering considerable damage; ethylene dibromide was the fumigant of choice for these two fruits.

Methyl bromide fumigation is relatively cheap, the chemical being only a fraction of the cost of application, and can be done in fairly simple facilities. The main consideration is that fumigation chambers do not measurably leak. Treatment times are short, most being done in 0.5–2 h. Chief disadvantages are that a number of tropical fruits do not tolerate the treatment, and organic shippers will not accept it.

Methyl bromide has been implicated as a significant stratospheric ozone-depleting substance, and is regulated under the Montreal Protocol. Post-harvest phytosanitary uses have been indefinitely exempted from restrictions. However, the price of the chemical has risen several-fold since regulation was initiated, and there is no guarantee that phytosanitary uses will be exempted forever. Users are researching alternatives, and funding has been directed toward that effort, leading to other treatment advances discussed in this chapter.

Methyl bromide is a versatile treatment when combined with other treatments, such as preceding or following cold (FAO 1984). The reasoning for this is that certain commodities may not tolerate either treatment alone to the degree necessary for complete control, but will tolerate reduced doses of both treatments applied sequentially. Another reason would be to apply a shorter than efficacious cold treatment in ship transit following a reduced methyl bromide fumigation when the transit time is insufficient for a full cold treatment. The main disadvantage is the complication in having to do two separate treatments.

21.3.2.5 Phosphine Fumigation

Phosphine has typically not been used on living plants or parts thereof, such as fresh fruits and vegetables, because of the damage it causes to these. Another disadvantage is that phosphine requires several days to achieve the complete kill necessary for phytosanitary purposes. The potential use of phosphine broadened with its formulation as a gas ($ECO_2FUME^{®}$) containing 2±0.2 % phosphine, with the remainder carbon dioxide. It shows promise for use on fresh produce, and has been registered as an interstate phytosanitary treatment for cut flowers in Australia. Williams et al. (2000) concluded that $ECO_2FUME^{®}$ fumigation (48 h) would be efficacious against fruit fly larvae and eggs in citrus fruits.

21.3.2.6 Sulfuryl Fluoride Fumigation

Sulfuryl fluoride penetrates wood more easily than does any other available fumigant, and is extensively used to kill termites and other wood-boring insects. It is approved to disinfest non-food items of ticks, and is registered for a number of food items in the US, such as meat, cheese, coconut, cottonseed, peanut, ginger, and legumes (EPA 2005). Sulfuryl fluoride did not control California red scale, *Aonidiella aurantii*, on lemons at a dose level tolerated by the fruit (Aung et al. 2001). It showed promise at killing larvae of Lepidoptera in walnuts and almonds, especially at low atmospheric pressure (Zettler and Leesch 2000).

21.3.2.7 Hot Water Immersion

Immersion of mangoes in 46.1 °C water for 65–110 min, depending on shape and weight, is used to disinfest nearly all mangoes imported by the US from Latin America of tephritid fruit fly eggs and larvae. This treatment replaced ethylene dibromide when it was banned in the mid-1980s. It is also approved for disinfesting longans and lychees of tephritids from Hawaii for export to the continental US.

Fresh commodities often tolerate heated air better than heated water (Shellie and Mangan 2000). Damage in heated water can be alleviated in some cases by gradual heating (McGuire 1991), or preconditioning the fruit with sublethal heating before the actual heat treatment (Jacobi et al. 2001). A unique problem with heated water treatments is that people have actually died from eating hot-water treated fruit! Mangoes absorb a small amount of water through the stem end upon heating, and a widely dispersed case of salmonella poisoning in the US in late 1999 was traced to an unsanitary hot water immersion facility. This problem should be avoidable with proper levels of chlorination of both the water used for heating, and any water used for cooling fruit after heating.

21.3.2.8 Pesticidal Dips or Sprays

Insecticides are often used as phytosanitary treatments for items not to be consumed, such as bulbs, seeds, other plant propagative materials, and dry plant material. Some insecticides may be allowed on tobacco; for example, the insect growth regulator, methoprene, considered rather safe for mammals, is permitted by some countries for control of cigarette beetle, *Lasioderma serricorne*, and tobacco moth, *Ephestia elutella*. Many countries do not permit post-harvest insecticide applications to fresh fruits or vegetables. Australia is an exception; fenthion or dimethoate at 400 ppm in water are used for interstate movement of quarantined fruit, but are not used for any fruit exported from that country (Heather 1994).

21.3.2.9 Ionizing Irradiation

Irradiation using an ionizing source to remove electrons from their normal orbits, resulting in free radicals that often recombine in ways different from the original, was postulated as a phytosanitary treatment in the 1920s (Hallman 2001). It was not until 1995 that it began to be used as a commercial phytosanitary treatment on a continuous basis. Unlike all other commercial phytosanitary treatments, irradiation does not provide acute mortality;

Fig. 21.1 Boxes of papayas two boxes deep on a conveyor system to be irradiated by X-rays for phytosanitary control of tephritid fruit flies in Hawaii before export

organisms may be alive for days after treatment. However, they will not complete development nor reproduce, because of the damage done to genetic macromolecules through faulty recombination after ionization.

On a commercial scale, irradiation is used to disinfest produce grown in Hawaii and Florida of a variety of quarantine insects for shipment to other US states (Fig. 21.1). Internationally, it has been used since late 2004 to treat Australian mangoes for shipment to New Zealand. The US has approved a default dose of 150 Gy for all tephritid fruit flies on all hosts, and 400 Gy for all insects except pupae and adults of Lepidoptera (APHIS 2006). Irradiation has great potential, in that it is the most widely tolerated phytosanitary treatment for several uses. For example, very few fruits are known to not tolerate the tephritid default dose of 150 Gy, given a dose uniformity ratio of 2 (Hallman 2001). Irradiation has the advantage of application after packing and palletizing.

21.3.2.10 Miscellaneous Treatments

A number of treatments have been used in unique cases, such as pressure, at times combined with phosphine fumigation, to kill Hessian fly, *Mayetiola destructor*, and cereal leaf beetle, *Oulema melanopus*, in baled hay shipped from the US to Japan and Canada (Yokoyama and Miller 2002, 2003). Cleaning and inspection may be used to remove surface pests. Cherimoya and limes from Chile may be washed and waxed as a phytosanitary treatment for the mite *Brevipalpus chilensis* for export to the US.

21.3.2.11 Researched but not yet Applied Treatments

Considerable research has been carried out on a number of phytosanitary treatment possibilities that have not yet reached commercial application (Hallman 2002). Atmospheres with very little oxygen and high levels of carbon dioxide have been used successfully to disinfest commodities of pests. One successful trial shipment of asparagus, held for 4.5 days at 0–1 °C and 60 % CO_2 to disinfest it of aphids and thrips, was exported from New Zealand to Japan (Carpenter and Potter 1994).

Low oxygen/high carbon dioxide treatments combine well with elevated temperatures. For example, diapausing spider mites, *Tetranychus urticae*, were killed in one-seventh of the time at 40 °C when treated in an atmosphere of 0.4 % O_2 and 20 % CO_2, compared with ambient atmosphere (Whiting and van den Heuvel 1995). Low oxygen/high carbon dioxide phytosanitary treatment research is frequently carried out at low temperatures, and often it is the low temperature alone, with apparently no benefit from the modified atmosphere, that causes pest mortality (Hallman 1994).

Radiofrequency heating has been researched as a phytosanitary treatment since the late 1920s, with no commercial application yet (Hallman 2002). The potential benefits are that radiofrequency could heat a commodity uniformly, as opposed to heated air and water that heat from the outside in, and that radiofrequency could heat commodities rapidly, in a matter of seconds and as part of a conveyor line, rather than during 1 to many hours per batch load, as is done with heated air and water. In practice, radiofrequency heating has had problems with uniformity when heating fresh produce. The drier nature of nuts and dried fruit might offer an advantage to radiofrequency heating, as this type of heating selectively heats water; thus, moist insects inside of dry products might be selectively heated (Wang et al. 2001).

21.4 Future Challenges

To satisfy the demand for world trade in products quarantined because of invasive species as well as reduce the temptations to smuggle these products, improvements in the efficacy, ease of application, and cost of phytosanitary methods should be sought. Because methyl bromide fumigation is a key phytosanitary treatment that is being more heavily regulated because of its role as a stratospheric ozone-depleting substance, alternatives to all of its phytosanitary uses should be developed. Furthermore, it is desirable to have more than one solution to any quarantine, in case some are rejected because of questions about efficacy, an unacceptable side effect, excessive cost, unavailability, or another unforeseen problem. To avoid creating unnecessary barriers to trade, researchers should be careful when constructing host lists of pests, and give extent of infestation and precise conditions under which a commodity has been found to be a host.

One way to reduce the risk of invasive species through trade is to reduce the amount of imported products, especially fresh ones that can support a greater quantity and variety of invasive species. Although it is argued that present regulatory controls preclude international trade from being a significant source of invasive species, it is not known by what route the great majority of invasive species became established, and the massive amount and variety of trade provide abundant avenues for invasion. Voucher specimens of pests used in host determinations and development of phytosanitary measures should be deposited in a curated collection for future questions on identification.

The energy banker and author Matthew Simmons, who predicts very high oil prices in the near future, says that off-season trade in fresh commodities "will become a thing of the past" (Ward 2006). He argues that this would be favorable because fresh produce that is shipped long distances is not of good quality, and that increased supply and canning of locally grown produce, picked at its optimum in flavor and nutrition, can more favorably fulfill our nutritional needs than is the case for fresh imported produce that is often picked well before its optimum stage of ripening in order to survive the long trip to market (Goldman et al. 1999). Higher transportation costs will make local farming more competitive by reducing competition from cheap imported food, and could paradoxically help the rural poor of the world, most of who are involved in farming (Lindskog 2005).

References

APHIS (2004) Mexican avocado import program. Federal Register 69:69747–69774

APHIS (2006) Treatments for fruits and vegetables. Federal Register 71:4451–4464

Armstrong JW (1986) Pest organism response to potential quarantine treatments. In: Proc Regional Conf Plant Quarantine Support for Agricultural Development, 1985, Serdang, Malaysia, pp 25–31

Armstrong JW (1994) Commodity resistance to infestation by quarantine pests. In: Sharp JL, Hallman GJ (eds) Quarantine treatments for pests of food plants. Westview Press, Boulder, pp 199–211

Aung LH, Leesch JG, Jenner JF, Grafton-Cardwell EE (2001) Effects of carbonyl sulfide, methyl iodide, and sulfuryl fluoride on fruit toxicity and insect mortality. Ann Appl Biol 139:93–100

Carpenter A, Potter M (1994) Controlled atmospheres. In: Sharp JL, Hallman GJ (eds) Quarantine treatments for pests of food plants. Westview Press, Boulder, pp 171–198
Coley PD (1999) Hungry herbivores seek a warmer world. Science 284:2098–2099
Cowley JM, Baker RT, Englberger KG, Langi TG (1991) Methyl bromide fumigation of Tongan watermelons against *Bactrocera xanthodes* (Diptera: Tephritidae) and analysis of quarantine security. J Econ Entomol 84:1763–1767
Cowley JM, Baker RT, Harte DS (1992) Definition and determination of host status for multivoltine fruit fly (Diptera: Tephritidae) species. J Econ Entomol 85:312–317
Denlinger DL, Yocum GD (1998) Physiology of heat sensitivity. In: Hallman GJ, Denlinger DL (eds) Temperature sensitivity in insects and application in integrated pest management. Westview Press, Boulder, pp 7–53
EPA (2005) Sulfuryl fluoride; pesticide tolerance. Federal Register 70:40899–40908
FAO (1984) International plant quarantine treatment manual. FAO, Rome, Plant Production and Protection Paper no 50
FAO (1996) Guidelines for pest risk analysis. FAO, Rome, ISPM Publ no 2
FAO (2002) Guidelines for regulating wood packaging materials in international trade. FAO, Rome, ISPM Publ no 15
FAO (2003) Pest risk analysis for quarantine pests including analysis of environmental risks. FAO, Rome, ISPM Publ no 11
FAO (2004) Glossary of phytosanitary terms. FAO, Rome, ISPM Publ no 5
Follett PA, McQuate GT (2001) Accelerated development of quarantine treatments for insects on poor hosts. J Econ Entomol 94:1005–1011
Goldman IL, Kader AA, Heintz C (1999) Influence of production, handling and storage on phytonutrient content of foods. Nutrition Rev 57:S46–S52
Hallman GJ (1994) Controlled atmospheres. In: Paull RE, Armstrong JW (eds) Insect pests and fresh horticultural products: treatments and responses. CAB International, Wallingford, pp 121–136
Hallman GJ (2001) Irradiation as a quarantine treatment. In: Molins R (ed) Food irradiation principles and applications. Wiley, New York, pp 113–130
Hallman GJ (2002) Quarantine treatments: achieving market access for agricultural commodities in the presence of invasive pests. In: Hallman GJ, Schwalbe CP (eds) Invasive arthropods in agriculture: problems and solutions. Science Publishers, Enfield, pp 271–291
Hallman GJ, Armstrong JW (1994) Heated air treatments. In: Sharp JL, Hallman GJ (eds) Quarantine treatments for pests of food plants. Westview Press, Boulder, pp 149–163
Hansen JD, Chan HT Jr, Hara AH, Tenbrink VL (1991a) Phytotoxic reaction of Hawaiian cut flowers and foliage to hydrogen cyanide fumigation. Hort Sci 26:53–56
Hansen JD, Hara AH, Chan HT Jr, Tenbrink VL (1991b) Efficacy of hydrogen cyanide fumigation as a treatment for pests of Hawaiian cut flowers and foliage after harvest. J Econ Entomol 84:532–536
Hardenburg RE, Watada AE, Wang CY (1986) The commercial storage of fruits, vegetables, and florist and nursery stocks. Frederick, MD, USDA-ARS Handb no 66
Heather NW (1994) Pesticide quarantine treatments. In: Sharp JL, Hallman GJ (eds) Quarantine treatments for pests of food plants. Westview Press, Boulder, pp 89–100
Hennessey KK, Baranowski RM, Sharp JL (1992) Absence of natural infestation of Caribbean fruit fly (Diptera: Tephritidae) from commercial Florida 'Tahiti' lime fruits. J Econ Entomol 85:1843–1845
Jacobi KK, MacRae EA, Hetherington SE (2001) Loss of heat tolerance in 'Kensington' mango fruit following heat treatment. Postharvest Biol Technol 21:321–330
Jang EB, Moffitt HR (1994) Systems approaches to achieving quarantine security. In: Sharp JL, Hallman GJ (eds) Quarantine treatments for pests of food plants. Westview Press, Boulder, pp 225–237

Jones WW, Holzman JJ, Galloway AG (1939) The effect of high-temperature sterilization on the Solo papaya. Hawaii Agric Exp Station Circ no 14

Landolt PJ, Chambers DL, Chew V (1984) Alternative to the use of probit 9 mortality as a criterion for quarantine treatments of fruit fly (Diptera: Tephritidae)-infested fruit. J Econ Entomol 77:285–287

Lindskog P (2005) How to cut world hunger in half. Science 310:1768

McGuire RG (1991) Market quality of grapefruit after heat quarantine treatments. Hort Sci 26:1393–1395

Parker JD, Burkepile DE, Hay ME (2006) Opposing effects of native and exotic herbivores on plant invasions. Science 311:1459–1461

Parmesan C, Ryerholm N, Stefanescu C, Hill JK, Thomas CD, Descimons H, Huntley B, Kaili L, Kullberg J, Thomas JA, Warren M (1999) Poleward shifts in geographical ranges of butterfly species associated with regional warming. Nature 399:579–583

Perrings C, Williamson M, Dalmazzone S (eds) (2000) The economics of biological invasions. Edward Elgar, Cheltenham

Robacker DC, Landolt PJ (2002) Importance and use of attractants. In: Hallman GJ, Schwalbe CP (eds) Invasive arthropods in agriculture: problems and solutions. Science Publishers, Enfield, pp 169–205

Schwalbe CP, Hallman GJ (2002) An appraisal of forces shaping the future of regulatory entomology. In: Hallman GJ, Schwalbe CP (eds) Invasive arthropods in agriculture: problems and solutions. Science Publishers, Enfield, pp 425–433

Shellie KC, Mangan RL (2000) Postharvest disinfestation heat treatments: response of fruit and fruit fly larvae to different heating media. Postharvest Biol Technol 21:51–60

Wang S, Ikediala JN, Tang J, Hansen JD, Mitcham E, Mao R, Swanson B (2001) Radio frequency treatments to control codling moth in in-shell walnuts. Postharvest Biol Technol 22:29–38

Ward S (2006) Twilight for oil? Barron's Magazine 86(1):30–32

Weller GL, van S Graver JE (1998) Cut flower disinfestation: assessment of replacement fumigants for methyl bromide. Postharvest Biol Technol 14:325–333

Whiting DC, van den Heuvel J (1995) Oxygen, carbon dioxide, and temperature effects on mortality responses of diapausing *Tetranychus urticae* (Acari: Tetranychidae). J Econ Entomol 88:331–336

Williams P, Hepworth G, Goubran F, Huhunthan M, Dunn K (2000) Phosphine as a replacement for methyl bromide for postharvest disinfestation of citrus. Postharvest Biol Technol 19:193–199

Wilson EO (1984) Biophilia. Harvard University Press, Cambridge, MA

Yokoyama VY, Miller GT (2002) Bale compression and hydrogen phosphide fumigation to control cereal leaf beetle (Coleoptera: Chrysomelidae) in exported rye straw. J Econ Entomol 95:513–519

Yokoyama VY, Miller GT (2003) Quarantine control of Hessian fly (Diptera: Cecidomyiidae) in exported hay: a new treatment for large-size, polypropylene fabric-wrapped bales and a 3-d fumigation for compressed standard bales. J Econ Entomol 96:1340–1344

Zettler JL, Leesch JG (2000) Sulfuryl fluoride: a disinfestation treatment for walnuts and almonds. Methyl Bromide Alternatives 2:5–6

22 Limits and Potentialities of Eradication as a Tool for Addressing Biological Invasions

Piero Genovesi

22.1 Introduction

22.1.1 Definition

Eradication is the complete and permanent removal of all wild populations of an alien plant or animal species from a defined area, by means of a time-limited campaign. This measure is therefore different from control, i.e. the reduction of population density and abundance in order to keep damage at an acceptable level, and containment, aimed at limiting the spread of a species by containing its presence within defined geographical boundaries (Bomford and O'Brien 1995). Following this definition, also the removal of very few individuals is an eradication, if these have the potentialities of reproducing and establishing in the wild (i.e. this does not include the removal of single animal individuals but includes removal of seeds or plant propagules in the wild, or of a few pairs of animals).

Eradication of unwanted alien species is an increasingly important tool for conservation of biological diversity. In fact, although the most effective way for mitigating the impacts caused by biological invasions is the prevention of new unwanted introductions, once prevention has failed and an alien species has invaded a new area, eradication is the best alternative, considering the costs and undesired effects related to permanent control or to a "do-nothing" policy.

This general approach has been identified as the key for action on invasive aliens species by the Convention on Biological Diversity which, with Decision VI/23 on *Alien Species that threaten ecosystems, habitats and species* (adopted at COPVI, The Hague, April 2002), has called parties to adopt a hierarchical approach for addressing biological invasions. Prevention of unwanted introduction of invasive alien species between and within states is the priority. If an

Ecological Studies, Vol. 193
W. Nentwig (Ed.)
Biological Invasions

invasive alien species has been introduced, then early detection and rapid eradication are crucial to prevent its establishment. Only when prevention has failed and an unwanted alien species has established into the wild is eradication the preferred response, when this option is feasible. If eradication does not appear to be feasible, then containment and long-term control measures should be implemented, if appropriate.

22.1.2 History and Recent Developments

With his arrival on earth, *Homo sapiens* has directly caused the extinction of many species, either from localised geographic areas or from the entire biosphere. The history of human-mediated extirpations ranges from mass extinctions of megafauna through overharvesting in the Pleistocene (Lyons et al. 2004) to more recent cases, such as the passenger pigeon *Ectopistes migratorius* which became extinct at the beginning of the last century (Blockstein and Tordoff 1985). Extermination of species has been in some cases the result of a clear commitment and policy by man, as in the case of all large carnivore species eradicated from western Europe in the 19th century because of predation on livestock, or of eradications carried out for health purposes. For example, smallpox has been successfully extirpated from earth, and mosquito species have been eradicated from many areas of the world to combat malaria (e.g. *Anopholes labranchiae* eradicated from Sardinia between 1946 and 1951; Hall 2004).

We humans have indeed a unique ability to exterminate organisms from all taxonomic groups, even if long-established or inhabiting very large areas. In most cases, this is simply a matter of time and perseverance. The fast-growing number of biological invasions calls upon us now to use this ability to preserve biological diversity, rather than reducing it.

The first eradications of alien species have been carried out for sanitary purposes; for example, *Anopheles gambiae* from over 30,000 km^2 of Brazil in the 1950s, to combat a yellow fever outbreak and prevent a spread of the disease to North America (Davis and Garcia 1989). Eradications carried out for conservation purposes started in the 1930s, and have become a routine management action only in the 1980s. Over 156 eradications have been successfully carried out so far in New Zealand to protect native species, 23 on islands of NW Mexico and 48 on islands of NW Australia. Most eradications have involved terrestrial vertebrates but there have also been many successful campaigns in other taxonomic groups, including freshwater fishes (e.g. Copp et al. 2005) and several terrestrial invertebrates, such as the fruit fly, successfully eradicated from Nauru (Allwood et al. 2002) or the screw-worm (*Cochiomyia hominivorax*) extirpated from the south-eastern United States, Central America and North Africa (Myers et al. 2002). Even some marine organisms have been eradicated (when invasion was still localised), such as a mussel (*Mytilop-*

sis sp.) inadvertently introduced in Cullen Bay (Australia; Bax et al. 2002) and a sabellid polychaete (*Terebrasabella heterouncinata*) successfully removed from a mariculture facility in California (Galil 2002). An eradication of the alien algae *Caulerpa taxifolia* is being completed in the lagoon of Agua Hedionda, in southern California (Anderson 2005). Within the context of the latter case, it should be noted that, despite the eradication of plants being much more challenging than that of animals, many plant eradications have been carried out worldwide. In most cases, these were for infestations which occurred in isolated areas or which were still in an early stage (Timmins and Braithwaite 2001; Rejmánek and Pitcairn 2002).

The recent increase in the number (Figs. 22.1, 22.2) and complexity of eradication projects is due to both an increased awareness of the need to mit-

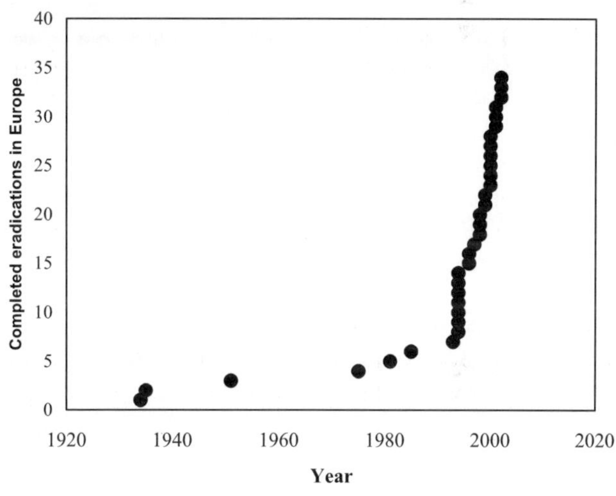

Fig. 22.1 Successfully completed eradications in Europe (Genovesi 2005)

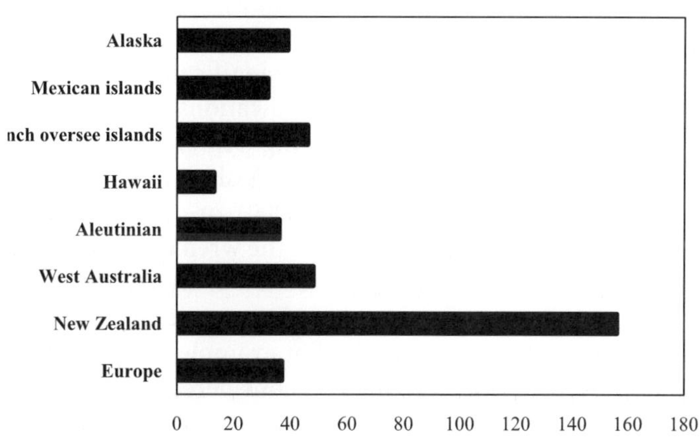

Fig. 22.2 Number of successfully completed eradications on islands (Pascal 1999; Burbridge and Morris 2002; Tershy et al. 2002; Courchamp et al. 2003; Genovesi 2005; D. Veitch, personal communication)

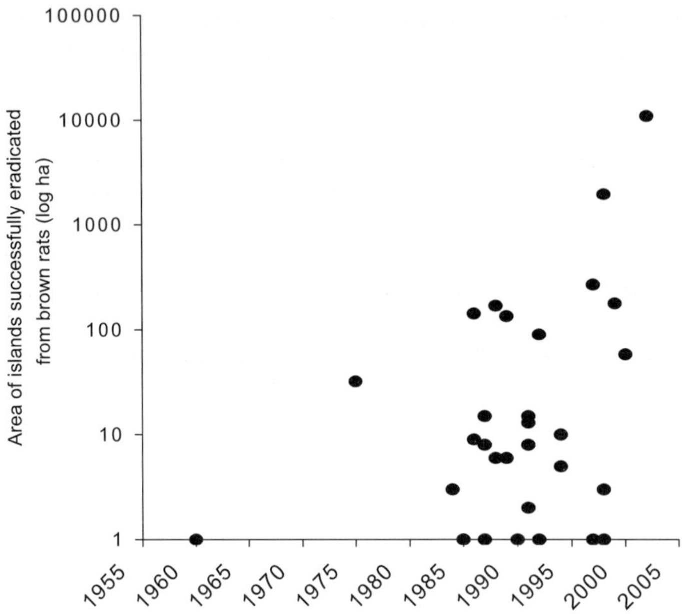

Fig. 22.3 Eradications of *Rattus rattus* in New Zealand: increasing size of islands successfully eradicated (modified from Veitch 1995)

igate the impacts of invasive alien species and also significant advances in technical aspects of eradications (Veitch and Clout 2002). For example, the development of second-generation anti-coagulant poisons has made it possible to increase the size of islands from which rats have been successfully eradicated. This was confirmed by the successful removal, completed in 2003, of the Norway rat (*Rattus norvegicus*) from sub-Antarctic Campbell Island (over 11,000 ha), by far the largest area in the world successfully eradicated from rats (Fig. 22.3).

22.1.3 Outcomes

Successful eradications can lead to significant effects in terms of recovery of native species and habitats. Rat eradications from islands have promoted the recovery of many colonial nesting seabirds, including the storm petrel (*Hydrobates pelagicus*) and the Cory's shearwater (*Calonectris diomedea*) in the Mediterranean (Martín et al. 2000), or of terrestrial bird species such as the dunnock (*Prunella modularis*), the wren (*Troglodytes troglodytes*) and the rock pipit (*Anthus petrosus*) (Kerbiriou et al. 2003).

In many cases, the removal of alien species is an essential prerequisite for subsequent recovery programs. For example, the recently completed rat eradication from Campbell Island was considered an essential condition for starting a recovery program for the endemic Campbell Island teal (*Anas nesiotis*), a flightless bird very vulnerable to rat predation. Cat eradication from Long Cay Island (Caicos Bank, British West Indies) was essential for carrying out a recovery program for the highly endangered iguana *Cyclura carinata* (Mitchell et al. 2002).

Eradications bring benefits not only in terms of preservation and recovery of biodiversity but also in economic terms, health and human wellbeing. Before successful eradication, *Anopheles gambiae*, introduced in northeast Brazil, was responsible for an outbreak of malaria which caused over 30,000 casualties. Eradication of three species of introduced fruit flies from Nauru Island allowed people to again eat mangoes and breadfruit, after a decade of collapse in fruit production (Allwood et al. 2002). Eradication of the coypu from East Anglia prevented severe impacts on wetland biological diversity but also severe economic losses (Panzacchi et al. 2006).

22.2 Key Elements of Eradications

22.2.1 Biological Aspects

In general, it is said that an eradication is biologically feasible when all reproductive individuals of the target population can be removed, and risk of reinvasion is zero (or close to zero). As both vulnerability to removal methods and ability to re-invade largely depend on the biological traits of the target species, eradication programs require a good understanding of the species' biology. In particular, the dispersal ability of the target species, its reproductive biology and, more generally, its life history should be carefully taken into account for evaluating the feasibility of eradication (Myers et al. 2000). In general, eradications are more complex when target species have a dormant life stage (e.g. soil seed bank), or for species with high dispersal capacity and reproductive rates. This is the main reason why plants are particularly difficult to eradicate once they are established; furthermore, eradication of plant species commonly requires many years of monitoring before completion can be verified. The complex challenges linked to plant eradications explain why there is, in fact, no known case of successful eradication of a well-established alien plant whereas many plant invasions detected early enough have been successfully eradicated from both island and mainland areas (e.g. Timmins and Braithwaite 2001; Rejmánek and Pitcairn 2002).

Successful eradication requires the complete removal of all reproductive individuals or, at least, the reduction of the population well below a threshold of viability (Liebhold and Bascompte 2003). This is obviously a very challenging aspect of eradication campaigns, as the failure to remove the very last individuals can promote the recovery of the target species, totally undermining all the removal efforts. The ability of the species to establish and expand from a nucleus of very few individuals is thus another aspect to carefully take into account in eradications. For example, eradication of squirrel species is complicated by the ability of many species to establish and expand from a nucleus of only two or three pairs (Bertolino and Genovesi 2005).

Nevertheless, analysing the target species' life history may not be enough, as ecological interactions with other species can play a major role in the final outcomes of the eradication. For example, the eradication of rabbits from two islands of the Kerguelen archipelago led to an overall increase in plant species richness but also in the decline of some native species (Chapuis et al. 2004). Eradication of two introduced brushtail possums and brushtail rock wallabies from Rangitoto and Motutapu islands also resulted in a proliferation of several weed species (Mowbray 2002). In general, when eradications are carried out in areas invaded by more than one alien species, as in the case of islands with domestic cats, rats, and rabbits, eradication of superpredators (cats) may result in an explosion of mesopredators (rats), with the risk of aggravating impacts on native species. Conversely, eradication of rats can cause a shift of the diet of cats towards native species. Multi-species interactions also have effects on the feasibility of eradication: Courchamp and Sugihara (1999) showed that the removal of alien prey is in some cases an essential condition to successfully eradicate alien predators, because abundant prey populations may substantially limit the efficacy of predator removal efforts. Therefore, eradication planning requires also an understanding of the ecological interaction of the target species with other alien and native species and, in some cases, concurrent control of more than one alien species can be the best alternative.

22.2.2 Lag Phase

Biological invasions are often characterized by a lag phase followed by a rapid expansion (Williamson 1996), and this common pattern calls for starting eradications at the earliest possible stage, when the chances of successfully removing the infestation are highest. Prompt removal has been the key to success in the few cases of eradications of marine organisms attempted to date as well as eradications of freshwater fishes (e.g. Jackson et al. 2004), of the Canadian beaver from France (Rouland 1995) and of alien plants (Rejmánek and Pitcairn 2002). On the contrary, when the start of control activities is delayed, eradication often becomes impractical (e.g. the American grey squirrel; Bertolino and Genovesi 2003).

The case of *Caulerpa taxifolia* is paradigmatic of the need to adopt a precautionary approach when addressing alien species. This alien alga was accidentally detected in 1984 in France when the invaded area covered only 1 m^2; by 1989, this had increased to 3 ha, and to 31 ha by 1991. Response was delayed by an academic debate within the French and the international academic world, concerning the origin of the alga, its impact and potential long-term effects. It was only in 1995 that a first formal recommendation was approved, calling affected states to control the proliferation of the alga. This decision arrived far too late because the so-called killer alga had by then expanded its range to many areas of the Mediterranean basin, making its eradication an unrealistic option (Meinesz 1999). The reaction to an analogous infestation by this alga in California was very different – in this case, containment and treatments of the infestation by *C. taxifolia* were initiated only 17 days after discovery, based on timely identification, proactive approach by competent authorities, availability of emergency funds, and involvement of local diver crews (Anderson 2005).

Prompt eradication is usually cost-beneficial even when costs are high; for example, the prompt eradication of Himalayan porcupines from Devon required ca. € 230,000 for the removal of only 12 animals but likely prevented much more severe economic losses in the long term, considering the potential impact of the species on crops (Smallshire and Davey 1989). The duration of the lag phase can vary strongly from case to case, also with regard to the potential increase of the species and its ability to spread and, consequently, the period after which eradication becomes unpractical varies, too. For example, the eradication of the Canadian beaver *Castor canadensis* from a river in France was carried out 9–10 years after its introduction into the wild, when the population was still very localised (24 individuals; Rouland 1985).

22.2.3 Removal Methods

Removal methods applied to eradications can be very diverse, including hand removal of plants, mechanical means such as traps, shooting, and the application of poisons and toxicants against plants. Also, biological control agents and introduced pathogens have been applied to eradications, although in general pathogens can only reduce population abundance, and total eradication requires the integrated application of other techniques (Courchamp and Sugihara 1999). More recently, the use of engineered viruses has been investigated, although this technique has yet not been applied to any actual eradication campaign and remains in an experimental phase.

The integration of different methods has sometimes proved to be a very effective strategy, and should be considered when planning an eradication. In domestic cat eradications, for example, toxins and biological controls are considered more effective at an early stage of the campaign whereas, once the population has been reduced, hunting and trapping are usually necessary to remove the remaining animals (Nogales et al. 2004).

Removal means, based either on single methods or integrated, should have the potentiality to affect all individuals of the target species. Therefore, the efficacy of removal techniques strongly depends on the species' biology; for example, the eradication of flies from Nauru Island was possible because the target species could be sterilized by irradiation.

A particularly complex aspect to take into account when implementing an eradication is that removal methods need to be effective also when the density of the target species decreases to a very low level. Predicting the effort required to complete an eradication can thus be difficult, as the removal of the very last individuals can require very significant and scarcely predictable efforts and resources.

Impact of control on non-target species is often a major concern, and the selection of removal methods should thus be carefully evaluated when planning an eradication campaign. This risk is evident when using toxicants or poison baits. In the case of rodent eradications, for example, the numerous successful programs completed in New Zealand have been facilitated by the absence of any native rodent species in that country, thereby enabling the use of aerial poison baiting. The situation is different in other regions of the world, where islands are often inhabited by both native and alien rodents. In these cases, eradications must often be carried out with the use of selective baiting stations, which prevent the access to bait for non-target species. For example, in planning the eradication of the house mice on Thevenard Island in Western Australia, where also the native short-tailed mouse *Leggadina lakedownensis* is present, a tailored poison-delivering station was developed in which the small entrance hole could be used only by the smaller alien species, access for young native individuals being prevented through a careful timing of baiting (Moro 2001). Poisoning can have secondary effects on scavengers or non-target species (e.g. Howald et al. 1999) which, in some cases, can be prevented by carefully selecting bait dispensers, the time of year of application, and the attractants used.

In other cases, it may be impossible to carry out an eradication without impacting on non-target species. One notable example is the rat eradication program in the Channel Islands National Park (California), where several endemic subspecies of *Peromyscus maniculatus* are also present. To avoid impacting the native taxa, the eradication plan included the translocation of *P. maniculatus* from treatment islands before dispersing poisoned bait, and subsequent reintroduction once the eradication had been completed (Pergams et al. 2000).

There are also cases in which foreseen impacts on non-target species can be considered acceptable. For eradications carried out in freshwater or marine environments, the only available removal method is generally the extensive use of toxicants. For example, when in 1999 an alien *Mytilopsis* species was recorded in the Cullen Bay Marina (Darwin, North-East Australia), in order to prevent severe impacts on aquaculture, fishery, tourism and industrial activities at the site, the local authorities decided to immediately treat the entire bay with chlorine and copper sulphate, managing to successfully eradicate the alien mollusc species. In all, 187 t liquid sodium hypochlorite and 7.5 t copper sulphate were pumped into the area. The decision to proceed with the treatment of the bay was taken considering that, once the eradication was completed, all native species could re-colonize the area from nearby coastal environments (Bax et al. 2002). It is interesting to note that even when using toxicants in marine environments, it may be possible to avoid the large-scale diffusion of chemicals. In the eradication of *C. taxifolia* from California (cf. above), chlorine was applied by using underwater tarps, anchored and sealed to the bottom, thereby much reducing the diffusion of the chemical into these waters.

Aside from environmental risks, it should be noted that the use of toxicants can also affect human wellbeing and health, in some cases causing serious concern to local residents. For the eradication of malaria from Sardinia, some 10,000 t of DDT mixture were doused on the island in 5 years, associated with risks to the local population and livestock. Still, the campaign eradicated this pathology from the island where, in the 1930s, over 70,000 Sardinians suffered from malaria (Hall 2004).

Not only toxicants can cause undesired effects. If not properly planned, also biological control agents or genetically engineered viruses can severely impact non-target species (Chaps. 17 and 23), and require careful risk evaluation before being released in the wild (Cory and Myers 2000).

Considering the risks of undesired effects of removal methods, an adequate monitoring program should always be carried out during and after the eradication, in order to facilitate the prompt detection of any impacts on non-target species, to assess the achievement of objectives, and to enable rapid response in case of reinvasion. This latter aspect is a particularly challenging element of eradications, as removing single reinvading individuals can be disproportionately difficult (Russell et al. 2005).

22.2.4 Costs

The economics of eradications has scarcely been investigated so far, although the cost/benefit ratio of removal campaigns is indeed a critical element for defining future policies on invasive alien species. In fact, in order for eradications to progress from an anecdotic to a routine management tool, it would be necessary to identify those resources required to ensure the ability of competent authorities to rapidly remove new, unwanted introduced species and to implement large-scale eradication campaigns.

Eradications are often viewed as extremely costly programs and, indeed, many campaigns (albeit not necessarily the most successful ones) have required huge monetary resources. In only 2 years, an attempt to eradicate the medfly from California required (converted into €) over € 80 million (Myers et al. 2000). Coypu eradication in East Anglia cost € 5 million in

11 years (Panzacchi et al. 2006). The Ruddy duck eradication campaign launched in the United Kingdom is predicted to cost € 5–13 million. Even small-scale eradications can sometimes be very costly: for example, the removal of only twelve Himalayan porcupines from Devon cost a staggering € 230,000.

One of the problems in assessing how much eradications have cost on average is that the available literature (either scientific or "grey") often does not report such data, and the results of prompt eradication projects (cf. removals carried out in the early stages of invasions) are often not published at all. If it is complex to assess costs, it is even more difficult to compare costs with benefits of eradications, because these depend on parameters which have very high levels of uncertainty, such as the probability of the target species, if not removed, to establish, expand and cause damage. The eradication of the Canadian beaver (*Castor canadensis*) from France required only one operator working for a limited amount of time but likely prevented very costly impacts if the species had been allowed to expand (Rouland 1985). In an attempt to compare costs of eradication vs. permanent control, Panzacchi and co-authors (2006) showed that the successful eradication of the coypu from East Anglia, costing about € 5 million in 11 years, may have prevented much more severe economic impacts in the long term, considering that permanent control of the species in Italy causes losses of over € 3.4 million per year and that future annual costs are predicted to exceed € 12 million (see also Chaps. 18 and 19).

In the case of the invading alga *C. taxifolia*, successful eradication from California required (converted into €) over € 2.5 million in 3 years, and additional costs will be necessary for medium-term monitoring. In Europe, when the species was initially detected, it could have been removed within a few days of work (cf. above) – nowadays, it is widespread, and permanent control in many areas of the Mediterranean basin is costing huge amounts of money. In the case of the black-striped mussel from Cullen Bay in Australia, the decision of eradicating the invasive – at a cost of (converted into €) over € 1.3 million "only" – was taken because of the risk that an expansion of the alien mollusc could have impacted the local € 24 million pearl fishery.

More comprehensive economic assessments of eradications are evidently needed, in order to provide state governments with critical data for revising their policies on the issue, and to assist the authorities competent of taking decision about when to start an eradication (see also Chaps. 18 and 19).

22.2.5 Legal and Organizational Constraints

Invasive alien species are a cross-cutting issue, involving many different aspects (such as agriculture, forestry, horticulture, aquaculture and hunting), and implementation of eradications is often regulated by different laws. For example, a common legal constraint to eradication is that several national and supra-national legislations automatically protect alien species, and often do not explicitly include eradication as a management tool (Shine et al. 2000).

This complex situation causes an unclear repartition of roles and responsibilities in eradication, and equivocal decision and authorisation processes. As a result of these legal and organizational constraints, eradications are often delayed for too long or, in many cases, never even start. For example, an attempt to eradicate the American grey squirrel from Italy failed also because of the unclear legal status of alien species under the Italian legal framework, and the inadequate repartition of responsibilities between the national and local institutions (Genovesi and Bertolino 2001).

One of the reasons for the successful eradication of *C. taxifolia* from California (cf. above) is the extraordinary commitment by all competent authorities in the invaded area. These formed a contingency body and ensured adequate powers and funds to the program, thereby enabling a prompt and effective response to the invasion. Similarly, the eradication of the coypu from East Anglia was also made possible by the strong commitment and the relevant resources mobilised by the decision makers.

More generally, eradications should be integrated into national strategies, and particular attention should be paid to the coordination of all involved departments and agencies. For example, the many eradications carried out in New Zealand are also the result of a coordinated policy, with a clear repartition of roles and authorisation process under the Hazardous Substances and New Organisms Act of 1996. The successful eradication of over 53 infestations of 16 weed species in California has also been made possible by the coordinated policy to prevent and control weeds, formalised in 2005 through the adoption of a Noxious and Invasive Weed Action Plan (Rejmánek and Pitcairn 2002; Shoenig 2005).

In order to enable the competent agencies to carry out eradications and, more generally, to mitigate the impacts caused by invasive alien species, state authorities should review national institutional and legal frameworks, and develop coherent, comprehensive and consistent policies on biological invasions. In particular, legal constraints to the application of the necessary mitigation measures should be removed, legal tools should adopt terminology consistent with international definitions, and the competent authorities should be equipped with adequate powers to take the appropriate actions (Shine et al. 2000; Genovesi and Shine 2004). Establishing a coordination body, with access to adequate contingency funds, can be a critical element for resolving the institutional fragmentation of competencies on the different issues related to eradications.

22.2.6 Human Dimensions

Biological invasions are a product of human action, and human dimensions are thus critical for effectively addressing the threats posed by invasive alien species (McNeely 2001). An analysis of the eradication programs so far attempted – or, even more important, non-attempted – shows how public support is essential for successful eradication projects, and how opposition to the control methods – or, more generally, to removing an alien population – has been the major factor obstructing the implementation of many eradication projects. For example, the eradication of the Barbary sheep from the Canary Islands or of the Tahr from New Zealand (Bomford and O'Brien 1995) and South Africa have been strongly opposed by hunters and the general public, who fought to maintain an important game or ornamental species. The eradication of the American grey squirrel from Italy did not fail for technical reasons but rather because of opposition from animal rights groups (Genovesi and Bertolino 2001). On the other hand, one of the main reasons for the success in eradicating *C. taxifolia* from California was the involvement of local key stakeholders.

In planning eradications, particular attention should be given to the conflicts which could arise in implementing the projects, and methods for managing these conflicts should be identified and used. Social science has shown us there can be different types of conflicts, including cognitive ones (cf. different knowledge leads to different conclusions), values conflicts (e.g. different importance given by different societal groups to the protection of biodiversity with respect to animal welfare), conflicts caused by different perceptions of the cost/benefits of the project and, last but not least, conflicts caused by a distrust of the authorities involved in the program (Bath 1999; see also Chaps. 19 and 20).

Before attempting an eradication, detailed analysis of the human dimensions involved in the presence and removal of the target species should be carried out, and key beliefs of all potentially affected societal sectors should be considered. If the results of these analyses show that there are gaps of information which can cause an opposition to the program, then information campaigns aimed at bridging these specific gaps should be launched. Also, when presenting eradication campaigns, it is important to emphasize the recovery scope of the programs, rather than simply focusing on the removal of an alien species – the language used to communicate this should be carefully chosen accordingly (Larson 2005).

All sectors directly affected by the target species should be involved in the decision-making process of how to manage the problem, and this at the earliest possible stage. For example, the cat eradication from Isla Isabel (Pacific coast of Mexico) was made possible by the involvement of hundreds of fishermen, also through a 2.5-year education program. The fishermen voluntarily not only helped in removing the cats and in the post-eradication monitoring but, more importantly, their involvement was critical for preventing further releases of domestic cats (Rodriguez et al. 2006).

Prompt, transparent and credible public information should be ensured. This is important not only for the eradication being planned but also to demonstrate the credibility of the responsible authorities, thereby creating a positive basis for any future program to be developed. In the Hebridean mink project, aimed at eradicating the American mink from some of the Western Islands (Scotland), a bulletin, both in Gaelic and English, is regularly and widely circulated to both experts and the general public, informing on all updates of the project, including the numbers of trapped minks and trapping areas.

22.3 Management Implications

22.3.1 How to Plan an Eradication

The many successful eradications carried out in recent years, and the increasing technical challenges overcome in terms of target taxa and eradication areas clearly indicate that, with proper planning, substantial funds, and adequate political and social support, it is possible to successfully eradicate many alien animal and plant species in all kinds of environment, also from non-isolated areas. However, eradication remains a challenging management option and should always be based on a rigorous assessment of feasibility, and a careful planning of key biological, technical, organizational, sociological and economic elements.

Many lessons can also be drawn from the numerous failures recorded to date, in particular the many cases of unwanted introduced species which could easily have been removed with only limited efforts but where this failed because of ignorance, scarce awareness by decision makers, gaps in the legal framework or authorisation process, or because of social opposition. Here, a synthetic overview of technical guidelines for increasing the ability of states, public agencies or others to plan, implement and evaluate eradication projects is provided.

22.3.2 Rapid Response to New Invasions

Prompt removal of newly detected invaders – before they become established – is the best option for preventing any future impact to biological diversity and human wellbeing. Considering the difficulty in predicting the chances of establishment and impacts of the alien species, eradication should not be delayed by lengthy assessments of such aspects – rather, as soon as a new incursion of a non-authorised alien species is detected, the eradication should be promptly started.

To facilitate prompt reaction to invasions, it is important to increase the ability to detect new alien species through an early warning system, focused on most harmful species and most vulnerable areas. Early warning systems, focusing especially on key areas, should be set up, also based on lists of most dangerous alien species. Also, the competent authorities should establish contingency plans for eradicating specific taxa or groups with similar characteristics (e.g. plants, invertebrates, marine organisms, freshwater organisms, freshwater fishes, reptiles, amphibians, birds, small mammals, large mammals). Information management is particularly critical for reducing the time lag before a new invasion is detected and, for this reason, it is important to set up mechanisms for collecting and circulating information, including identification keys for different taxonomic groups but also references to experts for the various taxa, control methods, etc.

The authorisation process should be streamlined, in order to allow rapid response; where urgent eradication action is needed, the use of emergency orders should be considered. Competent authorities should be equipped in advance with powers to take appropriate mitigation measures. The founding of a national coordination body should be considered, and adequate funds and equipment for rapid response to new invasions should be secured. Relevant staff should be trained to use the eradication methods (see also Chap. 21).

22.3.3 Planning the Eradication of Established Populations

Eradications of long-established alien species require proper planning, based on an assessment of key biological, technical and economical aspects, and the involved human dimensions. In many cases, public information and awareness are critical, and the development of specific education programs or public awareness campaigns should be considered. Participation of local communities should be encouraged, and relevant societal sectors (hunters, NGOs, foresters, landowners, fishermen, etc.) should be engaged in the eradication programs whenever possible.

22.3.4 Legal-Organizational Aspects

National legislation and local regulatory tools should be reviewed to ensure that the legal status of alien species is compatible with control measures, and that the roles and competencies of the relevant authorities are clearly defined. In order to improve coordination and reduce the time for initiating the response to an invasion, it can be very useful to establish an advisory committee, involving all the competent public agencies and the relevant private organizations.

22.3.5 Removal Methods

Removal methods, either single or integrated, should be selected primarily on the basis of their efficacy but also taking into account the selectivity and risks of causing unnecessary stress and pain to target animal species. To avoid undesired effects of eradications, a feasibility study should assess the ecology of the species to be removed, ecological interactions with native and any other alien species present in the area, and the functional role of the target alien species.

22.3.6 Eradication vs. Control

Compared with control or geographic containments, and as a general rule, eradication should be considered a better alternative because it prevents potential future impacts caused by the introduced species, and allows us to avoid permanent costs and undesired effects of removal methods. However, eradication is not always a practicable alternative and, thus, it is important to define explicit criteria in a decision-making process aimed at selecting between either eradication, sustained harvest, geographic containment or a "do-nothing" alternative. Decisions should take into account the success probability of eradication, required effort, impacts and economic costs of the invasion, and monitoring costs.

22.3.7 Monitoring

Monitoring during and after the eradication is essential to detect undesired effects of removal methods, assess results of eradication, detect cases of re-invasion, and collect information useful for preventing new invasions. Adequate funds should be secured for monitoring, which should be based on the best available scientific and other expertise. It may be useful that completion of eradication be verified by an independent body, not involved in the eradication. Eradication programs should be periodically reviewed on the basis of monitoring.

Acknowledgements. Funding of the European Union within the FP 6 by the specific targeted research project DAISIE (SSPI-CT-2003-511202) is gratefully acknowledged. Mick Clout, Maj De Poorter, Alan Saunders and Dick Veitch provided data, shared ideas and gave valuable insights on the topics of this chapter.

References

Allwood AJ, Vueti ET, Leblanc L, Bull R (2002) Eradication of introduced *Batrocera* (Diptera: Tephritidae) in Nauru using male annihilation and protein bait application techniques. In: Veitch CR, Clout MN (eds) Turning the tide: the eradication of invasive species. IUCN, Gland, pp 19–25

Anderson LWJ (2005) California's reaction to *Caulerpa taxifolia*: a model for invasive species rapid response. Biol Invasions 7:1003–1016

Bath A (1999) Human dimension in wildlife management – gaining public acceptance. Environ Encounters 41:121–125

Bax N, Hayes K, Marshall A, Parry D, Thresher R (2002) Man-made marinas as sheltered islands for alien marine organisms: establishment and eradication of an alien invasive marine species. In: Veitch CR, Clout MN (eds) Turning the tide: the eradication of invasive species. IUCN, Gland, pp 26–39

Bertolino S, Genovesi P (2003) Spread and attempted eradication of the grey squirrel (*Sciurus carolinensis*) in Italy, and consequences for the red squirrel (*Sciurus vulgaris*) in Eurasia. Biol Conserv 109:351–358

Bertolino S, Genovesi P (2005) The application of the European strategy on invasive alien species: an example with introduced squirrels. Hystrix Ital J Mammal 16:59–69

Blockstein DE, Tordoff HB (1985) A contemporary look at the extinction of the passenger pigeon. Am Birds 39:845–852

Bomford M, O'Brien P (1995) Eradication of Australia's vertebrate pests: a feasibility study. In: Grigg GC, Hale PT, Lunney D (eds) Conservation through sustainable use of wildlife. Centre for Conservation Biology, University of Queensland

Burbridge AA, Morris KD (2002) Introduced mammal eradications for nature conservation on Western Australian islands: a review. In: Veitch CR, Clout MN (eds) Turning the tide: the eradication of invasive species. IUCN, Gland, pp 64–70

Chapuis JL, Frenot Y, Lebouvien M (2004) Recovery of native plant communities after eradication of rabbits from the subantarctic Kerguelen Islands, and influence of climate change. Biol Conserv 117:167–179

Copp GH, Wesley KJ, Vilizzi L (2005) Pathways of ornamental and aquarium fish introductions into urban ponds of Epping Forest (London, England): the human vector. J Appl Ichthyol 21:1–12

Cory J, Myers J (2000) Direct and indirect ecological effects of biological control. Trends Ecol Evol 15:137–139

Courchamp F, Sugihara G (1999) Biological control of alien predator populations to protect native island prey species from extinction. Ecol Appl 9:112–123

Courchamp F, Pascal M, Chapuis JL (2003) Mammal invaders on islands, impact, control and control impact. Biol Rev 78:347–383

Davis JR, Garcia R (1989) Malaria mosquito in Brazil. In: Dahlsten DL, Garcia JR (eds) Eradication of exotic pests. Yale University Press, New Haven, pp 274–283

Galil BS (2002) Between serendipity and futility: control and eradication of aquatic invaders. Ballast Water News 11:10–12

Genovesi P (2000) Guidelines for eradication of terrestrial vertebrates: a European contribution to the invasive alien species issue. Council of Europe, T-PVSs65e:1–61

Genovesi P (2005) Eradications of invasive alien species in Europe: a review. Biol Invasions 7:127–133

Genovesi P, Bertolino S (2001) Human dimension aspects in invasive alien species issues: the case of the failure of the grey squirrel eradication project in Italy. In: McNeely JA (ed) The great reshuffling: human dimensions of invasive alien species. IUCN, Gland, pp 113–119

Genovesi P, Shine C (2004) European strategy on invasive alien species. Council of Europe Publishing, Strasbourg, Nature and Environment 137

Hall M (2004) Today Sardinia, tomorrow the world: killing mosquitoes. BardPolitik 5:21–28

Howald GR, Mineau P, Elliott JE, Cheng KM (1999) Brodifacoum poisoning of avian scavengers during rat control on a seabird colony. Ecotoxicology 8:431–447

Jackson JE, Raadik TA, Lintermans M, Hammer M (2004) Alien salmonids in Australia: impediments to effective impact management, and future directions. N Z J Mar Freshwater Res 38:447–455

Kerbiriou C, Pascal M, Le Viol I, Goroche J (2003) Conséquences sur l'avifaune terrestre de l'Ile de Trielen (Réserve naturelle d'Iroise, Bretagne) de l'éradication du rat surmulot (*Rattus norvegicus*). Rev Evol – Terre et Vie 59:319–329

Larson BMH (2005) The war of the roses: demilitarizing invasion biology. Frontiers Ecol Environ 3:495–500

Liebhold AM, Bascompte J (2003) The allee effect, stochastic dynamics and the eradication of alien species. Ecol Lett 6:133–140

Lyons SK, Smith FA, Brown JH (2004) Of mice, mastodons and men: human-mediated extinctions on four continents. Evol Ecol Res 6:339–358

Martín A (2002) Rabbit eradication on the islet of Montaña Clara (Lanzarote, Canary Islands). Proc Workshop Invasive Alien Species on European Islands and Evolutionary Isolated Ecosystems and Group of Experts on Invasive Alien Species. Council of Europe, T-PVS/INF 33:23–25

Martín JL, Thibault JC, Bretagnolle V (2000) Black rats, island characteristics and colonial nesting birds in the Mediterranean: consequences of an ancient introduction. Conserv Biol 14:1452–1466

McNeely JA (2001) The great reshuffling: human dimensions of invasive alien species. IUCN, Gland

Meinesz A (1999) Killer algae. University of Chicago Press, Chicago

Mitchell N, Haeffner R, Veer V, Fulford-Gardner M, Clerveaux W, Veitch CR, Mitchell G (2002) Cat eradication and the restoration of endangered iguanas (*Cyclura carinata*) on Long Cay, Caicos Bank, Turks and Caicos Islands, British West Indies. In: Veitch CR, Clout MN (eds) Turning the tide: the eradication of invasive species. IUCN, Gland, pp 206–212

Moro D (2001) Evaluation and cost benefits of controlling house mice *Mus domesticus* on islands: an example from Thevenard Island, Western Australia. Biol Conserv 99:355–364

Mowbray SC (2002) Eradication of introduced Australian marsupials (brushtail possum and brushtailed rock wallaby) from Rangitoto and Motutapu Islands, New Zealand. In: Veitch CR, Clout MN (eds) Turning the tide: the eradication of invasive species. IUCN, Gland, pp 226–232

Myers JH, Simberloff D, Kuris AM, Carey JR (2000) Eradication revisited: dealing with exotic species. Trends Ecol Evol 15:316–320

Nogales M, Martín A, Tershy BR, Donlan CJ, Veitch D, Puerta N, Wood B, Alonso J (2004) A review of feral cat eradication on islands. Conserv Biol 18:310–319

Panzacchi M, Bertolino S, Cocchi R, Genovesi P (2006) Population control of coypu in Italy vs. eradication in UK: a cost/benefit analysis. Wildlife Biol (in press)
Pascal M (1999) Eradication of mammals introduced in the islands. In: Proc Worksh Control and Eradication of Non-Native Terrestrial Vertebrates. Council of Europe, Environ Encounters 41:31–42
Pergams ORW, Lacy RC, Ashley MV (2000) Conservation and management of Anacapa Island deer mice. Conserv Biol 14:819–832
Rejmánek M, Pitcairn J (2002) When is eradication of exotic pest plants a realistic goal? In: Veitch CR, Clout M (eds) Turning the tide: the eradication of invasive species. IUCN, Gland, pp 249–253
Rodriguez C, Torres R, Drummond H (2006) Eradicating introduced mammals from a forested tropical island. Biol Conserv 130:98–105
Rouland P (1985) Les castors canadiens de la Puisaye. Bull Mens Office Natl Chasse 91:35–40
Russell JC, Towns DT, Anderson SH, Clout MN (2005) Intercepting the first rat ashore. Nature 437:1107
Schoenig S (2005) California noxious and invasive weed action plan. http://www.cdfa.ca.gov/phpps/ipc/noxweedinfo/pdfs/noxious_weed_plan.pdf
Shine C, Williams N, Gündling L (2000) A guide to designing legal and institutional frameworks on alien invasive species. IUCN, Gland
Smallshire D, Davey JW (1989) Feral Himalayan porcupines in Devon. Nature in Devon 10:62–69
Tershy BR, Donlan CJ, Keitt BS, Croll DA, Sanchez JA, Wood B, Hermosillo MA, Howald GR, Biavaschi N (2002) Island conservation in north-west Mexico: a conservation model integrating research, education and exotic mammal eradication. In: Veitch CR, Clout MN (eds) Turning the tide: the eradication of invasive species. IUCN, Gland, pp 293–300
Timmins SM, Braithwaite H (2002) Early detection of invasive weeds on islands. In: Veitch CR, Clout MN (eds) Turning the tide: the eradication of invasive species. IUCN, Gland, pp 311–318
Veitch CR (1995) Habitat repair: a necessary prerequisite to translocation of threatened birds. In: Serena M (ed) Reintroduction biology of Australian and New Zealand fauna. Surrey Beatty, Chipping Norton, pp 97–104
Veitch CR, Clout MN (2002) Turning the tide: the eradication of invasive species. IUCN, Gland
Williamson M (1996) Biological invasion. Chapman & Hall, London

23 Pros and Cons of Biological Control

DIRK BABENDREIER

23.1 Introduction

Biological control involves the deliberate introduction of natural enemies for the control of pest organisms, including insects, weeds and diseases. A general difference exists between augmentative releases where biological control agents are used periodically, i.e. once or several times within a season, and classical biological control where agents are released with the aim of establishment and, ideally, a permanent pest control. Whereas native candidates are generally given preference in augmentative biological control, in some cases exotic species have been used. By contrast, for classical biological control the rule is that exotic natural enemies are introduced for the control of exotic pests. Pest species may either interfere with agricultural production without being invasive per se or may be invasive on a larger scale, thereby threatening ecosystems and natural reserves. Whereas arthropod biological control generally applies to the former, it is in the weed control section that biological control agents are often released to control invasive species. This means that biological control finds itself in the unique position of being both an important strategy for the control of alien invasive species and also a route by which potentially damaging new alien species (i.e. the natural enemies) are themselves introduced and spread (Chap. 2).

This chapter will briefly review the positive aspects of biological control and will highlight a few examples. It will further review negative aspects of biological control introductions. One of the examples where biological control led to detrimental environmental effects was the introduction of the ladybeetle *Harmonia axyridis*, and this case will be outlined in more detail. This example will also be used to explore some of the population biology mechanisms which can contribute to the net effects of introduced natural enemies. Finally, some information on recent developments and improvements in risk assessment of biological control agents is provided.

Ecological Studies, Vol. 193
W. Nentwig (Ed.)
Biological Invasions

23.2 Pros of Biological Control

Since the spectacular success of the vedalia beetle *Rodolia cardinalis* which was released to control a scale insect in California more than 100 years ago, there is hardly any doubt that biological control can be a very effective strategy. One of the most striking advantages is that, if biological control works, then it virtually works forever. This means that it can be an extremely cost-effective pest control method, and the benefits may exceed the initial costs of the projects by several orders of magnitude (Hoddle 2004). A clear advantage of a successful biological control program is the saving of sometimes huge amounts of pesticides, of which many are known to be harmful for numerous non-target insects, vertebrates and even humans. For example, the biological control program against alfalfa weevil conducted in the US reduced pesticide use by 95 % from 1968 to 1983, and is saving farmers more than $ 100 million each year in insecticide and application costs. Similarly, the use of biological control agents against greenhouse pests altogether has saved substantial amounts of pesticides. More than 100 species of beneficial organisms are commercially available for control of nearly all important insect and mite pests. As an example, we can take the parasitoid *Encarsia formosa* which is successfully being used against whiteflies since more than 20 years. Special advantages in augmentative biological control are that agents can be used where pesticides are no longer efficient due to resistance, and also that the growers of vegetables can make use of bumblebees for pollination, which would be prohibited with the use of pesticides.

Probably the most impressive advantage of biological control is that it may be the single one solution for restoration of ecosystems which have been impacted by invasive species. For example, the invasion of purple loosestrife (*Lythrum salicaria*) into North American freshwater wetlands has altered decomposition rates and nutrient cycling, led to reductions in wetland plant diversity, reduced pollination and seed output of the native *Lythrum alatum*, and reduced habitat suitability for several specialized wetland bird species (Blossey et al. 2001). After years of research in Europe, it was determined that potential benefits outweigh risks, and four biocontrol agents were introduced in 1992 and 1994. These species are attacking flowers, leaves and roots, and this combination was predicted to enhance control. At some of the early release sites, the attack by the host-specific insects has resulted in dramatic declines of purple loosestrife and, often, the once monotypic stands of *L. salicaria* are replaced by a diverse wetland plant community (Blossey et al. 2001). There are clearly many more such success stories, of which only a very few have been touched upon here. For a comprehensive review of successes in all disciplines of biological control, I would like to refer to Gurr and Wratten (2000).

23.3 Cons of Biological Control

As mentioned above, biological control can work forever. Nevertheless, a quite common problem is that, in very many cases, biological control simply fails to control the pest sufficiently. In classical biological control of insects, for instance, only about one third of the introduced agents have been able to establish, from which again about one third is able to suppress pest populations. Though success rates are somewhat higher for weed biological control, this means that numerous exotic natural enemies have been added to the native fauna without noticeable benefits.

Despite the fact that both classical biological control and augmentative biological control were regarded safe for much of their history, concerns about detrimental effects of introduced exotic species on the native fauna have been increasingly expressed over the last two decades. This development led to the release of several papers which reviewed the impact of biological control agents on non-target species and, more importantly, reported that many effects may have passed unnoticed because no study had been conducted (Howarth 1991; Simberloff and Stiling 1996; Samways 1997; Lynch et al. 2001).

Potential risks concerning the introduction of exotic biological control agents include those to human health, to the economics and to the environment. No serious health risks are known for any macro-organisms, though some cases of allergy in the mass-production of predatory mites or nematodes may occur. Economic issues are dealt with in Chap. 18 and I thus would like to focus on environmental risks and non-target effects here. Although the term non-target effect is not very well defined, it clearly encompasses a large spectrum varying from very small effects, e.g. 2% parasitization of a parasitoid biological control agent on a non-target insect, to massive effects at the population or even ecosystem level. Until now, there is no general agreement on how to judge the magnitude of non-target effects, and whether these effects can be tolerated or are unacceptable. Clearly, the most serious negative aspect of biological control would be the displacement of non-target species on a large, geographical scale or even globally, and the change of complete ecosystems.

23.3.1 Weed Biological Control

In weed biological control, the most serious concern is that an exotic introduced herbivore would be able to feed on crop plants, thereby becoming a pest itself. As a consequence, the assessment of the candidate's host range became routine already several decades ago, and only herbivores with a narrow host range are considered for release. Compared to agents released against arthropods, agents released for weed control indeed generally have a better safety record, although in some cases non-target effects have materialized. One of the hotly debated cases is the flower head weevil *Rhinocyllus conicus* which was introduced to North America already in 1969, and later also into other countries for control of weedy thistles. This decision was taken at that time by the national authorities despite host range tests indicating that some non-target thistles may be attacked as well to some degree. In fact, feeding on the seeds of some non-target *Cirsium* spp. did occur shortly after introduction, and there is evidence from field experiments that populations of these thistles are decreasing due to lower seed production (Louda 2000; Louda et al. 2003). In addition, Louda (2000) showed that populations of a native floral-feeding tephritid decreased simultaneously, indicating that also indirect effects are associated with the introduced beetle. This case reflects the former lack of concern over non-target effects on non-economic species. More generally, it reflects a different perception of risks and benefits of the authorities at that time, as the predicted feeding on several non-economic plants was disregarded in light of the high pest pressure and the probability of solving this problem.

Another interesting case is the Argentine pyralid moth *Cactoblastis cactorum*. In 1926, this moth was introduced against *Opuntia* spp. in Australia where it is clearly one of the major success stories. In 1957, it was introduced into the Caribbean islands and also reduced populations of the target species. It was considered for introduction into the US but this was not permitted because feeding on native *Opuntia* spp. was suspected. However, *C. cactorum* did accidentally arrive in Florida by the end of the 1960s, and the moth was indeed found to feed on non-target *Opuntia* spp. Included in its diet were several endangered species, which led to increasing efforts to prevent these from going extinct. The most important conclusion here is that even a successful biological control agent with a narrow host range can cause serious non-target effects in specific geographical areas. The case further suggests to include adjacent regions within potential dispersal distance in pre-release risk assessments (Louda and Stiling 2004).

23.3.2 Arthropod Biological Control

In contrast to weed biological control, the potential risks for non-target organisms have only recently received attention and, consequently, most of the relevant studies have been published only within the last 10–15 years (Babendreier et al. 2005). In an attempt to quantify the number of cases where non-target effects have occurred and also the relevance of these effects, Lynch et al. (2001) screened the BIOCAT database. Of the 5,279 classical introductions of insects listed in BIOCAT, 80 were associated with one or more such non-target effect records. However, this includes all kinds of smaller effects such as low parasitism of a non-target species at a single location – indeed, these cases constitute the great majority. By contrast, the evidence for population reduction or extinction is fairly weak in many cases. Those cases where more serious effects were observed have virtually always happened on islands. For instance, the introduction of predatory land snails (especially *Euglandina rosea*) for control of the alien giant African snail into Hawaiian Islands in the 1950s – and later even into other countries – had disastrous consequences for the non-target mollusc fauna (Howarth 1991). There is good evidence that the extirpation of endemic tree snails is caused by this introduced predator. In addition to this, more cases of non-target effects were documented and reviewed by Howarth (1991) and Hoddle (2004). A common feature of these cases is that polyphagous predators are largely responsible for the observed effects, as demonstrated in several projects conducted early in the 20th century. Some of the most serious effects originated from the introduction of vertebrate predators (e.g. mongoose against rats) – conducted not by biological control experts but rather by other stakeholders. Nevertheless, a critical question still is whether these known negative reports represent only the tip of an iceberg – many non-target effects may simply have escaped our attention (Howarth 1991; Simberloff and Stiling 1996).

In an attempt to address this concern, several post-release studies were conducted recently with a focus on those biological control projects where population declines of non-target species have been observed or on projects where the potential for non-target effects have been judged high due to the polyphagous nature of the biological control agent. Obviously, such a selection is strongly biased and, thus, can not be representative. In one of these cases, Barron et al. (2003) evaluated parasitism by the introduced biological control agent *Pteromalus puparum* on the New Zealand red admiral butterfly *Bassaris gonerilla*, in comparison to other mortality factors. From an extensive dataset, Barron et al. (2003) constructed a partial life table and concluded that the level of mortality caused by *P. puparum* is low relative to egg parasitism by *Telenomus* sp., and also low in comparison to larval disappearance and pupal parasitism caused by the accidentally introduced ichneumonid *Echthromorpha intricatoria*.

Similarly, Benson et al. (2003) tested whether the introduced parasitoids *Cotesia glomerata* and *C. rubecula* may have been responsible for the decline of the native butterfly *Pieris virginiensis* in New York and Ontario. Although *P. virginiensis* is an acceptable and suitable host, Benson et al. (2003) concluded that populations of this butterfly do not appear to be at risk because both *C. glomerata* and *C. rubecula* do not forage in forested habitats, even when they are locally present in adjacent meadows.

Extensive studies have been carried out on potential risks of the polyphagous egg parasitoid *Trichogramma brassicae*, which is being mass-released against the European corn borer in many countries. Although *T. brassicae* did parasitize various butterfly species (including rare ones) under field cage conditions (Babendreier et al. 2003a), subsequent studies have demonstrated that parasitism of non-target species under field conditions is zero or restricted to a few meters from the release field (Orr et al. 2000; Babendreier et al. 2003a, 2003b). All these studies concluded that non-target effects of the biological control agent have been negligible, which is remarkable in light of the fact that such effects were presumed in these cases. Although still no comprehensive answer is possible on whether we have missed many non-target effects, evidence is increasing from the abovementioned studies and others that at least the more serious effects would have been detected.

23.4 *Harmonia axyridis*, a Case Study

Already in the section above it was mentioned that especially the polyphagous predators have the potential to cause serious non-target effects if introduced into foreign countries. The multicoloured Asian ladybeetle, *Harmonia axyridis*, is native to large parts of Asia and definitely is a polyphagous predator. It is a voracious feeder, preying mostly on aphids but also on immature stages of various other insects – psyllids, butterflies and aphid predators, including other ladybeetles. Moreover, cannibalism is often observed in adults and larvae of *H. axyridis*, which consume conspecific eggs and sometimes smaller larvae. Due to its effects on native coccinellids in the US and its recent invasion in Europe, *H. axyridis* has attracted quite some attention. Many studies have already been conducted and, thus, *H. axyridis* may serve as a case study to be examined in greater detail.

Since *H. axyridis* feeds voraciously on aphids and has strong dispersal capacities, it was released as a classical biological control agent in the United States already in 1916. However, these early introductions failed and periodic releases continued until the 1980s. First established populations were documented only in 1988 and, subsequently, the beetle spread rapidly across North America. Currently, *H. axyridis* occurs throughout much of the continental United States and also in southern Canada. The ability of *H. axyridis* to feed on many non-target herbivores, intraguild prey and conspecifics may contribute to its success to invade new ecosystems but also to control aphid numbers. After the introduction in North America, it provided control of pests in several systems. For instance, *H. axyridis* provides effective biological control of *Aphis spiraecola* in apple orchards (Brown 2004), and the biological control of several citrus pests may also be benefiting from the establishment of *H. axyridis* (Michaud 2002a). *Harmonia axyridis* has also been utilized successfully in augmentative biological control in Asia, Europe and North America. In Europe, releases were first undertaken in 1982, i.e. before the negative impact had emerged, but continued until only a few years ago.

However, the same attributes responsible for the effectiveness of *H. axyridis* against aphids may also cause less desirable effects. A number of recent studies indicated that the establishment of exotic ladybeetles may have adverse affects on native coccinellids. In South Dakota, the abundance of *Coccinella transversoguttata richardsoni* and *Adalia bipunctata* was approximately 20 times lower after the establishment of an exotic coccinellid, *C. septempunctata* (Elliott et al. 1996). Evidence is building to indicate that *H. axyridis* may be having similar adverse effects on native Coccinellidae. Over a 13-year period, Brown and Miller (1998) monitored the abundance of various coccinellid species in apple orchards. The abundance of native coccinellids decreased after the establishment and rapid rise to dominance of the exotics, *C. septempunctata* and *H. axyridis* (Brown and Miller 1998). A 9-year study of the abundance of various Coccinellidae in an agricultural landscape showed a decrease in the abundance of *Brachiacantha ursina*, *Cycloneda munda* and *Chilocorus stigma* after the establishment of *H. axyridis* (Colunga-Garcia and Gage 1998). Similarly, a field study conducted in the major citrus-producing regions of Florida over 5 years showed that the introduced *H. axyridis* has increased in abundance while the formerly dominant ladybeetle *Cycloneda sanguinea* has declined (Michaud 2002b). The author concluded that competitive displacement of *C. sanguinea* by *H. axyridis* may be in progress in this citrus ecosystem in Florida. Recently, a study conducted on the coccinellid community inhabiting potato crops in northern Maine over 31 years showed dramatic changes. Prior to the arrival of the exotic species, ladybeetle communities were comprised almost exclusively of the two native ladybeetles, *Coccinella transversoguttata* and *Hippodamia tredecimpunctata*. Starting 1980, the exotic *C. septempunctata* became permanently established and quickly began to dominate the ladybeetle community (Alyokhin and Sewell 2004). Two other exotic species, *H. axyridis* and *Propylea quatuordecimpunctata*, became prominent members of the ladybeetle community in 1995 and 1996. Altogether, these studies conducted in the US demonstrate the profound effects that the exotic natural enemy *H. axyridis* may have on the abundance of native coccinellid species.

To date, research has concentrated on the effect of common natural enemies of agricultural crops because of fears regarding the loss of biocontrol function. As a consequence, there is a paucity of studies on the potential adverse effects of *H. axyridis* on coccinellids in non-agricultural settings or generally on less common coccinellids. In addition, few studies have been conducted on other non-target insects. In one of those few studies, Koch et al. (2003) recently identified *H. axyridis* as a potential hazard to immature Monarch butterflies, *Danaus plexippus*. In laboratory and field-cage studies, eggs and larvae of *D. plexippus* incurred significant predation by *H. axyridis* adults and larvae. A follow-up study showed that this effect is mediated by the presence of aphids, the preferred prey of *H. axyridis*. However, although predation on the monarch decreased in the presence of aphids, some monarch larvae were nevertheless consumed (Koch et al. 2005).

In addition to the concerns for non-target effects, *H. axyridis* has been identified as a potential pest in the fruit and wine industry, as aggregations can occur on fruit during harvest. The beetles are particularly difficult to remove from grapes, and this can lead to tainted wine due to contamination by alkaloids from the ladybeetle (Koch 2003). Last not least, *H. axyridis* has started to disturb local communities in North America, since aggregations can also occur on house walls, leading not only to the staining of the outside walls but also to the invasion of houses. In the latter case, the reflex-bleeding of huge numbers of beetles can cause staining of furniture. Due to these characteristics, *H. axyridis* has become a significant pest in several regions of the United States.

Although several studies have unequivocally demonstrated effects on coccinellids in agricultural habitats, the addition of *H. axyridis* to an existing system apparently does not negatively affect aphid control. For instance, Lucas et al. (2002) showed that the addition of *H. axyridis* to the predator guild on apple trees did not hinder the suppression of *Aphis citricola* and *Tetranychus urticae*, which is consistent with the findings of other studies. Thus, it generally appears as if the main effect of the *H. axyridis* invasion in North America is not that of causing new problems with pest control but rather a decrease in populations of native ladybirds and potentially other non-targets.

Harmonia axyridis has also been released in Europe as a biological control agent. It was first introduced to France for the control of various aphid species in 1982, i.e. before the negative effects were known from the US. In the 1990s, *H. axyridis* was commercialized for the control of aphids in greenhouses and field crops in several European countries (e.g. Katsoyannos et al. 1997). Reports are now accumulating which show that this ladybeetle established feral populations in several countries. After the first sightings in Germany, Belgium and The Netherlands in the period 2000–2002, monitoring studies have indicated a rapid spread of populations in all three countries (Fig. 23.1). The beetle is now being found also in parts of France and the United Kingdom (www.harlequin-survey.org). *Harmonia axyridis* has also been released in Italy but so far no beetles have been found in the field.

Based on the most recent records of massive numbers of *H. axyridis* at several sites in The Netherlands, Belgium and Germany, and the rapid spread in these countries (Fig. 23.1), it is inevitable that the beetle will soon spread over larger parts of Europe. At present, however, it is impossible to predict what its precise impact might be. In contrast to many studies conducted in North America, so far hardly any data are available on the potential impact this beetle may have in Europe. To what extent data obtained in North America can be extrapolated remains unclear. One might presume that the situation in Europe will become comparable to that in North America, though it should be acknowledged that North America is characterized by a different set of native and also exotic coccinellid species. Besides *H. axyridis*, probably the most important exotic coccinellid in North America is *C. septempunctata*, which is

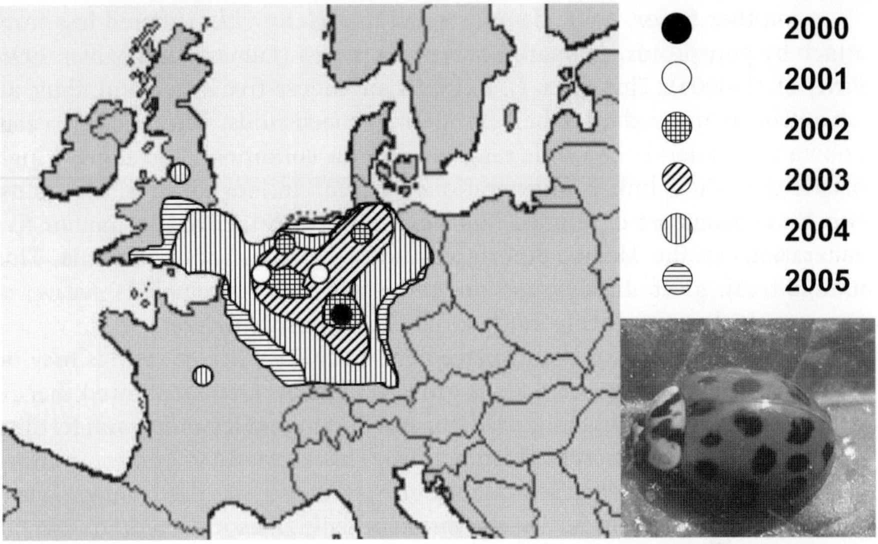

Fig. 23.1 Map showing the invasion of *Harmonia axyridis* in Europe from the first sightings in the year 2000. This ladybeetle has been used for the biocontrol of aphids in greenhouses and the field since 1982 (map extracted from Crop Protection Compendium and based on information provided by A.J.M. Loomans, Plant Protection Service NL, photograph A.J.M. Loomans)

native to Europe. The situation in the US therefore warrants being concerned about the invasion of *H. axyridis* but can not be used to predict the potential environmental impact in Europe with high certainty.

23.5 Why Has *H. axyridis* Become Invasive?

Since it is well known that, once established, only a small fraction of all exotics subsequently become invasive, the question posed here more generally relates to our ability to predict whether a species will become invasive or not. Despite the many studies conducted on *H. axyridis*, the reasons for the displacement of native coccinellids by *H. axyridis* and its general success in North America are still not well understood. Intraguild predation has been examined as a mechanism leading to displacement of native species by *H. axyridis*. Several studies showed that *H. axyridis* is intrinsically superior to many other coccinellids (e.g. Michaud 2002b), and results of Yasuda et al. (2004) suggest that intraguild predation by *H. axyridis* strongly influence aphidophagous guild structure. Thus, *H. axyridis* acts as a top predator in the guild of aphidophagous insects, which may be one factor for its success as an invader.

As another factor, it was demonstrated that *H. axyridis* suffered less from attack by parasitoids, compared to native species (Lamana and Miller 1996; Firlej et al. 2005). This gives *H. axyridis* an enemy-free space and, thus, an advantage compared to other competing coccinellids. The beetle is also known to be tolerant of a wide range of climatic conditions, and shows a high plasticity in the number of generations per year. In most temperate regions, two generations are developed but the number can increase to four or five generations in the Mediterranean, and even more in southern India. This undoubtedly gives *H. axyridis* an advantage over coccinellids native to Europe, which generally are limited to one generation per year.

Important factors for the displacement or coexistence of species may be specific life history traits. In citrus groves, Michaud (2002b) showed that *H. axyridis* was a more voracious predator and had higher fecundity and fertility than was the case for *Coccinella sanguinea*. This is clearly important in using resources efficiently when abundant but may become a disadvantage when prey is scarce. However, coccinellids and especially *H. axyridis* react to this situation of prey limitation through an increase in the number of infertile eggs, which are used as food for the hatched larvae. This allows the hatched larvae to consume a higher number of eggs, increasing the amount of energy available to search for other prey over greater distances (Perry and Roitberg 2005). Furthermore, the broad host range may allow *H. axyridis* populations to survive even if the main prey (aphids) is scarce. Last not least, *H. axyridis* is known to be a strong disperser which flies long distances to overwintering sites in autumn and to new breeding sites in spring. Many of these traits are correlated with body size and, thus, being one of the largest coccinellids might be an advantage in itself for *H. axyridis*. Altogether, these factors may allow *H. axyridis* to adapt to new environments, to invade new territories and to establish new populations (see also Chaps. 6 and 8).

23.6 How to Avoid '*Harmonia* Cases'?

There is now general agreement that the potential for non-target effects has to be evaluated before releasing biological control agents. As noted above, a rather stringent host range assessment procedure is already in place in weed biological control, and this since decades. This approach was shown to be effective because it allows us to reliably judge risks to native flora before introduction (Pemberton 2000) – and the overall need for doing so is now also acknowledged in arthropod biological control. A problem specific to the latter is that there are also companies involved which are selling biological control agents for augmentative releases. Within the last two decades, augmentative biological control has become a major success story, especially in fighting greenhouse pests, and it is feared that any strong obligations to conduct host range tests prior to putting new agents on the market would "kill" the business. However, also for all other forms of biological control, there is a risk that fewer projects will be developed and fewer agents will be approved if hurdles of host range testing or risk assessment in general are becoming too high. Thus, the question basically becomes how to prevent '*Harmonia* cases' without overly hampering biological control.

During the last 10 years, several international activities have addressed the assessment of non-target effects and the regulation of arthropod biological control agents. A starting point towards international regulation was marked by the FAO Code of Conduct for the Import and Release of Exotic Biological Control Agents; this was adopted in 1995 by the FAO Conference and published in 1996 as the International Standard for Phytosanitary Measures No. 3 (IPPC 1996). One objective of the Code was to provide a standard for those countries lacking adequate legislation and procedures to regulate importation and to analyze risks related to biological control agents. The revised version of this Code of Conduct has extended its range from classical biological control to inundative biological control, native natural enemies, microorganisms and other beneficial organisms, and it also includes evaluation of environmental impacts (IPPC 2005). This standard will certainly continue to provide guidance for countries which are developing their own legislative systems for biological control regulation.

Shortly after the Code's first publication, the European and Mediterranean Plant Protection Organization (EPPO) broadly endorsed the FAO Code but recommended that regulation should not slow the importation of import of biological control agents. Several workshops resulted in two guidance documents and a 'positive list' of organisms for safe use in EPPO countries (EPPO 1999, 2001, 2002). The documents concluded that a certification system should be put in place for Europe, rather than a registration procedure, to ensure a 'light' regulatory system with efficient and rapid mechanisms. The reasoning behind this decision was based on previous experience with the registration system for microbial biological control agents in Europe: the EU Directive and its implementation is so stringent that it is basically impossible to register a new microorganism in EU countries.

In 2000, the North American Plant Protection Organization (NAPPO) published its Guidelines for Petition for Release of Exotic Entomophagous Agents for the Biological Control of Pests (RSPM No. 12; NAPPO 2000). These guidelines are intended to assist researchers and companies in drafting a petition for the release of exotic entomophagous agents for biological control of pest insects and mites. It will also assist reviewers and regulators in assessing the risk of exotic introductions intended for biological control.

Island nations, such as Australia and New Zealand, have had serious problems with invasive species and, thus, have set up stringent rules to regulate the import of any organisms from abroad. The 1996 Hazardous Substances and New Organisms (HSNO) Act in New Zealand (http://www.legislation.govt.nz)

has attracted considerable attention internationally as very environmentally focused legislation, and its implementation by ERMA NZ has been observed with interest. In Australia, biological control agents are regulated by two agencies under three separate Acts, and have been similarly heralded as a thorough and biosafety-conscious approach. The two systems have some key differences in approach, and the notability of these approaches, particularly in the area of scope of the regulatory process, is in the opportunity for public participation, and the degree of risk-aversion of the regulatory agencies.

An initiative of OECD (Organization for Economic Co-operation and Development) countries resulted in the development of a guidance document for biological control agents. The document (OECD 2003) proposes guidance to member countries on information requirements for (1) the characterization and identification of the organism, (2) the assessment of safety and effects on human health, (3) the assessment of environmental risks and (4) the assessment of efficacy of the organism. In Europe, the biological control industry expressed concern when the OECD guidance document was published, as the information requirements were considered to be too stringent. As a result, a Commission of the IOBC/WPRS was established in 2003 and a meeting of scientists, together with the biocontrol industry and regulators, resulted in the production of a document which gives detailed guidance on regulation procedures for exotic and indigenous biocontrol agents (Bigler et al. 2005). Most recently, the European Commission released a call for project applications with the aim to develop a new, appropriate and balanced system for regulation of biological control agents (micro- and macro-organisms), semiochemicals and botanicals. It is expected that, in the foreseeable future, the EU members and other European countries may regulate invertebrate biological control agents under uniform principles (but see also Chap. 20).

From this overview on activities worldwide, it is becoming evident that the potential for non-target effects of biological control agents has become an important issue recently, and that important progress has been achieved. However, it was recognized that all these initiatives and guidelines generally highlight *what* should be done or what knowledge is required but they are not designed to provide detailed methods on *how* one should test for non-target effects. This gap was addressed by a guide to best practice in host range testing published by Van Driesche and Reardon (2004). In addition, a comprehensive review on the current methods used to assess potential risks of biological control agents was provided by Babendreier et al. (2005), and a recently published book attempts to go a step further by providing guidance on methods to be used to assess non-target effects of invertebrate biological control agents of insect pests (Bigler et al. 2006).

Though several different aspects may become relevant for risk assessment of arthropod biological control agents, including the potential for establishment (in augmentative biocontrol only) or hybridisation (Chap. 16), host range is the pivotal point in most cases. As Hoddle (2004) pointed out, the most promising agents are those which exhibit high levels of host and habitat fidelity, and maximal impact on the target species, ensuring at the same time minimal impact off target. Host range assessment is essential to all of the abovementioned documents, and many recent introductions have already been accompanied by appropriate host range tests. Therefore, there is reason to expect that polyphagous predators such as *H. axyridis* will no longer be released as biological control agents into foreign countries in the future or, if so, then only after a thorough assessment of possible risks.

23.7 Conclusions

The vast majority of non-native species pose no threat to native biota in the new environment. However, a small number are highly invasive. In parallel, by far most introductions of classical biological control agents of both weeds and arthropods have not caused any harm to the environment, only a small number leading to unwanted non-target effects. It should be stressed, however, that those negative cases which have been documented to date (1) mostly result from projects conducted relatively early in the history of biological control and (2) most often relate to the introduction of vertebrate top predators in arthropod biological control. Nowadays, biological control is conducted in a much more reasonable manner and, as Frank (1998) noted correctly, it may not be appropriate to criticize today's biological control for disastrous introductions of the distant past.

However, adding species to ecosystems (new environments) can have complicated consequences and, undoubtedly, a variety of factors can influence host use by biological control agents under field conditions. This means that precluding any negative effects of new biological control introductions will be notoriously difficult, if at all possible. To reduce this potential as far as possible, host range testing and general risk assessment procedures must be conducted. Fortunately, much progress has been made in terms of regulation of biological control agents and methods to be used for assessing potential effects before introduction into new environments. In fact, the predictive power of the outcome of biological control introductions (in terms of realized host range) is rather good, assuming that sound ecological studies have been conducted beforehand (Babendreier et al. 2005). Saying this, I would even like to suggest that the large experience and the many data available from the purposeful introduction of biological control agents could be used more intensively to the benefits of invasion biology in general.

In this chapter, I provided some examples of detrimental environmental effects stemming from introductions of biological control agents but also showed that many deliberate introductions for biological pest control have resulted in long-term ecological and economic benefits. Altogether, I suggest conducting careful and well-balanced analyses of potential risks and benefits for biological control projects in the future, keeping in mind that all plant protection methods as well as the 'doing nothing' strategy bear risks and benefits which all need to be evaluated and weighed up against each other. Clearly, the value of such analyses much depends on our knowledge on the potential for non-target effects as well as on the potential benefits. It further depends on our scale of rating of risks. In view of the current shift towards a 'risk-free world' and overly stringent regulations, there might even be a risk that promising agents will not be developed because the potential of non-target effects is judged too high.

References

Alyokhin A, Sewell G (2004) Changes in a lady beetle community following the establishment of three alien species. Biol Invasions 6:463–471

Babendreier D, Kuske S, Bigler F (2003a) Parasitism of non-target butterflies by *Trichogramma brassicae* Bezdenko (Hymenoptera: Trichogrammatidae) under field cage and field conditions. Biol Control 26:139–145

Babendreier D, Schoch D, Kuske S, Dorn S, Bigler F (2003b) Non-target habitat exploitation by *Trichogramma brassicae* (Hym. Trichogrammatidae): what are the risks for endemic butterflies? Agric Forest Entomol 5:199–208

Babendreier D, Bigler F, Kuhlmann U (2005) Methods used to assess non-target effects of invertebrate biological control agents of arthropod pests. BioControl 50:821–870

Barron MC, Barlow ND, Wratten SD (2003) Non-target parasitism of the endemic New Zealand Red Admiral Butterfly (*Bassaris gonerilla*) by the introduced biological control agent *Pteromalus puparum*. Biol Control 27:329–335

Benson J, Pasquale A, Van Driesche R, Elkinton J (2003) Assessment of risk posed by introduced braconid wasps to *Pieris virginiensis*, a native woodland butterfly in New England. Biol Control 26:83–93

Bigler F, Bale J, Cock M, Dreyer H, GreatRex R, Kuhlmann U, Loomans A, van Lenteren J (2005) Guideline on information requirements for import and release of invertebrate biological control agents in European countries. Biocontrol News Information, 10 pp

Bigler F, Babendreier D, Kuhlmann U (2006) Environmental impact of invertebrates for biological control of arthropods: methods and risk assessment. CABI, Wallingford, 299 pp

Blossey B, Skinner LC, Taylor J (2001) Impact and management of purple loosestrife (*Lythrum salicaria*) in North America. Biodiversity Conserv 10:1787–1807

Brown MW (2004) Role of aphid predator guild in controlling *Spirea* aphid populations on apple in West Virginia, USA. Biol Control 29:189–198

Brown MW, Miller SS (1998) Coccinellidae (Coleoptera) in apple orchards of eastern West Virginia and the impact of invasion by *Harmonia axyridis*. Entomol News 109:143–151

Colunga-Garcia M, Gage SH (1998) Arrival, establishment, and habitat use of the multicolored Asian lady beetle (Coleoptera: Coccinellidae) in a Michigan landscape. Environ Entomol 27:1574–1580

Elliott N, Kieckhefer R, Kauffman W (1996) Effects of an invading coccinellid on native coccinellids in an agricultural landscape. Oecologia 105:537–544

EPPO (1999) First import of exotic biological control agents for research under contained conditions. EPPO Bull 29:271–272
EPPO (2001) Import and release of exotic biological control agents. EPPO Bull 31:33–35
EPPO (2002) List of biological control agents widely used in the EPPO region. EPPO Bull 32:447–461
Firlej A, Boivin G, Lucas E, Coderre D (2005) First report of *Harmonia axyridis* Pallas being attacked by *Dinocampus coccinellae* Schrank in Canada. Biol Invasions 7:553–556
Frank JH (1998) How risky is biological control? Comm Ecol 79:1829–1834
Gurr G, Wratten S (2000) Biological control: measures of success. Kluwer, Hingham, MA, 429 pp
Hoddle MS (2004) Restoring balance: using exotic species to control invasive exotic species. Conserv Biol 18:38–49
Howarth FG (1991) Environmental impacts of classical biological control. Annu Rev Entomol 36:485–509
IPPC (1996) Code of conduct for the import and release of exotic biological control agents. FAO, Rome, International Standards for Phytosanitary Measures no 3, 21 pp
IPPC (2005) Guidelines for the export, shipment, import and release of biological control agents and other beneficial organisms. FAO, Rome, International Standards for Phytosanitary Measures no 3, 12 pp
Katsoyannos P, Kontodimas DC, Stathas GJ, Tsartsalis CT (1997) Establishment of *Harmonia axyridis* on citrus and some data on its phenology in Greece. Phytoparasitica 25:183–191
Koch R (2003) The multicoloured Asian lady beetle, *Harmonia axyridis*: a review of its biology, uses in biological control, and non target impacts. J Insect Sci 3:1–16
Koch RL, Hutchison WD, Venette RC, Heimpel GE (2003) Susceptibility of immature monarch butterfly, *Danaus plexippus* (Lepidoptera: Nymphalidae: Danainae), to predation by *Harmonia axyridis* (Coleoptera: Coccinellidae). Biol Control 28:265–270
Koch RL, Venette RC, Hutchison WD (2005) Influence of alternate prey on predation of monarch butterfly (Lepidoptera: Nymphalidae) larvae by the multicolored Asian lady beetle (Coleoptera: Coccinellidae). Environ Entomol 34:410–416
Lamana ML, Miller JC (1996) Field observations on *Harmonia axyridis* Pallas (Coleoptera: Coccinellidae) in Oregon. Biol Control 6:232–237
Louda SM (2000) Negative ecological effects of the musk thistle biological control agent, *Rhinocyllus conicus*. In: Follett PS, Duan JJ (eds) Nontarget effects of biological control. Kluwer, Norwell, MA, pp 215–243
Louda SM, Stiling P (2004) The double-edged sword of biological control in conservation and restoration. Conserv Biol 18:50–53
Louda SM, Pemberton RW, Johnson MT, Follett PA (2003) Nontarget effects – the Achilles' heel of biological control? Retrospective analyses to reduce risk associated with biocontrol introductions. Annu Rev Entomol 48:365–396
Lucas E, Gagne I, Coderre D (2002) Impact of the arrival of *Harmonia axyridis* on adults of *Coccinella septempunctata* and *Coleomegilla maculata* (Coleoptera: Coccinellidae). Eur J Entomol 99:457–463
Lynch LD, Hokkanen HMT, Babendreier D, Bigler F, Burgio G, Gao ZH, Kuske S, Loomans, A, Menzler-Hokkanen I, Thomas MB, Tommasini G, Waage JK, vanLenteren JC, Zeng QQ (2001) Insect biological control and non-target effects: a European perspective. In: Wajnberg E, Scott JK, Quimby PC (eds) Evaluating indirect ecological effects of biological control. CABI, New York, pp 99–125
Michaud JP (2002a) Biological control of Asian citrus psyllid, *Diaphorina citri* (Hemiptera: Psyllidae) in Florida: a preliminary report. Entomol News 113:216–222

Michaud JP (2002b) Invasion of the Florida citrus ecosystem by *Harmonia axyridis* (Coleoptera: Coccinellidae) and asymmetric competition with a native species, *Cycloneda sanguinea*. Environ Entomol 31:827–835

NAPPO (2000) Guidelines for petition for release of exotic entomophagous agents for the biological control of pests. Secretariat of North American Plant Protection Organization, Ottawa

OECD (2003) Guidance for information requirements for regulations of invertebrates as biological control agents. OECD Environment, Health and Safety Publications, Series on Pesticides 21:1–22

Orr DB, GarciaSalazar C, Landis DA (2000). *Trichogramma* nontarget impacts: a method for biological control risk assessment. In: Follett PA, Duan JJ (eds) Nontarget effects of biological control. Kluwer, Norwell, MA, pp 111–125

Pemberton RW (2000) Predictable risk to native plants in weed biological control. Oecologia 125:489–494

Perry JC, Roitberg BD (2005) Ladybird mothers mitigate offspring starvation risk by laying trophic eggs. Behav Ecol Sociobiol 58:578–586

Samways M (1997) Classical biological control and biodiversity conservation: what risks are we prepared to accept? Biodiversity Conserv 6:1309–1316

Simberloff D, Stiling P (1996) How risky is biological control? Ecology 77:1965–1974

Van Driesche RG, Reardon R (2004) Assessing host ranges for parasitoids and predators used for classical biological control: a guide to best practice. Forest Health Technology Enterprise Team, Morgantown, WV

Van Lenteren JC, Bale J, Bigler F, Hokkanen HMT, Loomans AJM (2006) Assessing risks of releasing exotic biological control agents of arthropod pests. Annu Rev Entomol 51:609–634

Yasuda H, Evans EW, Kajita Y, Urakawa K, Takizawa T (2004) Asymmetric larval interactions between introduced and indigenous ladybirds in North America. Oecologia 141:722–731

24 General Conclusion, or what Has to be Done now?

Wolfgang Nentwig

Now, after numerous experts have given their detailed and well-founded statements on various aspects of biological invasions in this book, it is time to settle back for a moment and to rethink this flood of information. Two points come immediately to mind:

1. The number of alien and invasive species is still increasing, no change is yet visible, many habitat modifications are irreversible, and earlier pristine conditions of nature can no longer be restored.
2. There are many recommendations and conclusions which should be followed as consequentially as possible – and as soon as possible – to mitigate the impact of biological invasions. Most of these recommendations can be attributed to one of the following five categories: scientific research, management, technical solutions, legislation and administration, and socio-economy, including education of the public.

24.1 Need for more Research

Though alien species and biological invasions have become very topical in recent years, it is astonishing (or, should I say, embarrassing) how limited our knowledge still is. We lack information on the specific characteristics (traits) of those species which become invasive, and we do not really know how successful invaders differ from closely related non-invaders. This makes plausible prognoses extremely difficult (if these will ever be possible) but it has also to be stressed that explanation and prediction are two different things. We know that it is unrealistic to expect to detect single characteristics, or even syndromes, which fully explain invasiveness. Biological invasions are to a large extent idiosyncratic, triggered by a huge variety of events and driven by specific factors. A far more realistic approach, therefore, focuses on a complex interaction of many factors, including species traits, environmental aspects and human influence, which need to be more fully unravelled. Today, it is

Ecological Studies, Vol. 193
W. Nentwig (Ed.)
Biological Invasions

widely accepted that it is futile to manage particular invasions without attending to ill-managed landscapes.

We have only limited information on the spreading capabilities of species, their pathways to the invaded habitats, and on the differences in invasibility of ecosystems. Apart from expressing concern, we can often offer too few reliable, concrete data on the impact of an alien species in an invaded ecosystem. This point is particularly important because more precise information would enable us to prioritize a given set of aliens, and to focus at first on those species with the most detrimental impact. In addition, high-quality inventories of alien species and of experts working on these species are rare.

Thus, one of our utmost urgent needs is to intensify research on alien species and biological invasions. More experimental research is needed at the ecosystem level, notably on how invasive species alter ecosystem services. We also need to learn more about the effects of invasive species in conjunction with global change, including climate and land use, and other key factors such as the cycling and enrichment of N, P or other critical substances, hydrological changes and impacts of fire. In particular, supporting and regulating services presently feature only low levels of research, although both are highly significant elements of our combat against biological invasions. Recognition of the value of ecosystem services would, among others, raise the awareness of the general public.

For the development of appropriate conservation strategies, detailed knowledge of the ecology of alien species is required. Currently, in many cases negative impacts of invasive species on native species have been deduced from correlative evidence, since experimental studies on interactions between invasive and native species are largely lacking. Also, the management and conservation of native taxa threatened by hybridization with invading taxa need more studies which integrate ecological (including genetic) information at the level of local populations.

One special group of alien species are genetically modified organisms, which will become increasingly part of our future environment. Similarly to alien species, their introduction into the natural environment is also irreversible, potentially causing hazard to ecosystems. We need to understand the mechanisms and impacts of introduced genes as well as their potential impact on biodiversity or ecosystem functions.

24.2 Management from Detection to Eradication or Control

Greater efforts have to be made to eradicate several key alien species in certain regions. Experience clearly shows that, with proper planning, substantial financial support and adequate political and social assistance, it is indeed possible to successfully eradicate alien species. However, there are numerous cases of alien species which could easily have been removed but where this has not occurred because of ignorance, little awareness by decision makers, gaps in the legal framework or authorization process, or because of opposition by society.

Since our best strategy is the immediate removal of newly detected invaders before they become established, early detection of any new alien species through an early warning system is essential. This includes paying special attention to key areas, based also on lists of most dangerous alien species (e.g. warning lists). To act quickly, and as a precautionary measure, the authorities responsible need to have established contingency plans for the eradication of specific taxa. This rests heavily not only on an appropriate information management but also on the ready availability of expertise for the identification of relevant taxa as well as alternative control methods.

Eradication of an alien species is always better than its control because the latter implies the persistence of the alien, and affords no full guarantee against potential future ecological and economical impacts. In cases where eradication is no longer possible, biological control may be a suitable solution, at least to minimize the impact of some alien species. Within the last century, a large body of evidence has accumulated showing that the principle of biological control is sound and that the approach is effective – its high value in controlling alien species needs to be acknowledged. Current levels of biological control and associated risk assessment analyses need, however, to be intensified.

24.3 Technical Solutions

Overall, it is obvious that screening systems and border controls of traded goods and travellers have to be expanded and intensified. This includes technologies to prevent invasive organisms from being transported via containers, ships, planes, or other introduction vectors. These control costs are well-invested money, and prevent much higher ecological and economical follow-up costs. For the methods existing to date, further improvements in efficacy, ease of application, and cost-benefit tradeoffs have to be found. Additionally, it is necessary to foresee more than one solution for any particular problem, in the event of some measures having to be abandoned because of unexpected turns of event.

Since waterborne transport will continuously increase in future, a rise in the number of alien species translocated with ships could be the unavoidable result. For aquatic organisms, the most prominent invasion vectors are ballast water and hull fouling of ships. Convincing solutions need to be found for both these problematical aspects. Important pathways are the waterway networks in Europe or maritime channels, and both need also technical solutions to minimize the spread of aliens. This may be achieved by the installation of barriers such as deterrent electrical systems, and chloride or pH-altered locks.

24.4 Legislation and Administration

As one of the products of the 1992 Rio Earth Summit, the Convention on Biological Diversity has been signed by most countries of the world (among the few exceptions are Andorra, Iraq, Somalia and the United States of America). Article 8 (h) of the Convention states that contracting parties should, as far as possible and wherever appropriate, "prevent the introduction of, control or eradicate those alien species which threaten ecosystems, habitats or species". In Annex I of the Convention, the parties recommend adherence to "the precautionary approach" and, in Annex II, alien species are defined as "one of three priority issues" (www.biodiv.org). As parties of this Convention, national governments have to adapt this framework regulation to national law. Though there has been some progress in doing so, overall there is a considerable delay in fulfilling the commitments to this international treaty.

On a global scale, our level of protection against invasive species is weakened most by those countries with the poorest regulations against alien species (i.e. either no regulations or no execution of existing regulations). Since the budget of many developing countries engaged in prevention is small or negligible, it is meaningful to support these countries financially, thereby enabling them to better implement those measures and actions necessary to reduce the risk of dispersal of alien species.

National legislation and local regulations have to be reviewed to ensure that the legal status of alien species does not obstruct optimal management measures. The competencies of all involved authorities and various stakeholders need to be clearly defined. The authorization process should be streamlined, in order to facilitate rapid response. It is recommendable to establish a national coordination body and an independent advisory committee. It is also advisable to include into these bodies and structures not only relevant state agencies but also regional and local entities as well as industry, non-governmental organisations and other interest groups.

The management of alien species also requires effective international cooperation of neighbouring states. This includes exchange of information, cooperation and coordination among governmental agencies, non-governmental organisations and the private sector, e.g. trading companies and the hardly supervised aquaculture industries.

Subject Index

B

E

F

G

P

T

U

V

W

Ecological Studies

Volumes published since 2002

Volume 154
Geoecology of Antarctic Ice-Free Coastal Landscapes (2002)
L. Beyer and M. Bölter (Eds.)

Volume 155
Conserving Biological Diversity in East African Forests: A Study of the Eastern Arc Mountains (2002)
W.D. Newmark

Volume 156
Urban Air Pollution and Forests: Resources at Risk in the Mexico City Air Basin (2002)
M.E. Fenn, L. I. de Bauer, and T. Hernández-Tejeda (Eds.)

Volume 157
Mycorrhizal Ecology (2002, 2003)
M.G.A. van der Heijden and I.R. Sanders (Eds.)

Volume 158
Diversity and Interaction in a Temperate Forest Community: Ogawa Forest Reserve of Japan (2002)
T. Nakashizuka and Y. Matsumoto (Eds.)

Volume 159
Big-Leaf Mahogany: Genetic Resources, Ecology and Management (2003)
A. E. Lugo, J. C. Figueroa Colón, and M. Alayón (Eds.)

Volume 160
Fire and Climatic Change in Temperate Ecosystems of the Western Americas (2003)
T. T. Veblen et al. (Eds.)

Volume 161
Competition and Coexistence (2002)
U. Sommer and B. Worm (Eds.)

Volume 162
How Landscapes Change: Human Disturbance and Ecosystem Fragmentation in the Americas (2003)
G.A. Bradshaw and P.A. Marquet (Eds.)

Volume 163
Fluxes of Carbon, Water and Energy of European Forests (2003)
R. Valentini (Ed.)

Volume 164
Herbivory of Leaf-Cutting Ants: A Case Study on *Atta colombica* in the Tropical Rainforest of Panama (2003)
R. Wirth, H. Herz, R.J. Ryel, W. Beyschlag, B. Hölldobler

Volume 165
Population Viability in Plants: Conservation, Management, and Modeling of Rare Plants (2003)
C.A Brigham, M.W. Schwartz (Eds.)

Volume 166
North American Temperate Deciduous Forest Responses to Changing Precipitation Regimes (2003)
P. Hanson and S.D. Wullschleger (Eds.)

Volume 167
Alpine Biodiversity in Europe (2003)
L. Nagy, G. Grabherr, Ch. Körner, D. Thompson (Eds.)

Volume 168
Root Ecology (2003)
H. de Kroon and E.J.W. Visser (Eds.)

Volume 169
Fire in Tropical Savannas: The Kapalga Experiment (2003)
A.N. Andersen, G.D. Cook, and R.J. Williams (Eds.)

Volume 170
Molecular Ecotoxicology of Plants (2004)
H. Sandermann (Ed.)

Volume 171
Coastal Dunes: Ecology and Conservation (2004)
M.L. Martínez and N. Psuty (Eds.)

Volume 172
Biogeochemistry of Forested Catchments in a Changing Environment: A German Case Study (2004)
E. Matzner (Ed.)

Volume 173
Insects and Ecosystem Function (2004)
W. W. Weisser and E. Siemann (Eds.)

Subject Index

B

E

F

G

P

U

V

W

Ecological Studies

Volumes published since 2002

Volume 154
Geoecology of Antarctic Ice-Free Coastal Landscapes (2002)
L. Beyer and M. Bölter (Eds.)

Volume 155
Conserving Biological Diversity in East African Forests: A Study of the Eastern Arc Mountains (2002)
W.D. Newmark

Volume 156
Urban Air Pollution and Forests: Resources at Risk in the Mexico City Air Basin (2002)
M.E. Fenn, L. I. de Bauer, and T. Hernández-Tejeda (Eds.)

Volume 157
Mycorrhizal Ecology (2002, 2003)
M.G.A. van der Heijden and I.R. Sanders (Eds.)

Volume 158
Diversity and Interaction in a Temperate Forest Community: Ogawa Forest Reserve of Japan (2002)
T. Nakashizuka and Y. Matsumoto (Eds.)

Volume 159
Big-Leaf Mahogany: Genetic Resources, Ecology and Management (2003)
A. E. Lugo, J. C. Figueroa Colón, and M. Alayón (Eds.)

Volume 160
Fire and Climatic Change in Temperate Ecosystems of the Western Americas (2003)
T. T. Veblen et al. (Eds.)

Volume 161
Competition and Coexistence (2002)
U. Sommer and B. Worm (Eds.)

Volume 162
How Landscapes Change: Human Disturbance and Ecosystem Fragmentation in the Americas (2003)
G.A. Bradshaw and P.A. Marquet (Eds.)

Volume 163
Fluxes of Carbon, Water and Energy of European Forests (2003)
R. Valentini (Ed.)

Volume 164
Herbivory of Leaf-Cutting Ants: A Case Study on *Atta colombica* in the Tropical Rainforest of Panama (2003)
R. Wirth, H. Herz, R.J. Ryel, W. Beyschlag, B. Hölldobler

Volume 165
Population Viability in Plants: Conservation, Management, and Modeling of Rare Plants (2003)
C.A Brigham, M.W. Schwartz (Eds.)

Volume 166
North American Temperate Deciduous Forest Responses to Changing Precipitation Regimes (2003)
P. Hanson and S.D. Wullschleger (Eds.)

Volume 167
Alpine Biodiversity in Europe (2003)
L. Nagy, G. Grabherr, Ch. Körner, D. Thompson (Eds.)

Volume 168
Root Ecology (2003)
H. de Kroon and E.J.W. Visser (Eds.)

Volume 169
Fire in Tropical Savannas: The Kapalga Experiment (2003)
A.N. Andersen, G.D. Cook, and R.J. Williams (Eds.)

Volume 170
Molecular Ecotoxicology of Plants (2004)
H. Sandermann (Ed.)

Volume 171
Coastal Dunes: Ecology and Conservation (2004)
M.L. Martínez and N. Psuty (Eds.)

Volume 172
Biogeochemistry of Forested Catchments in a Changing Environment: A German Case Study (2004)
E. Matzner (Ed.)

Volume 173
Insects and Ecosystem Function (2004)
W. W. Weisser and E. Siemann (Eds.)

Volume 174
Pollination Ecology and the Rain Forest: Sarawak Studies (2005)
D. Roubik, S. Sakai, and A.A. Hamid (Eds.)

Volume 175
Antarctic Ecosystems: Environmental Contamination, Climate Change, and Human Impact (2005)
R. Bargagli

Volume 176
Forest Diversity and Function: Temperate and Boreal Systems (2005)
M. Scherer-Lorenzen, Ch. Körner, and E.-D. Schulze (Eds.)

Volume 177
A History of Atmospheric CO_2 and its Effects on Plants, Animals, and Ecosystems (2005)
J.R. Ehleringer, T.E. Cerling, and M.D. Dearing (Eds.)

Volume 178
Photosynthetic Adaptation: Chloroplast to Landscape (2005)
W.K. Smith, T.C. Vogelmann, and C. Chritchley (Eds.)

Volume 179
Lamto: Structure, Functioning, and Dynamics of a Savanna Ecosystem (2006)
L. Abbadie et al. (Eds.)

Volume 180
Plant Ecology, Herbivory, and Human Impact in Nordic Mountain Birch Forests (2005)
F.E. Wielgolaski (Ed.) and P.S. Karlsson, S. Neuvonen, D. Thannheiser (Ed. Board)

Volume 181
Nutrient Acquisition by Plants: An Ecological Perspective (2005)
H. BassiriRad (Ed.)

Volume 182
Human Ecology: Biocultural Adaptations in Human Cummunities (2006)
H. Schutkowski

Volume 183
Growth Dynamics of Conifer Tree Rings: Images of Past and Future Environments (2006)
E.A. Vaganov, M.K. Hughes, and A.V. Shashkin

Volume 184
Reindeer Management in Northernmost Europe: Linking Practical and Scientific Knowledge in Social-Ecological Systems (2006)
B.C. Forbes, M. Bölter, L. Müller-Wille, J. Hukkinen, F. Müller, N. Gunslay, and Y. Konstantinov (Eds.)

Volume 185
Ecology and Conservation of Neotropical Montane Oak Forests (2006)
M. Kappelle (Ed.)

Volume 186
Biological Invasions in New Zealand (2006)
R.B. Allen and W.G. Lee (Eds.)

Volume 187
Managed Ecosystems and CO_2: Case Studies, Processes, and Perspectives (2006)
J. Nösberger, S.P. Long, R. J. Norby, M. Stitt, G.R. Hendrey, and H. Blum (Eds.)

Volume 188
Boreal Peatland Ecosystem (2006)
R.K. Wieder and D. H. Vitt (Eds.)

Volume 189
Ecology of Harmful Algae (2006)
E. Granéli and J.T. Turner (Eds.)

Volume 190
Wetlands and Natural Resource Management (2006)
J.T.A. Verhoeven, B. Beltman, R. Bobbink, and D.F. Whigham (Eds.)

Volume 191
Wetlands: Functioning, Biodiversity Conservation, and Restoration (2006)
R. Bobbink, B. Beltman, J.T.A. Verhoeven, and D.F. Whigham (Eds.)

Volume 192
Geological Approaches to Coral Reef Ecology (2007)
R.B. Aronson (Ed.)

Volume 193
Biological Invasions (2007)
W. Nentwig (Ed.)